Advances in Intelligent Systems and Computing

Volume 628

Series editor

Janusz Kacprzyk, Polish Academy of Sciences, Warsaw, Poland
e-mail: kacprzyk@ibspan.waw.pl

About this Series

The series "Advances in Intelligent Systems and Computing" contains publications on theory, applications, and design methods of Intelligent Systems and Intelligent Computing. Virtually all disciplines such as engineering, natural sciences, computer and information science, ICT, economics, business, e-commerce, environment, healthcare, life science are covered. The list of topics spans all the areas of modern intelligent systems and computing.

The publications within "Advances in Intelligent Systems and Computing" are primarily textbooks and proceedings of important conferences, symposia and congresses. They cover significant recent developments in the field, both of a foundational and applicable character. An important characteristic feature of the series is the short publication time and world-wide distribution. This permits a rapid and broad dissemination of research results.

More information about this series at http://www.springer.com/series/11156

M. Sreenivasa Reddy · K. Viswanath
Shiva Prasad K.M.
Editors

International Proceedings on Advances in Soft Computing, Intelligent Systems and Applications

ASISA 2016

Editors
M. Sreenivasa Reddy
R.L. Jalappa Institute of Technology
Doddaballapur, Bengaluru, Karnataka
India

Shiva Prasad K.M.
R.L. Jalappa Institute of Technology
Doddaballapur, Bengaluru, Karnataka
India

K. Viswanath
R.L. Jalappa Institute of Technology
Doddaballapur, Bengaluru, Karnataka
India

ISSN 2194-5357 ISSN 2194-5365 (electronic)
Advances in Intelligent Systems and Computing
ISBN 978-981-10-5271-2 ISBN 978-981-10-5272-9 (eBook)
https://doi.org/10.1007/978-981-10-5272-9

Library of Congress Control Number: 2017948217

Printed on acid-free paper

This Springer imprint is published by Springer Nature
The registered company is Springer Nature Singapore Pte Ltd.
The registered company address is: 152 Beach Road, #21-01/04 Gateway East, Singapore 189721, Singapore

Preface

Opening Remarks

The International Conference on Advances in Soft Computing, Intelligent Systems and Applications (ASISA 2016) has become one of the major events of the year in the field of in soft computing, intelligent systems, and applications. ASISA 2016 conference is jointly organized by R.L. Jalappa Institute of Technology, ARPVGSI, and IEEE Robotics and Automation Society.

The world is going through a rapid change with the advances in soft computing and intelligent systems. This HMI advantage should be extended to all: literate and illiterate, able-bodied and physically challenged users, urban and rural people. In this respect, our world, in fact, is divided into two: privileged and underprivileged. There is a need to bridge the gap between underprivileged users and cyber society. With this aim in our mind, we have chosen the theme of ASISA 2016 conference.

The conference has been planned so that students from academic institutions, experts from industries, and academia take part in the conference and share their experience. Participation in the conference has been increased significantly over the years, and the conference has become truly international. We have received 199 submissions. Out of these, the number of submissions from industries is 11, and from academics and research institutions are 188. Moreover, out of all submissions, the number of submissions from India is 102 and rest from the outside of India. We categorized all papers into eight tracks. All accepted papers were evaluated by the International Advisory Committee of ASISA 2016 comprising of 55 experts in the field of Advances in Soft Computing, Intelligent Systems and Applications and Research from India & abroad.

We ranked all papers submitted to the conference numerically based on grade 5. Of this grade, 20% of it was determined by technical quality, 20% of it novelty, 5% by writing style, 5% by the confidence of reviewer, 20% of the relevance to the conference, and finally 30% by the overall feeling of the paper. Each paper had been reviewed with double-blind peer-review process and finally by the track chair of the corresponding track. We rejected superficial reviews and paid careful attention to reviews stating reasons for the accept/reject decision. We have accepted 39 papers with the acceptance rate of 43.3% for the oral presentation and inclusion in the conference proceedings.

On behalf of the ASISA 2016 organizing committee, we thank all referees, track chairs, volunteers, and the paper authors. We are also very grateful to the 2016 executive committee of IEEE Robotics and Automation Society and the authority of R.L. Jalappa Institute of Technology, Kodigehalli, Doddaballapur, Bangalore, Karnataka, India, to help us all the way to organize such a big event.

Our special thanks to Er. Ch. Nagaraja, Secretary; Sri Devaraj Urs Educational Trust, Bangalore, India; Er. J. Nagendra Swamy, CEO, R.L. Jalappa Institute of Technology, Doddaballapur, Bangalore, India; Er. J. Rajendra, Director, R.L. Jalappa Institute of Technology, Doddaballapur, Bangalore, India; Prof. M. Sreenivasa Reddy, Principal, R.L. Jalappa Institute of Technology, Doddaballapur, Bangalore, India; Prof. N. Manjunath, HOD, Computer Science, R.L. Jalappa Institute of Technology, Doddaballapur, Bangalore, India; Prof. Vishwanath, R.L. Jalappa Institute of Technology, Doddaballapur, Bangalore, India; Dr. Khan M.S., Dean Academics, R.L. Jalappa Institute of Technology, Doddaballapur, Bangalore, India, for their invaluable guidance in organizing the conference. Last but not least, we express our heartfelt thanks to speakers of keynote speeches and invited talks, who in spite of their busy schedule manage their times and have kindly agreed to deliver highly stimulating talks.

We feel that this is an exceptionally high-quality technical event, but you must judge that for yourselves. We do hope that you will enjoy reading the collection of papers presented.

Doddaballapur, Bengaluru, India

M. Sreenivasa Reddy
K. Viswanath
Shiva Prasad K.M.

Conference Committee

Chairman, RLJIT

Er. Ch. Nagaraja, Secretary, Sri Devaraj Urs Educational Trust, Bangalore, India

Chairman, ASISA 2016

Er. J. Nagendra Swamy, CEO, R.L. Jalappa Institute of Technology, Doddaballapur, Bangalore, India

Co-Chairman, ASISA 2016

Er. J. Rajendra, Director, R.L. Jalappa Institute of Technology, Doddaballapur, Bangalore, India

General Chair

Dr. Khan M.S., Dean Academics, R.L. Jalappa Institute of Technology, Doddaballapur, Bangalore, India

Steering Committee Chair

Prof. M. Sreenivasa Reddy, Principal, R.L. Jalappa Institute of Technology, Doddaballapur, Bangalore, India
Prof. N. Manjunath, HOD, Computer Science, R.L. Jalappa Institute of Technology, Doddaballapur, Bangalore, India
Prof. Vishwanath, R.L. Jalappa Institute of Technology, Doddaballapur, Bangalore, India

Program Committee

Prof. Qinggang Meng, Department of Computer Science, Loungborough University, UK.
Prof. Chrisos N. Schizas, Department of Computer Science, University of Cyprus, Nicosia
Schahram Dustdar, Technical University of Vienna
Sebastián Ventura Soto, University of Cordoba, Spain
Andreas Rausch, Clausthal University of Technology, Germany
Lefteris Angelis, Aristotle University of Thessaloniki, Greece
Andrew L. Nelson, Androtics LLC, Tucson, Arizona, USA
Janos Botzheim, Tokyo Metropolitan University, Japan
Heder Soares Bernardino, Federal University of Juiz de Fora, Brazil
Frantisek Zboril jr., Brno University of Technology, Czech Republic
Bernadetta Kwintiana Ane, University of Stuttgart, Germany
Toshiyuki Maeda, Hannan University, Japan
Virgilijus Sakalauskas, Vilinius University, Lithuania
Garenth Lim king Hann, Curtin University, Malaysia
Ing. Jan Samek, Brno University of Technology, Czech Republic
José Raúl Romero Salguero, University of Cordoba, Spain
Kuruvilla Varghese, Indian Institute of Science, Bangalore
Dalia Kriksciuniene, Vilinius University, Lithuania
Jimson Mathew, nm2 Logic, UK
Wolfgang Reif, University of Augsburg, Germany
Giancarlo Mauri, University of Milano-Bicocca, Italy
Francisco Javier Cabrerizo Lorite, UNED, Spain
Ender Özcan, University of Nottingham, UK
Sung-Bae Cho, Yonsei University, Korea
Roy P. Paily, Indian Institute of Technology, Guwahati
Hana Rezankova, University of Economics, Czech Republic
Boris Tudjarov, Technical University of Sofia, Bulgaria
Wei-Chiang Hong, Hangzhou Dianzi University, China
José Valente de Oliveira, Universidade do Algarve, Portugal
Juan Jose Flores Romero, University of Michoacan, Mexico
Rozita Jamili Oskouei, Institute for Advanced Studies in Basic Science (IASBS), Iran
Antonio Cicchetti, Mälardalen University, Sweden
Chanchal Roy, University of Saskatchewan, Canada
Sergio Segura, University of Seville, Spain
C. Samuel, Indian Institute of Technology, Varanasi
Frédéric Mallet, University Nice Sophia Antipolis, France
Marjan Mernik, University of Maribor, Slovenia
Sasikumar Punnekkat, Mälardalen University, Sweden
Carmine Gravino, University of Salerno, Italy

Helio Perroni Filho, University of Tsukuba, Japan
Juan M. Carrillo de Gea, University of Murcia, Spain
Abdel Obaid, University of Quebec at Montreal, Canada
Andrea Corradini, Kolding School of Design, Denmark
Giovanni Acampora, Nottingham Trent University, UK

Advisory Committee

Prof. Parvinder Singh Sandhu, Rayat & Bahra Institute of Engineering & Bio-Technology, India
Prof. Chris Hinde, Emeritus Professor of Computational Intelligence, Loughborough University, UK
Prof. S. Arockiasamy, Head Information Systems, CEMIS, University of Nizwa, Sultanate of Oman
Prof. Andrew Kusiak, Department of Mechanical and Industrial Engineering, University of Lowa, Lowa
Prof. Kimberly Newman, University of Colorado at Boulder, Boulder
Prof. Chandratilak De Silva Liyanage, Faculty of Science, University of Brunei, Brunei
Prof. Sohair Aly Aly Hassan, University of Kuala Lumpur, Malaysia
Prof. Kien Wen Sun, National Chiao Tung University, Taiwan
Prof. U.S.N. Raju, Department of Computer Science and Engineering, NIT, Warangal, India
Prof. Raghavendra Rao Chillarige, School of Computer and Information Sciences, University of Hyderabad, India
Prof. I.B. Turksen, Department of Industrial Engineering, TOBB-ETU, Ankara, Turkey
Dr. E. Evangelista Ordoño, Mapua Institute of Technology, Philippines
Dr. R.K. Sharma (ARO, CWC), Lucknow
Dr. Preeti Mishra (Mpec.), Kanpur
Dr. B.S. Chandel (DSC.), DBS Kanpur
Dr. Animesh Tripathi, Lucknow University, Lucknow
Dr. Manoj Tripathi, CSJM University, Kanpur
Sri. Swaminath Ram, Chief Environmental Officer, UPPCB, Lucknow
Dr. M.K. Shrma, PCB, Roorkee
Dr. P.N. Panday, Awadh University, Faijabad
Dr. Devendra Kumar, Agra University, Agra
Dr. Animesh Tripathi, Lucknow University, Lucknow
Dr. Manoj Tripathi, CSJM University, Kanpur
Sri. Swaminath Ram, Chief Environment Officer, UPPCB, Lucknow
Dr. M.K. Shrma, PCB, Roorkee
Dr. P.N. Panday, Awadh University, Faijabad

Dr. Devendra Kumar, Agra university, Agra
Prof. Madav Kumar S., R.L. Jalappa Institute of Technology, Doddaballapur, Bangalore, India
Prof. Anandreddy G.M., R.L. Jalappa Institute of Technology, Doddaballapur, Bangalore, India
Prof. Shankar N.B., R.L. Jalappa Institute of Technology, Doddaballapur, Bangalore, India
Prof. Basavaraj S. Pol, R.L. Jalappa Institute of Technology, Doddaballapur, Bangalore, India
Prof. Kamalamma K.V., R.L. Jalappa Institute of Technology, Doddaballapur, Bangalore, India
Prof. Gana K.P., R.L. Jalappa Institute of Technology, Doddaballapur, Bangalore, India
Prof. Madhukar H., R.L. Jalappa Institute of Technology, Doddaballapur, Bangalore, India
Prof. Veena K., R.L. Jalappa Institute of Technology, Doddaballapur, Bangalore, India
Prof. Vinay Kumar Y.B., R.L. Jalappa Institute of Technology, Doddaballapur, Bangalore, India
Prof. Mamatha E., R.L. Jalappa Institute of Technology, Doddaballapur, Bangalore, India
Prof. Shilpa M.N., R.L. Jalappa Institute of Technology, Doddaballapur, Bangalore, India
Prof. Sunil Kumar R.M., R.L. Jalappa Institute of Technology, Doddaballapur, Bangalore, India
Prof. Manjunath B.N., R.L. Jalappa Institute of Technology, Doddaballapur, Bangalore, India
Prof. Sowmya A.V., R.L. Jalappa Institute of Technology, Doddaballapur, Bangalore, India
Prof. Kishore Kumar R., R.L. Jalappa Institute of Technology, Doddaballapur, Bangalore, India
Prof. M.S. Rekha, R.L. Jalappa Institute of Technology, Doddaballapur, Bangalore, India
Prof. Prasanna Lakshmi, R.L. Jalappa Institute of Technology, Doddaballapur, Bangalore, India
Prof. Sukruth Gowda M.A., R.L. Jalappa Institute of Technology, Doddaballapur, Bangalore, India
Prof. Niranjan murthy S.V., R.L. Jalappa Institute of Technology, Doddaballapur, Bangalore, India
Prof. R. Lokesh, R.L. Jalappa Institute of Technology, Doddaballapur, Bangalore, India
Prof. Shivaprasad K.M., R.L. Jalappa Institute of Technology, Doddaballapur, Bangalore, India
Prof. Harish S., R.L. Jalappa Institute of Technology, Doddaballapur, Bangalore, India

Prof. Madhuchandra G., R.L. Jalappa Institute of Technology, Doddaballapur, Bangalore, India
Prof. Nataraja P.R., R.L. Jalappa Institute of Technology, Doddaballapur, Bangalore, India
Prof. Anil Kumar C., R.L. Jalappa Institute of Technology, Doddaballapur, Bangalore, India
Prof. K. Vishwanath, R.L. Jalappa Institute of Technology, Doddaballapur, Bangalore, India
Prof. Umashankar, R.L. Jalappa Institute of Technology, Doddaballapur, Bangalore, India
Prof. Nayana N.D., R.L. Jalappa Institute of Technology, Doddaballapur, Bangalore, India
Prof. Gaure Gundappa, R.L. Jalappa Institute of Technology, Doddaballapur, Bangalore, India
Prof. Darshan B.D., R.L. Jalappa Institute of Technology, Doddaballapur, Bangalore, India
Prof. Aarthi V., R.L. Jalappa Institute of Technology, Doddaballapur, Bangalore, India
Prof. Vijay Kumar, R.L. Jalappa Institute of Technology, Doddaballapur, Bangalore, India
Prof. Nandish N., R.L. Jalappa Institute of Technology, Doddaballapur, Bangalore, India
Prof. Manjesh N., R.L. Jalappa Institute of Technology, Doddaballapur, Bangalore, India
Prof. Lavanya Vaishnavi D.A., R.L. Jalappa Institute of Technology, Doddaballapur, Bangalore, India
Prof. Suchendra K.R., R.L. Jalappa Institute of Technology, Doddaballapur, Bangalore, India
Prof. Raghavendra Prasad, R.L. Jalappa Institute of Technology, Doddaballapur, Bangalore, India
Prof. Harish R.S., R.L. Jalappa Institute of Technology, Doddaballapur, Bangalore, India
Prof. Lokesh Yadhav B.R., R.L. Jalappa Institute of Technology, Doddaballapur, Bangalore, India
Prof. Hanumanthe Gowda, R.L. Jalappa Institute of Technology, Doddaballapur, Bangalore, India
Prof. Bharath H.N., R.L. Jalappa Institute of Technology, Doddaballapur, Bangalore, India
Prof. Sunil Kumar K., R.L. Jalappa Institute of Technology, Doddaballapur, Bangalore, India
Prof. Nagesh H.`C., R.L. Jalappa Institute of Technology, Doddaballapur, Bangalore, India
Prof. Vinay A.N., R.L. Jalappa Institute of Technology, Doddaballapur, Bangalore, India

Prof. Ravikumar M., R.L. Jalappa Institute of Technology, Doddaballapur, Bangalore, India
Prof. Jagannatha Reddy, R.L. Jalappa Institute of Technology, Doddaballapur, Bangalore, India
Prof. Vijay Praveen P.M., R.L. Jalappa Institute of Technology, Doddaballapur, Bangalore
Prof. Bharath L., R.L. Jalappa Institute of Technology, Doddaballapur, Bangalore, India
Prof. Gowri Shankar T.P., R.L. Jalappa Institute of Technology, Doddaballapur, Bangalore
Prof. Lakshminarayana T.H., R.L. Jalappa Institute of Technology, Doddaballapur, Bangalore
Prof. Shilpa T.V., R.L. Jalappa Institute of Technology, Doddaballapur, Bangalore, India
Prof. Sandeep Khelga, R.L. Jalappa Institute of Technology, Doddaballapur, Bangalore, India
Prof. Bharath G., R.L. Jalappa Institute of Technology, Doddaballapur, Bangalore, India

Financial Chair

Dr. Khan M.S., Dean Academics, R.L. Jalappa Institute of Technology, Doddaballapur, Bangalore, India

Publicity Chair

Mr. A. Yateesh, Head HRD, R.L. Jalappa Institute of Technology, Doddaballapur, Bangalore, India

Website Chair

Mr. A. Waseemulla, R.L. Jalappa Institute of Technology, Doddaballapur, Bangalore, India

Proceedings Chair

Dr. Khan M.S., Dean Academics, R.L. Jalappa Institute of Technology, Doddaballapur, Bangalore, India

Contents

About the Editors

M. Sreenivasa Reddy has been associated as Principal with R.L. Jalappa Institute of Technology (RLJIT), Bangalore, India, since 2001. He did his Ph.D. from Jawaharlal Nehru Technological University, Hyderabad; B.E. from University BDT College of Engineering, Karnataka; and M.Tech from B.M.S. College of Engineering. Dr. Reddy has 17 years of teaching, 2 years of industrial, 8 years of administration, and 5 years of research experience. He has several publications published in journals and conference proceedings. He is currently supervising 5 Ph.D. students.

K. Viswanath is currently working as associate professor at RLJIT, Bangalore, India. He did his Ph.D. from Pondicherry University, Pondicherry, and M.Tech in VLSI Design and B.Tech from Jawaharlal Nehru Technological University, Hyderabad. He has 12 years of teaching experience and 2 years of industrial. His areas of research interest are low-power VLSI, biomedical image processing, and artificial neural network. He has several publications to his credit.

Shiva Prasad K.M. is currently working as associate professor and HoD, ECE, at RLJIT, Bangalore, India. He is pursuing his Ph.D. from Jain University, Bangalore. His areas of interest are speech processing, image processing, electronic measurements and instrumentation, advanced computer architecture, wireless communication, real-time embedded system, synthesis and optimization of digital networks, optical networks, fiber optic communication, sensor and transducer engineering, biomedical signal processing, analog/digital communication engineering. He has several publications published in journals and conference proceedings.

Dynamic Stability Margin Evaluation of Multi-machine Power Systems Using Genetic Algorithm

I.E.S. Naidu, K.R. Sudha and A. Chandra Sekhar

Abstract This paper presents a method to find the dynamic stability margin of power system using genetic algorithm. Power systems are subjected to wide range of operating conditions. Modern power systems are equipped with fast-acting protective devices for transient stability problems. Hence, power systems are operated above the transient stability limit. The dynamic behaviour of system can be evaluated using small signal stability analysis. The maximum loading to which the system can be subjected can be obtained by observing the eigenvalue variations of the system under different loading conditions. The loading for which system exhibits a pair of imaginary eigenvalues is the maximum loading limit. Beyond this limit, the system will become unstable. The loading for which the power system exhibits imaginary eigenvalues is evaluated by using genetic algorithm. The dynamic stability margin is evaluated for a 3-machine 9-bus system. The efficacy of the proposed method is tested for the power system including conventional power system stabilizers (CPSS).

Keywords Dynamic stability · Eigenvalue analysis · Genetic algorithm
Power system stabilizer

I.E.S. Naidu (✉) · A. Chandra Sekhar
GITAM University, Visakhapatnam, India
e-mail: iesnaidu@gmail.com

A. Chandra Sekhar
e-mail: acs@gitam.edu

K.R. Sudha
A.U.C.E.(A), Andhra University, Visakhapatnam, India
e-mail: kramasudha.ee@auvsp.edu.in

M.S. Reddy et al. (eds.), *International Proceedings on Advances in Soft Computing, Intelligent Systems and Applications*, Advances in Intelligent Systems and Computing 628, https://doi.org/10.1007/978-981-10-5272-9_1

1 Introduction

Power system stability is the key aspect in operation and design of power system [1]. Small signal stability analysis at an operating point under steady state is important among several aspects of stability [2]. An uncontrolled power systems show slightly damped electromechanical oscillations to various operating conditions. The slightly damped electromechanical modes result in increase of the synchronizing torque in unstable oscillations. Significant efforts are made to identify, predict and control such stability problems [3]. Many methods have been developed to damp the oscillatory behaviour of power systems [4]. The power system nonlinear equations are linearized around an equilibrium point to derive the dynamic behaviour of a power system [5].

The stability of power system is decided by the eigenvalues at each operating point. For a change in operating point if the system eigenvalues crosses from left to right of the imaginary axis, then the system exhibits undamped oscillations. To improve the stability of the power system and to increase loadability margin, these oscillations must be damped out over a wide range of operating conditions [6]. The effective devices used for damping of these sustained oscillations are CPSS. The parameters of the CPSS can be tuned by understanding the characteristics of the power system under study and by providing proper stabilization signals over a wide range of operating conditions [7]. The traditional methods to design CPSS suffer from limitations such as fixed parameters and long processing time at an operating point. The designs of CPSS require the system eigenvalues and the damping ratio at each of the system modes in all operating conditions [8]. Novel computational methods can be employed to tune the CPSS parameters [9, 10].

The enhanced power transfer capability of power system with CPSS can be obtained by finding the maximum loadability limit (boundary conditions or limit cycles) [11]. Many investigations have been done on the methods to find the limit cycles of a power system [12]. The power system exhibits undamped oscillations beyond the maximum loadability limit [13, 15]. Hopf bifurcation (HB) theory can be used to find the limit cycle around an operating point [14]. At HB, a conjugate pair of eigenvalues of the linearized system crosses the imaginary axis [16, 17]. The HB appears due to either the variation of damping, exciter gain, line reactance or load [18]. Identification of these limit cycles is complex and tedious. These complex problems can be solved by using novel soft computing technique like genetic algorithm.

A novel computational method is proposed to find dynamic stability margin using genetic algorithm for multi-machine power system (MMPS). A case study considering 3-machine 9-bus system is presented. The dynamic stability margin for the given system without controller is obtained using the proposed genetic algorithm approach. The stability margin of the MMPS is enhanced by incorporating a CPSS at each of the three machines. The system state matrix is reconstructed by including stabilizer. The limit cycles are obtained for the modified state matrix using GA.

2 Power System Model

The power system under study consists of well-known 3-machine 9-bus system is given in Fig. 1. The set of state variables used to describe the behaviour of each machine of MMPS is represented using state-space approach. The state-space model of MMPS can be represented using following equations [4].

$$x' = \mathrm{A}\mathrm{x} + \mathrm{B}\mathrm{u} + \Gamma p(t) \tag{1}$$

$$x' = \begin{bmatrix} \Delta\omega_1 & \Delta\delta_1 & \Delta E'_{q1} & \Delta E_{fd1} & \Delta\omega_2 & \Delta\delta_2 & \Delta E'_{q2} & \Delta E_{fd2} & \Delta\omega_3 & \Delta\delta_3 & \Delta E'_{q3} & \Delta E_{fd3} \end{bmatrix} \tag{2}$$

$$\Delta u = \begin{bmatrix} \Delta u_1 & \Delta u_2 & \Delta u_3 \end{bmatrix} \tag{3}$$

$$p(t) = \begin{bmatrix} \Delta T_{m1} & \Delta T_{m2} & \Delta T_{m3} \end{bmatrix} \tag{4}$$

$$\Delta\omega_i = \frac{[\Delta T_{mi} - \Delta T_{ei}]}{M_i s + D_i} \tag{5}$$

$$\Delta\delta'_i = 2\Pi f \Delta\omega_i \tag{6}$$

$$\Delta E_{q_i} = \left[\frac{K_{3_{ii}}}{1 + sT'_{d0i}}\right] * \left[-K_{4_{ii}}\Delta\delta_i - K_{4_{ij}}\Delta\delta_j - \frac{1}{K_{3_{ij}}}\Delta E'_{q_j} + \Delta E_{fd}\right] \tag{7}$$

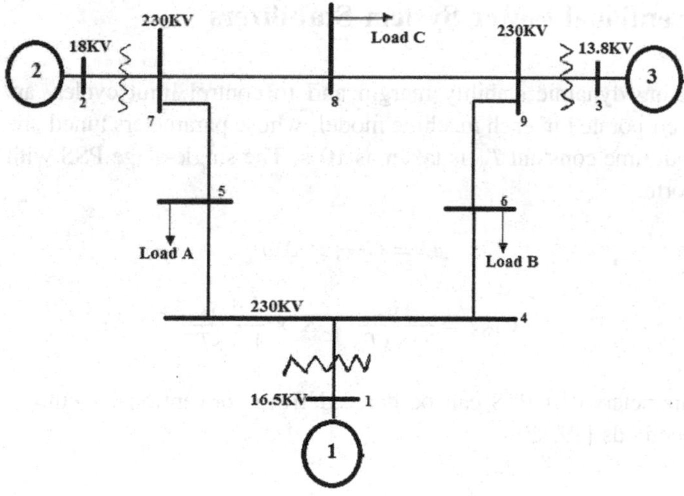

Fig. 1 Single line diagram of 3-machine 9-bus system

$$\Delta E_{fdi} = \left(\frac{-K_{A_i}}{1 + sT_{A_i}}\right)[\Delta v_i - u_i] \tag{8}$$

$$\Delta E'_{fdi} = \frac{-1}{T_{A_i}}\Delta E_{fdi} - \frac{K_{5_{ii}}K_{A_i}\Delta\delta_i}{T_{A_i}} - \frac{K_{6_{ii}}K_{A_i}\Delta E'_{N_i}}{T_{A_i}} - \frac{K_{5_{ij}}K_{A_i}\Delta\delta_j}{T_{A_i}} - \frac{K_{6_{ij}}K_{A_i}\Delta E'_{N_j}}{T_{A_i}} + \frac{K_{A_i}}{T_{A_i}}u_i \tag{9}$$

The above equations can be linearized around an operating point for small deviations and are considered from [3] resulting in 'A' matrix

$$\begin{bmatrix}
-\frac{D_1}{M_1} & \frac{-K_{1_{11}}}{M_1} & \frac{-K_{2_{11}}}{M_1} & 0 & 0 & \frac{-K_{1_{12}}}{M_1} & \frac{-K_{2_{12}}}{M_1} & 0 & 0 & \frac{-K_{1_{13}}}{M_1} & \frac{-K_{2_{13}}}{M_1} & 0 \\
2\Pi f & 0 & 0 & 0 & 0 & 0 & 0 & 0 & 0 & 0 & 0 & 0 \\
0 & \frac{-K_{4_{11}}}{T'_{d01}} & -\frac{1}{T'_{d01}K_{3_{11}}} & \frac{1}{T_{d01}} & 0 & \frac{-K_{4_{12}}}{T'_{d01}} & -\frac{1}{T'_{d01}K_{3_{12}}} & 0 & 0 & \frac{-K_{4_{13}}}{T'_{d01}} & -\frac{1}{T'_{d01}K_{3_{13}}} & 0 \\
0 & \frac{-K_{A_1}}{T_{A_1}}K_{5_{11}} & \frac{-K_{A_1}}{T_{A_1}}K_{6_{11}} & -\frac{1}{T_{A_1}} & 0 & \frac{-K_{A_1}}{T_{A_1}}K_{5_{12}} & \frac{-K_{A_1}}{T_{A1}}K_{6_{12}} & 0 & 0 & \frac{-K_{A_1}}{T_{A_1}}K_{5_{13}} & \frac{-K_{A_1}}{T_{A_1}}K_{6_{13}} & 0 \\
0 & \frac{-K_{1_{21}}}{M_2} & \frac{-K_{2_{21}}}{M_2} & 0 & -\frac{D2}{M2} & -\frac{K_{1_{22}}}{M_2} & \frac{-K_{2_{22}}}{M_2} & 0 & 0 & \frac{-K_{1_{23}}}{M_2} & \frac{-K_{2_{23}}}{M_2} & 0 \\
0 & 0 & 0 & 0 & 2\Pi f & 0 & 0 & 0 & 0 & 0 & 0 & 0 \\
0 & \frac{-K_{4_{21}}}{T'_{d02}} & -\frac{1}{T'_{d02}K_{3_{21}}} & 0 & 0 & \frac{-K_{4_{22}}}{T'_{d02}} & -\frac{1}{T'_{d02}K_{3_{22}}} & \frac{1}{T_{d02}} & 0 & \frac{-K_{4_{23}}}{T'_{d02}} & -\frac{1}{T'_{d02}K_{3_{23}}} & 0 \\
0 & \frac{-K_{A_2}}{T_{A_2}}K_{5_{21}} & \frac{-K_{A_2}}{T_{A_2}}K_{6_{21}} & 0 & 0 & \frac{-K_{A_2}}{T_{A_2}}K_{5_{22}} & \frac{-K_{A_2}}{T_{A_2}}K_{6_{22}} & -\frac{1}{T_{A_2}} & 0 & \frac{-K_{A_2}}{T_{A_2}}K_{5_{23}} & \frac{-K_{A_1}}{T_{A_1}}K_{6_{23}} & 0 \\
0 & \frac{-K_{1_{31}}}{M_3} & \frac{-K_{2_{31}}}{M_3} & 0 & 0 & \frac{-K_{1_{32}}}{M_3} & \frac{-K_{2_{32}}}{M_3} & 0 & \frac{-D_3}{M_3} & \frac{-K_{1_{33}}}{M_3} & \frac{-K_{2_{33}}}{M_3} & 0 \\
0 & 0 & 0 & 0 & 0 & 0 & 0 & 0 & 2\Pi f & 0 & 0 & 0 \\
0 & \frac{-K_{4_{31}}}{T'_{d03}} & \frac{-1}{T'_{d03}K_{3_{31}}} & 0 & 0 & \frac{-K_{4_{32}}}{T'_{d02}} & \frac{-1}{T'_{d03}K_{3_{32}}} & 0 & 0 & \frac{-K_{4_{33}}}{T'_{d03}} & \frac{-1}{T'_{d03}K_{3_{33}}} & \frac{1}{T_{d03}} \\
0 & \frac{-K_{A_3}}{T_{A_3}}K_{5_{31}} & \frac{-K_{A_3}}{T_{A_3}}K_{6_{31}} & 0 & 0 & \frac{-K_{A_3}}{T_{A_3}}K_{5_{32}} & \frac{-K_{A_3}}{T_{A_3}}K_{6_{32}} & 0 & 0 & \frac{-K_{A_3}}{T_{A_3}}K_{5_{33}} & \frac{-K_{A_3}}{T_{A_3}}K_{6_{33}} & \frac{-1}{TA_3}
\end{bmatrix}$$

3 Conventional Power System Stabilizers

For increasing dynamic stability margin and to control limit cycles, an adaptive CPSS is incorporated in each machine model, whose parameters tuned are K, T_1, T_2 and washout time constant T_w is taken as 10 s. The single-stage PSS with washout is in the form:

$$U = G_{PSS} \times \Delta\omega \tag{10}$$

$$G_{PSS} = \frac{sT_w}{1 + sT_w} \times K \times \frac{1 + sT_1}{1 + sT2} \tag{11}$$

The parameters of CPSS can be derived from conventional methods or meta-heuristic methods [19, 20].

4 Problem Formulation

4.1 System Model

Power systems are often modelled by a set of algebraic differential equations.

$$\dot{x} = f(x, y, p) \tag{12}$$

$$0 = g(x, y, p) \tag{13}$$

where

'*x*' is a state vector having the state variables of the power system whose derivatives appear in the system equations such as the machine excitation voltage, speeds and machine angles.
'*y*' is a state vector containing algebraic variables whose derivatives will not appear in power system equations such as network voltages and currents.
'*p*' is a state vector having operating parameters of the system like load demands.
'*f*' and '*g*' are functions whose dimensions are equal to that of '*x*' and '*y*', respectively.

The load parameters are subjected to vary over a wide range. As the '*p*' changes, the operating point changes with respect to Eqs. (12) and (13) by running load flow solutions.

For the given operating point (x_0, y_0, p_0), the dynamic stability of the system is analysed by linearizing it around the given operating point. The eigenvalues are computed at that operating point. If the real part of all eigenvalues is negative, then the power system is dynamically stable. With variation of parameter '*p*', if the real part of the eigenvalue drifts to zero, then the system becomes marginally unstable. This happens in three ways:

(i) A real part of the eigenvalue approaches origin.
(ii) A real part of eigenvalue crossing from $-\infty$ to $+\infty$.
(iii) A pair of complex eigenvalues crossing imaginary axis.

This existence of eigenvalues on imaginary axis refers to Hopf bifurcation. This Hopf bifurcation is associated with a limit cycle. These Hopf bifurcations are evaluated using a widely used soft computing technique genetic algorithm [20].

4.2 Computational Intelligence Algorithm

In the present paper, the application of genetic algorithm is proposed to find the dynamic stability margin of MMPS. The proposed method ensures that the given MMPS exhibits undamped oscillations associated with a pair of imaginary eigenvalues at that operating point. The solution obtained is optimal. The proposed method has been tested on MMPS with CPSS.

Genetic algorithms are robust optimization and search methods, which are evolved from the principles of nature to obtain its search directed for an optimal elucidation. The genetic algorithm uses population of solutions in every repetition as an alternative of a single solution. With the given problem as a unique optimization solution, then all the solutions of populations are expected to converge to one optimum solution. The genetic algorithm may be used for the problems with multiple solutions which will finally converge to a unique solution in its final population space [18].

The genetic algorithm uses the input population as random strings in the form of variables. Every string is measured using an objective function to find the fitness value of the string. The new populations are created from the string using three important GA operators known as reproduction, crossover and mutation. As the time goes on, the operators of GA are emerged enormously, and it gives feasibility of applying theses operators in the problem space.

Reproduction is a selection operator in GA which is used to gather the strings from the above average of the input population. New offsprings are created using crossover operation which is done by using mutual information exchange among the given input strings of the mating pool.

5 Dynamic Stability Margin Analysis Using Genetic Algorithm Approach

5.1 Algorithm

Step 1: Define the power system variables.

Step 2: Identify the variables on which the power system operating point depends (*P*, *Q*).
where

P Real power demand
Q Reactive power demand

Step 3: Generate the initial population.

Step 4: Set the initial individual.

Step 5: Find the final state values by running load flow solutions.

Step 6: Evaluate the initial conditions using results of load flow solutions.
Step 7: Form the reduced *Y* matrix treating the loads as admittances.
Step 8: Find the *K* matrices and the system state matrix '*A*'
Step 9: Find the eigenvalues of the state matrix '*A*' and obtain the

$$\text{Maximum}\{\text{Real}(\lambda_{i,k})\} \in \{\forall \text{Imag}(\lambda_{i,k} \neq 0)\}$$

Step 10: Repeat *Step 5* to *Step 9* for all individuals.
Step 11: Obtain the objective function

$$\underset{p,q}{\text{Maximize}}\ \text{J} = [\text{Maximum}\{\text{Real}(\lambda_{i,k})\} \in \{\forall \text{Imag}(\lambda_{i,k} \neq 0)\}]^{-1}$$

Step 12: Check for the convergence if yes, end; if no, generate new population by selection, recombination and mutation.
Step 13: Obtain the eigenvalues and dynamic response of the system for the obtained *P* and *Q* values.

The above objective function finds *P* and *Q* that results in Max J such that real part of eigenvalue becomes zero.

5.2 Flow Chart

The sequential steps to be followed for finding dynamic stability margin of power system is presented using flow chart given in Fig. 2.

6 Results and Analysis

In the present paper, a new powerful tool is developed to find the dynamic stability margin of multi-machine power system. The dynamic stability margin is evaluated by finding the eigenvalues of the power system '*A*' matrix under different loading conditions. The loading at which the system exhibits imaginary pair of eigenvalues is the dynamic stability margin of the power system. The load at bus 6 is considered as a variable load (*P*6 and *Q*6).

Table 1 shows the eigenvalues of the power system without PSS under different operating conditions. From Table 1, it is observed that the system is stable for loading condition LC1. The system exhibits a pair of imaginary eigenvalues at

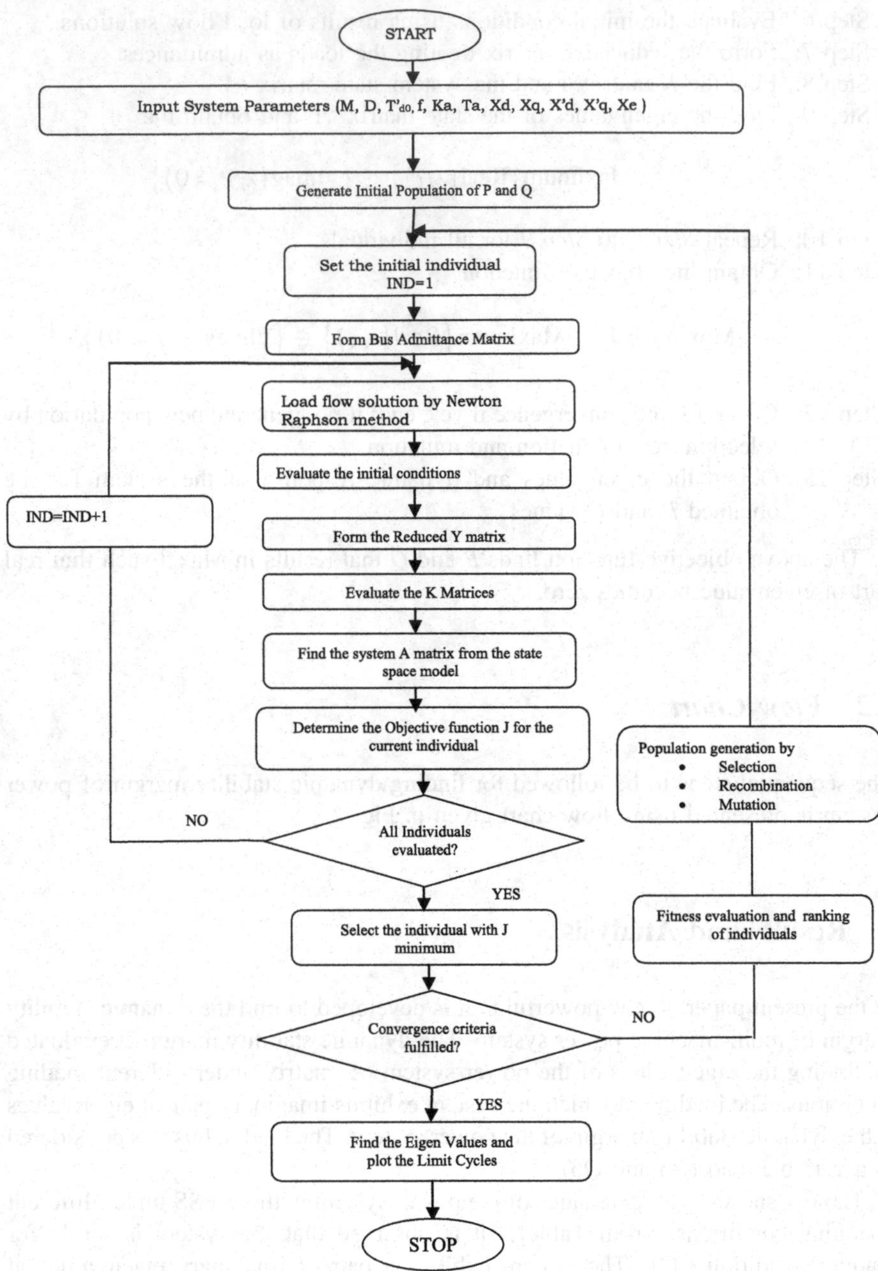

Fig. 2 Flow chart for finding dynamic stability margin by genetic algorithm

Table 1 Eigenvalues of MMPS without controller

Generation	Demand	Eigenvalues	System condition	
$P1 = 0.569$; $Q1 = 0.269$ $P2 = 1.63$; $Q2 = 0.039$ $P3 = 0.85$; $Q3 = -0.135$	$P5 = 1.25$; $Q5 = 0.5$ $P6 = 0.85$; $Q6 = 0.3$ $P7 = 0.9$; $Q7 = 0.3$	$-3.9110 \pm 8.6940i$ $-1.3532 \pm 7.9170i$ $-5.1806 \pm 5.2977i$ $-3.6202 \pm 5.0927i$ $-0.6230 \pm 3.4708i$ -1.3574 -3.7243	Stable	LC1
$P1 = 0.67$; $Q1 = 0.278$ $P2 = 1.63$; $Q2 = 0.041$ $P3 = 0.85$; $Q3 = -0.126$	$P5 = 1.25$; $Q5 = 0.5$ $P6 = 0.949$; $Q6 = 0.306$ $P7 = 0.90$; $Q7 = 0.3$	$-3.4490 \pm 8.6658i$ $-1.0841 \pm 7.5655i$ $-6.8642 \pm 4.7120i$ $\pm 3.1329i$ $-3.4336 \pm 5.0554i$ $-2.4042 \pm 2.7304i$	LS	LC2
$P1 = 0.956$; $Q1 = 0.272$ $P2 = 1.63$; $Q2 = 0.046$ $P3 = 0.85$; $Q3 = -0.116$	$P5 = 1.25$; $Q5 = 0.5$ $P6 = 1.019$; $Q6 = 0.308$ $P7 = 1.071$; $Q7 = 0.3$	$6.2475 \pm 10.9243i$ $-12.2386 \pm 8.3039i$ $-3.1574 \pm 10.8661i$ $-2.8957 \pm 5.9723i$ -0.5700 -2.8730 $-3.4709 \pm 2.3348ii$	Unstable	LC3

loading of LC2 and is exhibiting limit cycles at this operating point. The loading is further increased to LC3, a pair of system eigenvalues is imaginary with positive real part, and the system is unstable.

The rotor angle deviation of machine1 for the above operating conditions is shown in Fig. 3a. The dynamic stability margin is evaluated using genetic algorithm. The system exhibits undamped oscillations at $P6 = 0.949$ and $Q6 = 0.306$. The rotor angle oscillation between machines 1 and 2 is shown in Fig. 3b.

The variation of rotor angle deviation of machine1 versus rotor angle deviation of machine2 is shown Fig. 4a. The fitness value variation is shown in Fig. 4b.

6.1 Application of Conventional Power System Stabilizers to Increase the Dynamic Stability Margin

The dynamic stability of power system is improved by incorporating CPSS at each of the three machines. Table 2 shows the eigenvalues of the power system with CPSS under different operating conditions. From Table 2, it is observed that the

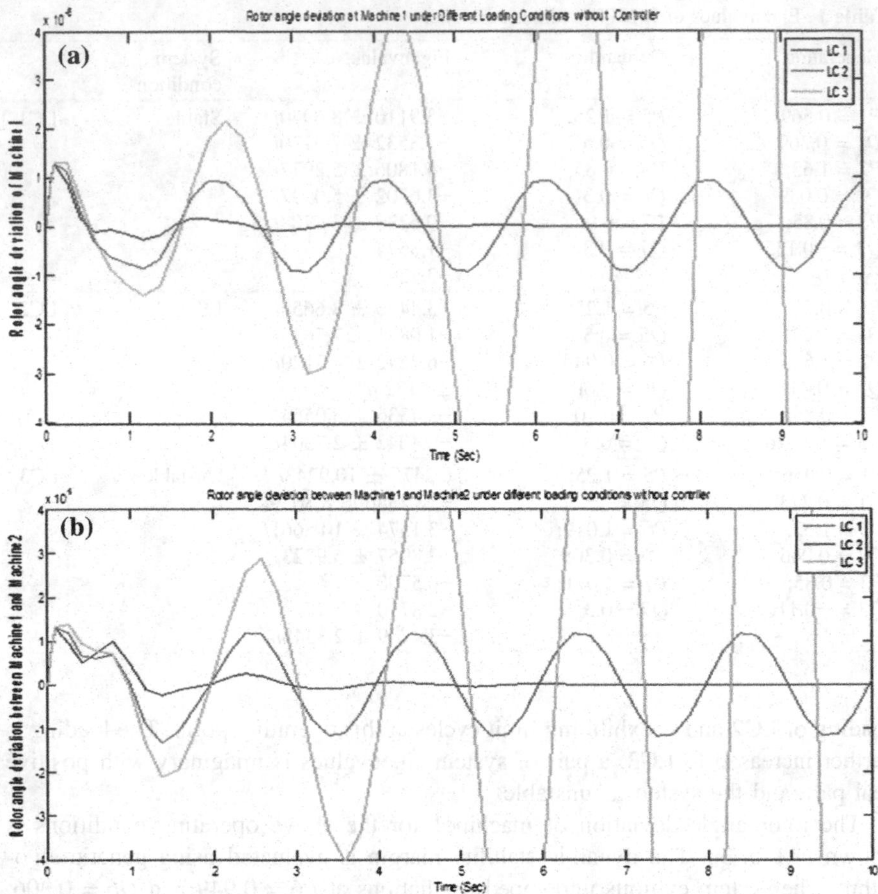

Fig. 3 **a** Rotor angle deviation at machine1 under different loading conditions without controller and **b** rotor angle deviations between machines 1 and 2 (δ_{12}) under different loading conditions without controller

system is stable for loading condition LC4. The system exhibits a pair of imaginary eigenvalues at loading of LC5 and is exhibiting limit cycles at this operating point. The loading is further increased to LC6, a pair of system eigenvalues is imaginary with positive real part, and the system is unstable.

The rotor angle deviation of machine1 for the above operating conditions is shown in Fig. 5a. The dynamic stability margin is again evaluated by using genetic algorithm for the modified system matrix including CPSS. The system exhibits

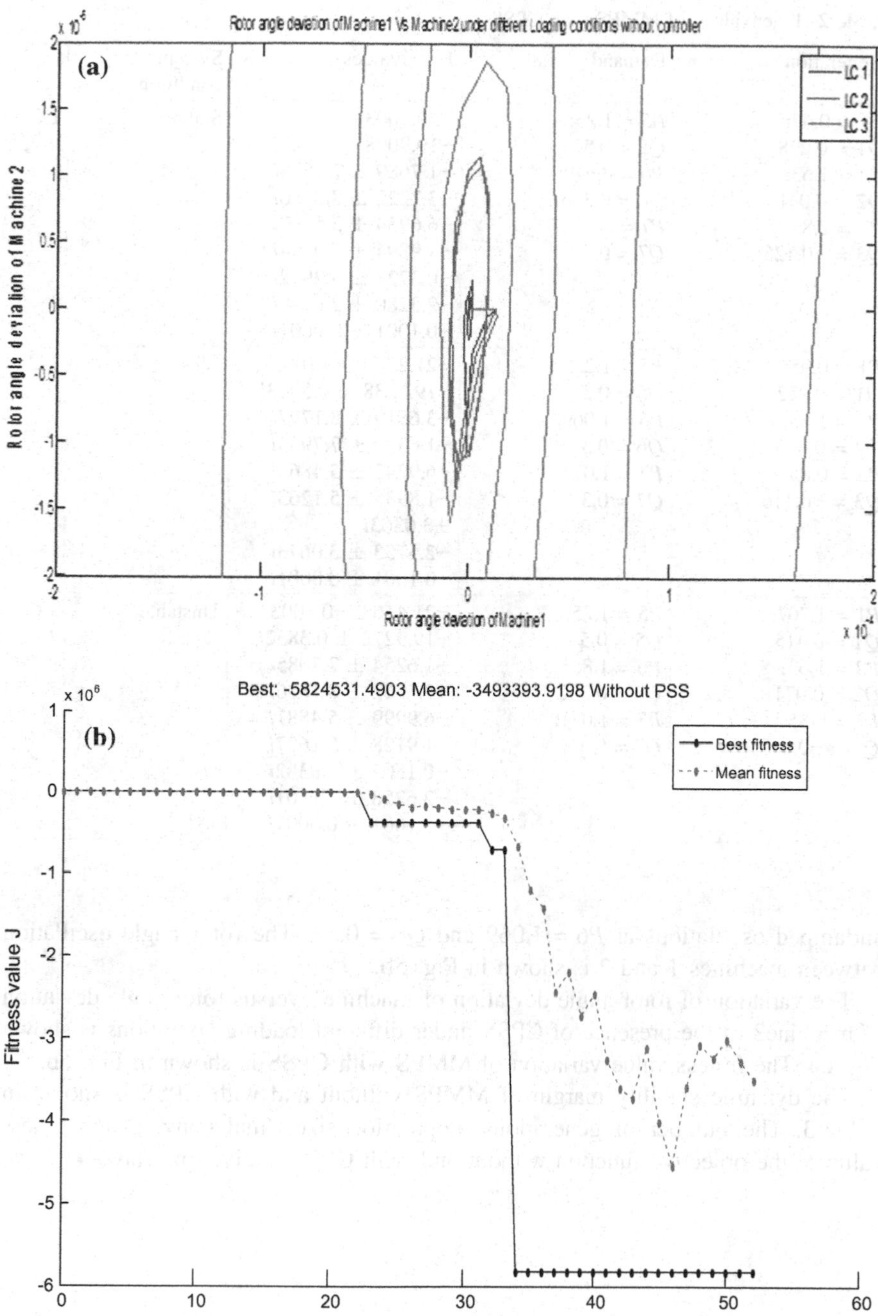

Fig. 4 **a** Rotor angle deviation of machine1 (δ_1) versus machine2 (δ_2) under different loading conditions without controller and **b** fitness value variations for MMPS without controller

Table 2 Eigenvalues of MMPS with CPSS

Generation	Demand	Eigenvalues	System condition	
$P1 = 0.67$; $Q1 = 0.278$ $P2 = 1.63$; $Q2 = 0.041$ $P3 = 0.85$; $Q3 = -0.126$	$P5 = 1.25$; $Q5 = 0.5$ $P6 = 0.949$; $Q6 = 0.306$ $P7 = 0.90$; $Q7 = 0.3$	−20.9003, −0.1002 $-19.9098 \pm 0.4271i$ $-1.7689 \pm 7.9595i$ $-3.7222 \pm 8.2386i$ $-6.6934 \pm 5.5540i$ $-1.9894 \pm 5.0754i$ $-0.3723 \pm 3.2922i$ $-2.3280 \pm 2.6138i$ $-0.1001 \pm 0.0001i$	Stable	LC 4
$P1 = 0.956$; $Q1 = 0.272$ $P2 = 1.63$; $Q2 = 0.046$ $P3 = 0.85$; $Q3 = -0.116$	$P5 = 1.25$; $Q5 = 0.5$ $P6 = 1.069$; $Q6 = 0.32$ $P7 = 1.071$; $Q7 = 0.3$	−21.2750, −0.1003 $-19.9238 \pm 0.3923i$ $-3.6519 \pm 8.1797i$ $-1.6350 \pm 7.7953i$ $-6.9247 \pm 5.4865i$ $-1.8935 \pm 5.1263i$ $\pm 3.0363i$ $-2.5753 \pm 3.0643i$ $-0.1001 \pm 0.0001i$	LS	LC 5
$P1 = 1.707$; $Q1 = 0.415$ $P2 = 1.63$; $Q2 = 0.074$ $P3 = 0.85$; $Q3 = -0.049$	$P5 = 1.25$; $Q5 = 0.5$ $P6 = 1.8$; $Q6 = 0.332$ $P7 = 1.071$; $Q7 = 0.3$	−21.4163, −0.1003 $-19.9272 \pm 0.3852i$ $-1.6254 \pm 7.7085i$ $-3.5962 \pm 8.1836i$ $-6.9999 \pm 5.4881i$ $-1.9128 \pm 5.1627i$ $-0.1507 \pm 3.0382i$ $-2.6258 \pm 3.1601i$ $-0.1001 \pm 0.0001i$	Unstable	LC 6

undamped oscillations at $P6 = 1.069$ and $Q6 = 0.32$. The rotor angle oscillation between machines 1 and 2 is shown in Fig. 5b.

The variation of rotor angle deviation of machine1 versus rotor angle deviation of machine2 in the presence of CPSS under different loading conditions is shown Fig. 6a. The fitness value variation of MMPS with CPSS is shown in Fig. 6b.

The dynamic stability margin of MMPS without and with CPSS is shown in Table 3. The number of generations, population size, final convergence fitness value of the objective function without and with CPSS is given in Table 4.

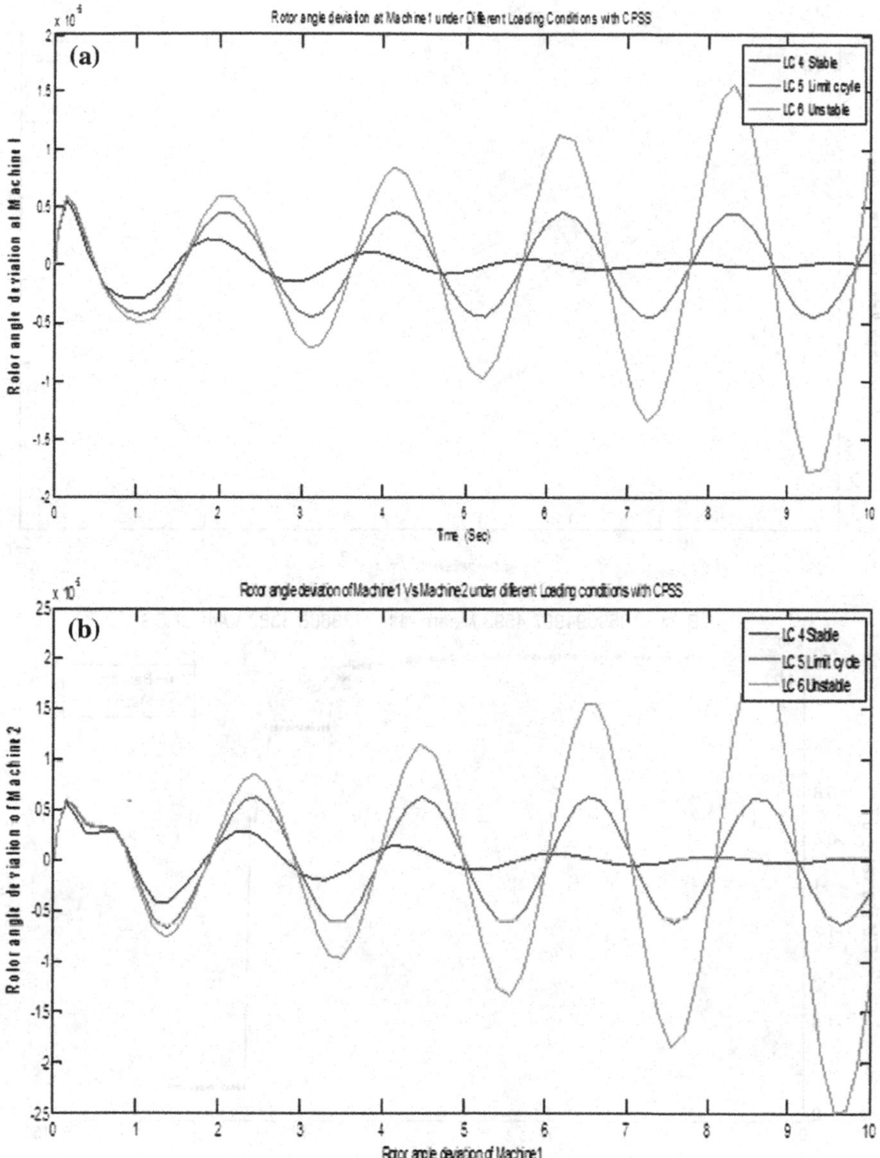

Fig. 5 **a** Rotor angle deviation at machine1 under different loading conditions with CPSS and **b** rotor angle deviations between machines 1 and 2 (δ_{12}) under different loading conditions with CPSS

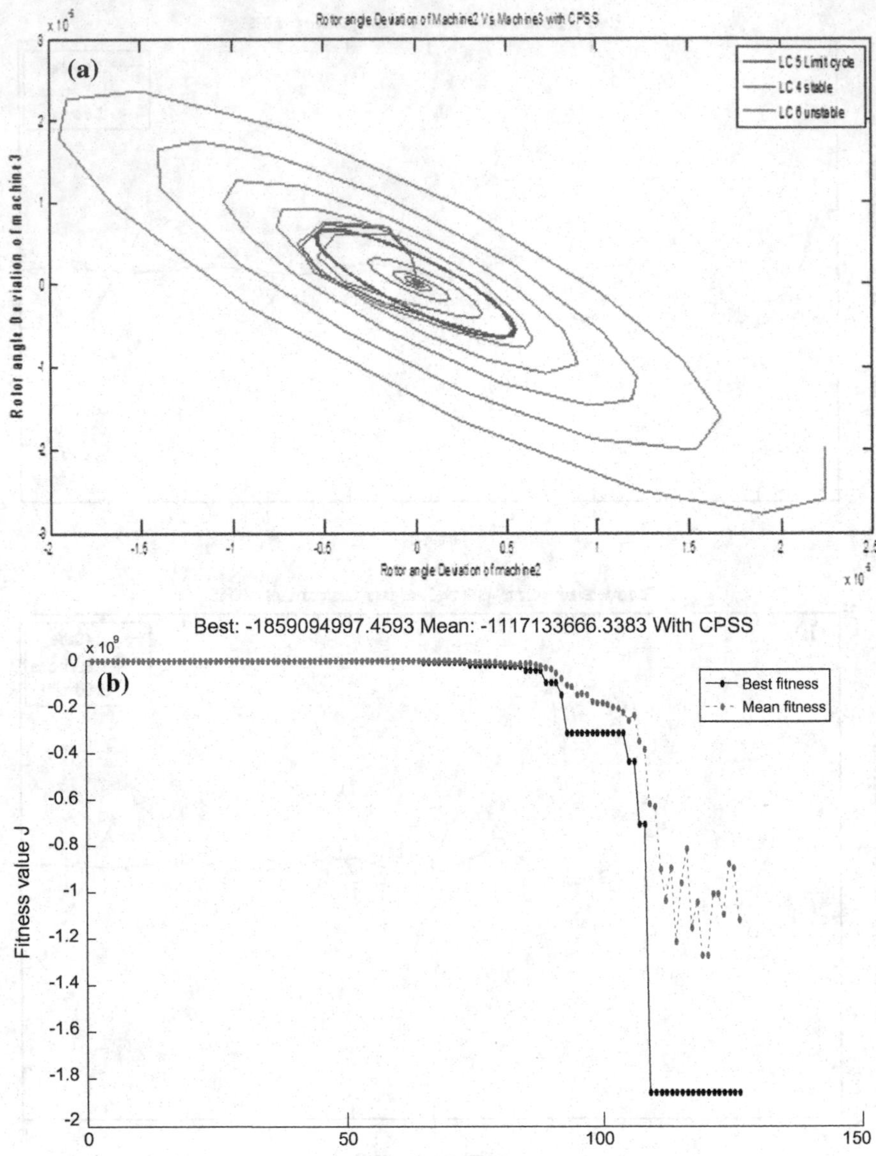

Fig. 6 **a** Rotor angle deviation of machine1 (δ_1) versus machine2 (δ_2) under different loading conditions with CPSS and **b** fitness value variations for MMPS with CPSS

Table 3 Comparison of dynamic stability margin of MMPS without and with CPSS

Power system model	Loading condition where Hopf bifurcation takes place	Eigenvalues associated with Hopf bifurcations
MMPS without controller	$P6 = 0.949$; $Q6 = 0.306$	$\pm 3.1329i$
MMPS with CPSS	$P6 = 1.069$; $Q6 = 0.32$	$\pm 3.0363i$

Table 4 Genetic algorithm results for MMPS without and with CPSS

Power system model	Number of generations	Objective function 'J' value
MMPS without controller	54	−5623526.4202
MMPS with CPSS	128	−1126183686.3263

7 Conclusion

A new computational method using genetic algorithm is presented in this paper to find the dynamic stability margin. The method is effectively implemented for a 3-machine 9-bus system. Using the proposed method, the loading condition at which the MMPS is exhibiting undamped oscillations is evaluated. The system performance is obtained for the derived load values and is observed that it is resulting in undamped oscillations at that value. The efficacy of the proposed technique is also tested for MMPS in the presence of CPSS for the modified system matrix and is observed that the method is working effectively in finding dynamic stability margin.

References

1. Kundur, P.S. 1994. *Power System Stability and Control*. New York: McGraw Hill Publications.
2. Awadallah, M., and H. Soliman. 2009. A Neuro-fuzzy Adaptive Power System Stabilizer Using Genetic Algorithms. *Electrical Power Components and Systems* 37 (2): 158–173.
3. Sauer, P.W., and M.A. Pai. 1997. *Power System Dynamics and Stability*. Prentice Hall.
4. Padiyar, K.R. 2008. *Power System Dynamics; Stability and Control*. B.S. Publications.
5. Soliman, H.M., and Saqr, M.M.F. 1988. Wide-Range Power System Pole Placer. *IEE Proc. Pt. C* 135 (3): 195–201.
6. Peres, Edimar, José De Oliveira, João Alberto Passos Filho, and Ivo Chaves Da Silva Junio. 2015. Coordinated Tuning of Power System Stabilizers Using Bio-inspired Algorithms. *Electrical Power and Energy Systems* 64: 419–428.
7. Saoudi, K., and M.N. Harmas. 2014. Enhanced Design of an Indirect Adaptive Fuzzy Sliding Mode Power System Stabilizer for Multi-machine Power Systems. *Electrical Power and Energy Systems* 54: 425–431.
8. Ross, Timothy J. 2004. *Fuzzy Logic with Engineering Applications*, 2nd ed. Willey.

9. Sambariya, D.K., and R. Prasad. 2014. Robust Tuning of Power System Stabilizer for Small Signal Stability Enhancement Using Metaheuristic Bat Algorithm. *Electrical Power and Energy Systems* 61: 229–238.
10. Sudha, K.R. 2004. *Design of Fuzzy Logic Power System Stabilizers: A Systematic Approach*. Dissertation, Andhra University.
11. Kovacic, Zdenko, and Stejpan Bogelan. 2006. *Fuzzy Controller Design: Theory and Applications*. CRC Press.
12. Howell, Frederic, and Vaithianathan Venkatasubramanian. 1999. Transient Stability Assessment with Unstable Limit Cycle Approximation. *IEEE Transactions on Power Systems* 14 (2): 667–677.
13. Mithulananthan, N., A.C. Claudio, and John Reeve. 2000. *Indices to Detect Hopf Bifurcations in Power Systems*, NAPS-2000, Waterloo, 1–7 Oct 2000.
14. IEEE Recommended Practice for Excitation System Models for Power System stability Studies, March 2005.
15. Demello, Francisco P., and Charles Concordia. 1969. Concepts of Synchronous Machine Stability as Affected by Excitation Control. *IEEE Transactions on Power Apparatus and Systems* 88 (4): 316–329.
16. Reddy, P., and I. Hiskens. 2007. *Limit-Induced Stable Limit Cycles in Power Systems*. In: Proceedings of St. Petersburg Power Tech, St. Petersburg, Russia, Paper No. 603.
17. Duan, Xianzhong, Jinyu Wen, and Cheng Shijie. 2009. Bifurcation Analysis for an SMIB Power System with Series Capacitor Compensation Associated with Sub-synchronous Resonance. *Science in China Series E: Technological Sciences* 52 (2): 436–441.
18. Hassan, L.H., et al. 2014. Optimization of power system stabilizers using participation factor and genetic algorithm. *Electrical Power and Energy Systems* 55: 668–679.
19. Vakula, V.S., and K.R. Sudha. 2012. Design of Differential Evolution based Fuzzy Logic Power System Stabilizer with Minimum Rule Base. *IET Generation, Transmission & Distribution* 6 (2): 121–132.
20. Andreiou, Adrian. 2002. *Genetic Algorithm Based Design of Power System Stabilizers*. Dissertation, Chalmers University of Technology.

Short-Term Estimation of Transmission Reliability Margin Using Artificial Neural Networks

V. Khatavkar, D. Swathi, H. Mayadeo and A. Dharme

Abstract This paper proposes a novel approach for estimation of transmission reliability margin (TRM) by using artificial neural networks (ANN). The laws of deregulated electricity industry have made mandatory to publish the hourly available transfer capability (ATC) values for planning, reliability, and secure system operation. Hence the accurate determination of ATC values is of utmost importance, but is challenging as well. ATC comprises of marginal values such as TRM and capacity benefit margin (CBM) with the existing commitments (EC). Since TRM itself is composed of factors like network parameters, load changes, outages; it not only helps in accurate determination of ATC, but also plays a vital role in system congestion management. Therefore, this paper emphasizes on a method of determination of TRM. The tool used is ANN with back propagation algorithm (BPA) radial basis function (RBF). This work is based on the data of Alberta electric system operator, Canada.

Keywords Artificial neural networks · Deregulated power system operation
Transmission reliability margin · Congestion management

V. Khatavkar
Bharati Vidyapeeth Deemed University College of Engineering, Pune, India
e-mail: vrushali_4@rediffmail.com

V. Khatavkar · D. Swathi
PES Modern College of Engineering, Pune, India
e-mail: renuka.swathipr@gmail.com

H. Mayadeo (✉)
Black & Veatch (P.) Ltd., Pune, India
e-mail: mheramb@gmail.com

A. Dharme
Govt. College of Engineering, Pune, India
e-mail: aad.elec@coep.ac.in

M.S. Reddy et al. (eds.), *International Proceedings on Advances in Soft Computing, Intelligent Systems and Applications*, Advances in Intelligent Systems and Computing 628, https://doi.org/10.1007/978-981-10-5272-9_2

1 Introduction

The deregulated and restructured environment of the power system is enabling power transactions from various generating stations to diverse customers. This facility helps the power producers to generate more energy from the cheaper source for greater profit margins. It also helps customers for choosing more reliable and cheaper power. Although the power producers and customers are different, both of them utilize the same transmission network. Because of all the power transactions from point of production to point of utilization take place simultaneously cause contingency and congestion on transmission network. Power wheeling transactions within the same zone is slightly easier for forecasting and planning as compared to interzone transactions. For making interzone transactions safe and secure the timely posting of cross-border capacities become indispensable. Cross-border capacities of an interconnected system can be labeled as available transfer capabilities in interconnected system.

Available transfer capability (ATC) is made available to all utilities through open access information system (OASIS), so that further committed loads can be interfaced ensuring the system security. This provides a signal to the market participants about the capability of transmission network that is available for the period under consideration.

ATC is defined additional amount of inter-area power that can be transferred without violating the system security [1, 2]. ATC inherently contains some marginal values in it. It is mathematically defined as, ATC=TTC-TRM-CBM-EC [3, 4]. The TRM accounts for uncertainty in transmission, and CBM accounts for uncertainty in generation. So, appropriate quantification of TRM is essential for accurate determination of ATC.

TRM can be defined as the amount of transfer capability that is necessary for secure operation of interconnected system considering uncertainties at a particular operating condition. Network parameters, load changes, planned and unplanned outages, fluctuations in the prices are the some of the contributing factors of the uncertainty. Gamut of methods is available for calculation and estimation of TRM. J. Zhang discusses a formula which can determine the TRM based on the sensitivities of transfer capabilities and probabilistic uncertainties [5]. The method proposed by the M.M. Othman for determining the TRM with large uncertainty in transfer capability uses the parametric bootstrap technique [6]. In [7], a new concept is devised to estimate the TRM on border values. In this paper, initially the relevant uncertainties are considered and their respective independent time series is derived. Later, each uncertainty is statistically analyzed and their probability density functions are derived. Finally, convoluting all these PDFs the respective TRM values are obtained. All these methods provide accurate and efficient results. But as the power transactions have to be updated periodically and frequently, they fail in fast computation and repetitive analysis. These methods consume excessive computational time and use complex energy management system (EMS) computers.

These can be avoided by using ANN. The complex structure of the problem, inherent uncertainty, and number of factors affecting the system make it difficult to analyze using stochastic methods. This makes ANN more suitable for the problem due to its inherent ability of pattern recognition. This work has been carried out using back propagation algorithm (BPA) and radial basis function (RBF) and validated accordingly.

The paper is organized as follows. Section 1 introduces the paradigm of the problem. Section 2 speaks about problem formulation. Section 3 provides brief review of artificial neural network and related algorithms. Section 4 has the test case and results. Section 5 is about the analysis of results obtained. Ultimately, Sect. 6 concludes the paper with concluding remarks on the work done and possible future scope.

2 Problem Formulation

2.1 *Problem Description*

Due to simultaneous use of transmission network by all participants, overloading and congestion on the transmission network occur. This causes the violation of transmission limits such as thermal, voltage, and stability limits which in turn cause the hindrance in the system security and reliability. Therefore, it is necessary to determine the available transfer capability accurately so that the system doesn't lose its reliability while operating on the multilateral transactions. As TRM accounts for considerable part in ATC, it is essential to determine it dynamically [8]. TRM is the amount transmission margin that is set aside for the uncertain conditions that transmission network has to deal with. When the uncertain condition such as line outage occurs this margin is added to that relevant ATC value to prevent jeopardizing the system security. Generally all independent system operators (ISO) allocate certain percentage of margin, i.e., 2–5% of total transfer capability as per NERC regulations [9]. In this paper, an attempt has been made to estimate the TRM at continuous time intervals using the real-time data. And this estimated data is verified against the real-time data.

2.2 *Methodology*

The TRM in real-time application is estimated using the artificial neural networks by back propagation algorithm. Considering the BC-Alberta system bilateral transactions (i.e., historical data), different inputs (uncertainties) are mapped to output to form training pattern. Then this pattern is tested using ANN functions and finally validated using real-time data. The uncertainties considered are planned and unplanned outages and load variations.

3 Artificial Neural Networks

ANN has evolved as a great promising statistical tool in solving power engineering problems. It has made it possible to obtain solutions in complex environments in the power systems area where the speed of system security and simultaneously, accurate analysis are ultimate objectives. ANNs have the capability of learning from the large input data, forming a pattern of relationship between the input and the target outputs. The two different multilayer perceptron models applied here are BPA and RBF

3.1 *Back Propagation Algorithm (BPA)*

The back propagation algorithm is the best method for the feed forward networks. There are two passes allowed in BPA, a forward pass and backward pass. The forward pass evaluates the input layer and hidden layer results and backward pass compares the target outputs and estimated outputs.

The activation function used in BPA can be given as,

$$f(x) = (1 + e^{-x})^{-1}. \tag{1}$$

Here, the input signal is given by the X_i (i = 1, 2, 3… n), which is multiplied by the weight W_{ij}. These are operated using the activation function $f(x)$ to reproduce the weights of the hidden layer (b_j). Where

$$b_j = f\left(\sum_{i=1}^{n} X_i W_{ij}\right). \tag{2}$$

These are forward pass operations. Similarly backward pass operation is also performed.

Here, a training procedure is followed such that the differences between the desired values and the outputs of hidden layers are minimized. Certain part of input data (70%) is reserved for training the network, and the remaining part of it (30%) is used for validation and testing.

The schematic diagram for back propagation algorithm is shown in Fig. 1. This network consists of five inputs neurons, one target output neuron, and hidden layer of 10 intermediate neurons. The five typical inputs taken into consideration are line outages, British Columbia Configuration (BC-conf), converter tests, ISO limits, and demand. Line outages, BC-conf comes under the unplanned outages and converter tests, ISO limits comes into category of forced outages [10]. The components of the neural network are as follows,

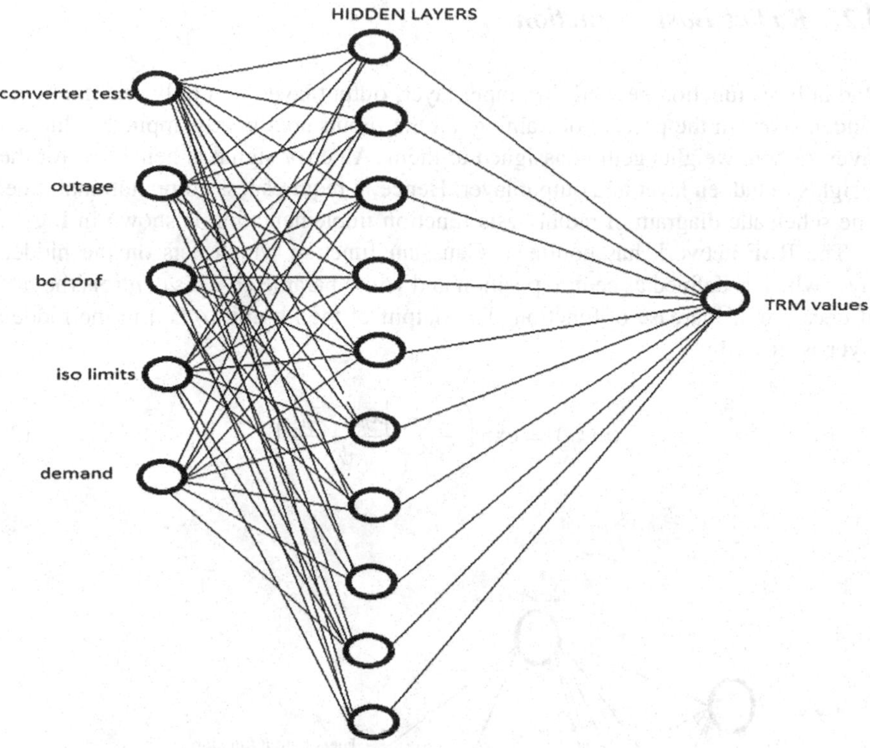

Fig. 1 Back propagation algorithm architecture

1. Input Vector:

Inputs	Limits applied	Limited unapplied
ISO limits	1	0
BC-CONF	1	0
Converter test	1	0
Outages	1	0
Demand conditions	The actual demand at every time interval is considered above the base case transfer	

2. Output Vector: The output vector is TRM for stipulated time intervals between sending area and receiving area.
3. Network architecture: As mentioned earlier, 10 hidden layers are used. The sample size of the data is 750. The activation function is used here is Levenberg-Marquardt optimization.

3.2 Radial Basis Function

Radial basis function network has input layer, output layer, and only one nonlinear hidden layer. In the process of training, all the inputs are directly applied to hidden layer without weights getting assigned to them. An error signal is generated for the weights of hidden layer and output layer. Hence, it requires less computational time. The schematic diagram of radial basis function implementation is shown in Fig. 2.

The RBF network has nonlinear Gaussian function which acts on the hidden layer which is defined as central position and width parameter. This controls the rate of decrease or increase of function. The output of the *i*th unit a_i (x_p) in the hidden layer is given by,

$$a_i(x_p) = \exp\left(-\sum_{j=1}^{r}\frac{|x_{jp} - \bar{x}_{ji}|}{\psi_i^2}\right), \tag{3}$$

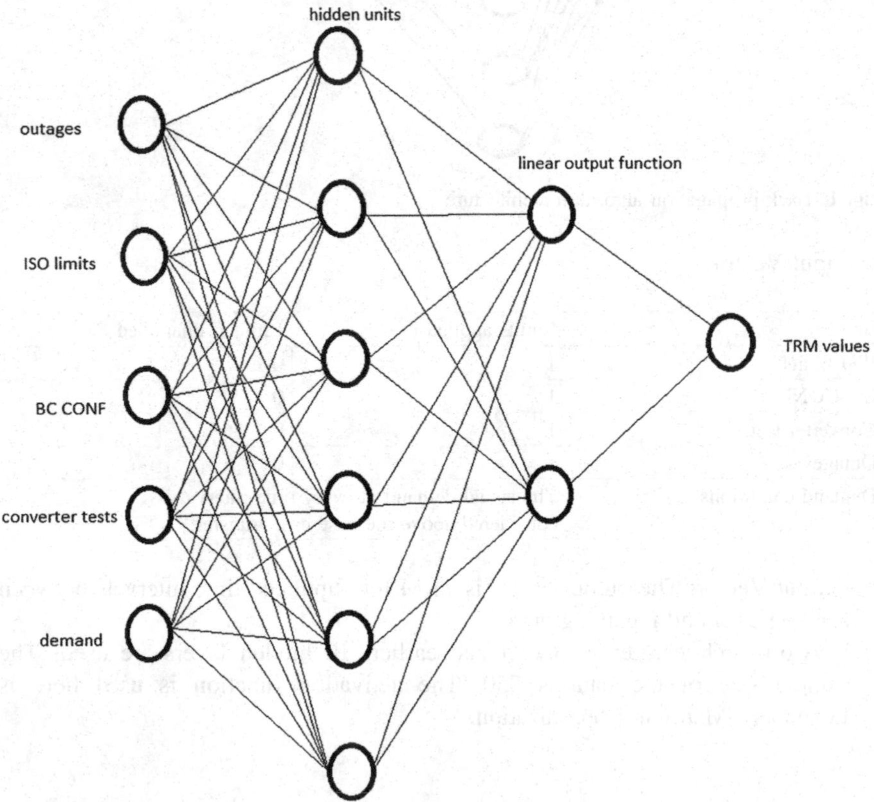

Fig. 2 Radial basis function architecture

where X_{ij} is center of the ith RBF unit of input variable j, ψ_i is the width of ith RBF unit, X_{jp} is jth variable of input pattern p, and r is dimension of input vector. The output value O_{qp} of the qth output node for pth is given by,

$$O_{qp} = \sum_{i=1}^{H} W_{qi} a_i(X_p) + W_{qo}, \tag{4}$$

where W_{qi} is the weight between ith RBF unit and qth output node, W_{qo} is the biasing term, and H is the number of hidden layer (RBF) nodes [8].

The inputs and their values remain same as considered in BPA. The only difference lies in the activation function.

4 Test Case and Results

For testing purpose, historical data of Alberta electric system Operator (AESO), Canada, is taken. The system chosen represents a modern power consuming society. Further description about the system is given in subsequent sections.

4.1 Test Case

Alberta electric system is taken as the test case. It is a commercial hub and represents peculiar load characteristics of a modern commercial load center. For purpose of testing, the proposed method is performed on AESO import and export TTC values. The system is divided into two areas. Dominant power flow is from British Columbia to Alberta. TRM is estimated between these two areas (Fig. 3).

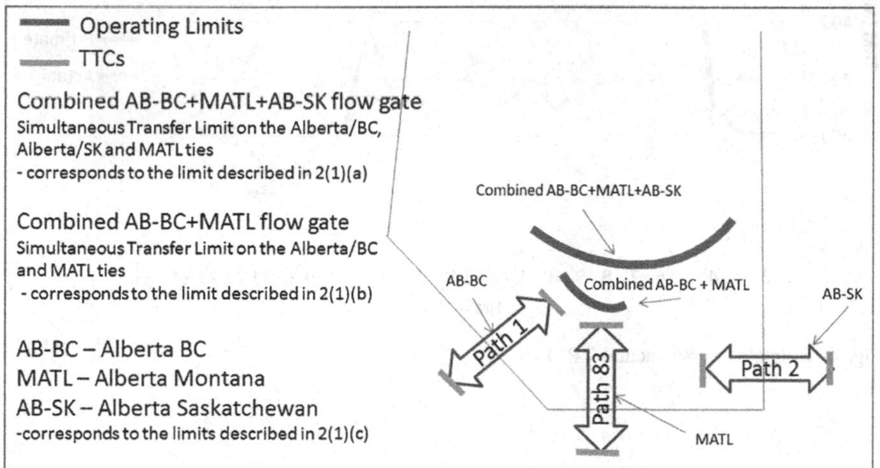

Fig. 3 Test system topology [9]

Testing patterns of neural network are representative of BC and Alberta path transaction. The planned, unplanned, and system parameters are taken into consideration of an independent system operator. Hence, individual transmission line status cannot be updated. So the uncertainties considered here are for the overall system transfer capabilities [10].

4.2 Results

Estimated TRM for every hour is obtained in 24 h window. Here, the TRM comprises of both the marginal values, i.e., CBM and TRM. Since CBM remains constant throughout, the TRM is the only variable; therefore, the whole marginal value is considered as TRM. Figures 4 and 5 give comparison of estimated values and real-time values of TRM for both the algorithms used.

To evaluate the performance of both the algorithms, we have considered relative error as the parameter of comparison. The performance of both the algorithms is given in Table 1.

The comparison of both the algorithms from computational efficiency point of view is given in Table 2.

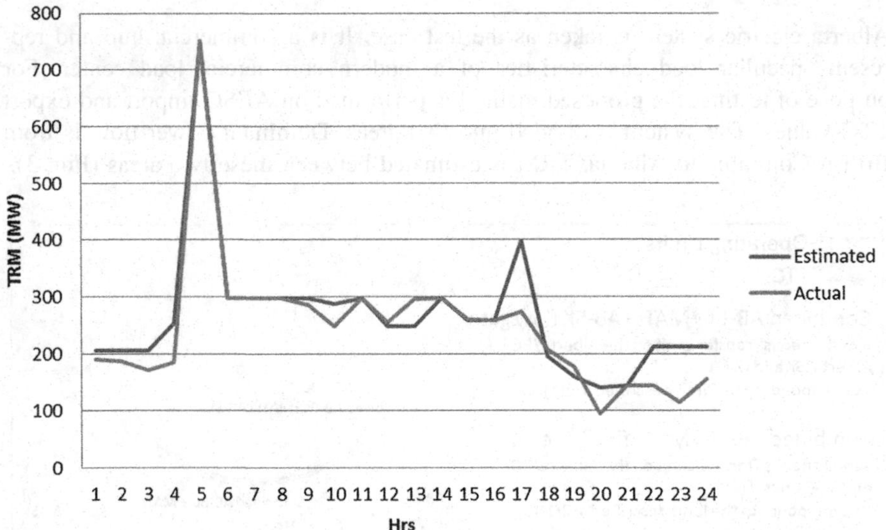

Fig. 4 Estimated versus actual TRM (BPA)

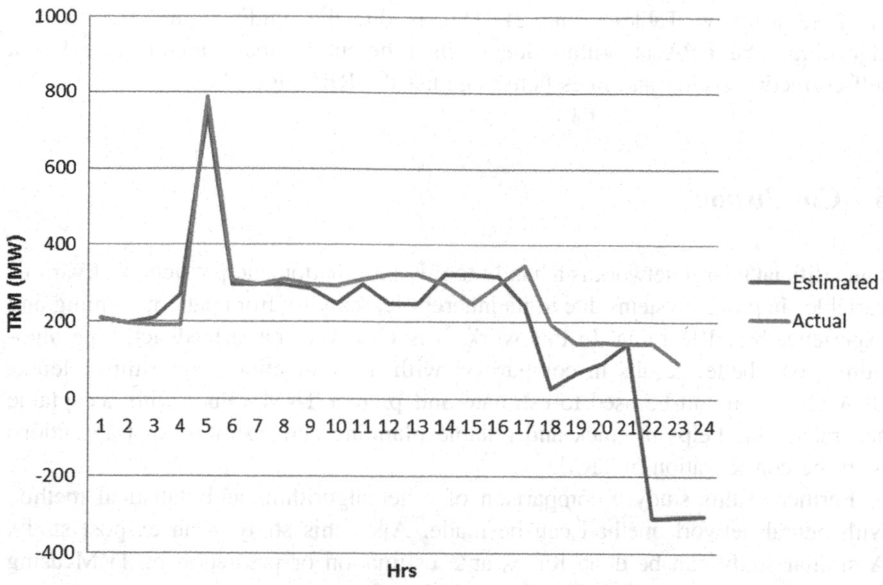

Fig. 5 Estimated versus actual TRM (RBF)

Table 1 Performance comparison

Statistical parameter of ϵ	BPA	RBF
Mean	−0.1337	0.5593
Standard deviation σ	0.2360	1.2784
Variance ν	0.0557	1.6344

Table 2 Comparison of computational efficiency

Parameter	BPA	RBF
EPOCH	7 iterations-1000	7 iterations-1000
Performance	193–32.098	256–32.098
Gradient	0.98e−05	0.98e−05
MU	100e+11	100e+10
Validation check	6	6

5 Discussion

From Figs. 4 and 5, it is evident that the vagaries in the difference between estimated TRM and actual TRM are prominent toward the end of the 24 h window. While using RBF algorithm, the estimated values tend to stray away from the actual values after 6th hour mark, but similar phenomenon is observed for BPA from 16th hour. The comparison of performance of the two algorithms shows that BPA is front runner in giving accurate solution to the problem under consideration. The RBF algorithm has much higher relative error especially in the trailing end of the time window (Tables 1 and 2). This is due the unidirectional flow of the algorithm. The BPA algorithm due to its inherent feedback architecture has a self-correcting ability and fares better against the RBF algorithm.

6 Conclusion

The artificial neural network is a handy tool for estimation and prediction of various variablcs in power systems due to its inherent learnability from pattern forming and experience-based learning. In this work, it is observed that a feedback type algorithm gives better results in comparison with a unidirectional algorithm. Hence, BPA algorithm can be used to estimate and predict TRM values with acceptable accuracy. This helps in quick and reliable planning and operation of transactions with the consideration of TRM.

Further to this study a comparison of other algorithms and statistical method with neural network method can be made. Also, this study is an ex-post study. A similar study can be done for ex-ante estimation or prediction of TRM using both, the neural network tool and statistical tool.

Acknowledgements The authors would like to thank Alberta Electric System Operator (AESO), Canada, for their invaluable support in providing the data for research purpose. Without their strong support this work could not have been completed. The authors would like to thank 'Bharati Vidyapeeth Deemed University, Pune', for providing research opportunities and facilities.

References

1. Available Transfer Capability Definitions and Determination. 1996. Internet: http://www.nerc.com.
2. Othman, M.M., A. Mohamed, and A. Hussain. 2004. A Neural Network Based ATC Assessment Incorporating Novel Feature Selection and Extraction Methods. *International Journal Electric Power Components and Systems* 32 (11): 1121–1136.
3. Sauer, P.W. 1997. Technical Challenges of Computing Available Transfer capability (ATC) in Electric Power System. In *Proceedings of 30th Annual Hawaii International Conference on Systems Sciences*.
4. Othman, M.M., A. Mohamed, and A. Hussain. *Available Transfer Capability Assessment Using Evolutionary Programming Based Capacity*.
5. Zhang, J., I. Dobson, and F.L. Alvarado. 2004. Quantifying Transmission Reliability Margin. *International Journal of Electrical Power & Energy Systems* 26 (9): 697–702.
6. Othman, M.M., A. Mohamed, and A. Hussain. 2008. Determination of Transmission Reliability Margin Using Parametric Bootstrap Technique. *IEEE Transactions on Power Systems* 23(4).
7. Bammert, J. 2015. Calculating Transmission Reliability Margins Based on Statistical Data. In *International Journal of European Energy Market (EMM), 2015. 12th International Conference*, 1–5.
8. Vinod Kumar, D.M., Narayan Reddy, and Ch. Venkaiah. Available Transfer Capability (ATC) Determination Using Intelligent Techniques.

9. Operating Policies and System Procedure–304. 2013. Internet: http://www.aeso.ca/rulesprocedures/9073.html.
10. Available Transfer Capability and Transfer Path Management. 2013. Internet: http://www.aeso.ca/downloads/2013-03-19_ID_2011-001R.

How Expert is EXPERT for Fuzzy Logic-Based System!

Kalyani Bhole, Sudhir Agashe and Jagannath Wadgaonkar

Abstract Anesthesia, an utmost important activity in operation theater, solely depends upon anesthesiologist, an expert. In the case of absence of expertise, drug dosing may go under-dose or overdose. To overcome this problem, an expert-based system can be designed to guide newcomers in the field of anesthesia. This structure is called as decision support system. As this system is dependent on experts' knowledge base, its performance depends on the expert's expertise which can be validated by comparison with other expert's knowledge base and finding maximum correlation among them. This paper demonstrates the application of prehistoric Gower's coefficient to validate the expert's expertise for fuzzy logic-based experts' system. Database is collected from ten experts. For the 80% level of confidence, eight experts are classified into one group leaving two aside. Database of these eight experts is used for the design of decision support system. A set of 270 results noted from decision support system is validated from the expert. Out of 270, expert declines 3 decisions accepting 98.88% result.

Keywords Intravenous anesthesia · Expert-based system · Fuzzy logic Gower's coefficient · Decision support system

1 Introduction

In the framework of complex unclear processes that can be controlled by trained human operators, modeling an uncertainty is a great challenge. Within the processes, the exact level of accuracy can be expressed by the characterization and

K. Bhole (✉) · S. Agashe · J. Wadgaonkar
Department of Instrumentation and Control, COEP, Pune, India
e-mail: kab.instru@coep.ac.in

S. Agashe
e-mail: sda.instru@coep.ac.in

J. Wadgaonkar
e-mail: jagannathvw14.instru@coep.ac.in

M.S. Reddy et al. (eds.), *International Proceedings on Advances in Soft Computing, Intelligent Systems and Applications*, Advances in Intelligent Systems and Computing 628, https://doi.org/10.1007/978-981-10-5272-9_3

quantification of uncertainty. Imprecision and inexactness of information that we have to characterize heave with complexity of processes. It is pragmatic to have an indistinguishable relationship between precision, information, and complexity. However, for most of the processes, we can achieve a better control in accepting some level of imprecision. Mapping of imprecision and uncertainty with the help of experts to control a complex process is the wreath of fuzzy logic [1]. Fuzzy logic-based expert's system for intravenous anesthesia is designed and proved its usefulness in our previous publications [2–4]. This system is dependent on information provided by experts. To validate the knowledge base provided by experts, different similarity measures are used which includes metric-based, set theoretic-based, and implication-based measurements. Different distance-based measurements such as Hamming, Euclidean, and Raha's coefficients are used to find similarity among quantitative measures [5–7]. Lazlo Koczy with his fellows has proposed a methodology for distance measurement between two fuzzy sets [8]. Koczy's methodology generates an output which is denoted by an interval and it is not able to cover qualitative properties. Gower's coefficient is published in 1971. Gower has proven its effectiveness for the use of qualitative as well as quantitative similarity measures for different applications through different publications [9].

In this sequel, a sincere attempt has been made to use classify experts' database using Gower's coefficient to extract effective knowledge base for the expert-based system and use the strong expert's database only for further development. The proposed methodology is applied to designing and development of decision support system for intravenous anesthesia.

This paper is divided into five sections. Section 2 discusses fuzzy modeling where the structure of fuzzy model is explained. Section 3 focuses on similarity measure, describing the use of Gower's coefficient. Section 4 explains results and discusses them with the case study of design and implementation of decision support system for intravenous anesthesia. Section 5 concludes the work.

2 Fuzzy Modeling

Fuzzy modeling has wide applications and proven to be useful, especially in control engineering. There are two approaches to model using fuzzy logic: (a) based on data and (b) based on knowledge acquisition. Fuzzy model for intravenous anesthesia is built using experts' knowledge base. This knowledge is stored in terms of membership functions and rule base. Membership functions are nothing but the human interpretation of variable in linguistic term. For example, we say that today, temperature is low. In this case, temperature is variable and low is linguistic term. In country like India, range of low temperature may be around 7–12 °C. This interpretation is represented in terms of mathematics using membership function. Figure 1 shows the block diagram of fuzzy model.

Main nuts and bolts of fuzzy logic system are fuzzification which is based on antecedent membership functions, inference mechanism which is driven by rule

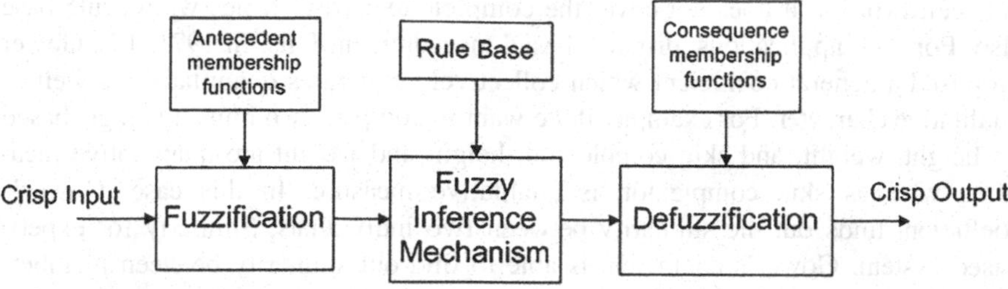

Fig. 1 Block diagram of fuzzy model

base, and defuzzification which is driven by consequence membership functions. Fuzzification is the process of representing human interpretation in terms of mathematical equations. This representation is called as membership function. Membership functions pursue different shapes such as triangular, trapezoidal, sigmoid, Gaussian and rely on the believability of occurrence of event/linguistic representation (e.g., when we say, antecedent is low, the plausibility of antecedent at typical condition will be highest which is considered as 1 whereas it decreases with upgradation or degradation of the antecedent). Membership functions are calculated for all linguistic variables. Next component of FLS is rule base where consequence mapping is stored in terms of if–then rules. This consequence mapping can be of expert-based or experiment-based. If system response is known, then consequence part of fuzzy rules is calculated from fuzzy patches. Fuzzy inference mechanism drives fuzzy rules for occurred antecedents' value and calculates fuzzified consequence which is defuzzified later by defuzzification method. Most famous and accurate defuzzification method is calculation of the center of mass or center of gravity (CG).

3 Similarity Measure

Similarity coefficient measures the likeness between two individuals based on characters or distinct kind of information. These characters can be of two types, quantitative or qualitative. Quantitative characters can be arranged into ordered set for comparison but qualitative characters do not form an ordered set though they may have many levels. The similarity measures are classified into three categories: (1) metric-based measures, (2) set theoretic-based measures, and (3) implicator-based measures. While dealing with distance-based measure, every distance axiom is clearly violated by dissimilarity measures, particularly the triangle inequality and consequently the corresponding similarity measure disobeys transitivity. Even in case of set theoretic measures, the perceived similarities do not follow transitivity. Experts' knowledge base consists of membership functions as well as rule base. Membership functions can be compared by simply finding out correlation factor between two, but it does not cover the complete expertise. It lies within rule base also. For such applications, distance-based measurement fails. In 1971, J.C. Gower proposed a general coefficient which collectively compares quantitative as well as qualitative character. For example, if we want to compare two human beings, based on height, weight, and skin complexion, height and weight are quantitative measures whereas skin complexion is qualitative measure. In this case, Gower's coefficient finds out the similarity between two individuals. Similarly for expert-based system, Gower's coefficient is able to find out similarity between membership functions as well as rule base.

(a) Gower's coefficient

If two individuals i and j *are* compared for character k, then the assigned a score (Gower's coefficient) is S_{ijk}.
For qualitative characters $S_{ijk} = 1$, if the two individuals i and j agree in the kth character and $S_{ijk} = 0$ if differs.
For quantitative characters,

$$S_{ijk} = 1 - \frac{|X_i - X_j|}{R_k},$$

$R_k \rightarrow$ Range of character k.

The similarity between i and j is defined as the average score taken over all possible comparisons.

$$S_{ij} = \frac{\sum_{k=1}^{v} S_{ijk}}{\sum_{k=1}^{v} \delta_{ijk}},$$

where δ_{ijk} is the possibility of making comparison.

$\delta_{ijk} = 1$ when character k can be compared for i and j.
$\delta_{ijk} = 0$ when character k cannot be compared for i and j.

(b) Properties of Fuzzy Relation [10]

An equivalence relation is the relation that holds between two individuals if and only if they are members of the same set that has been partitioned into different subsets such that every element of the set is a member of one and only one subset of the partition. The intersection of any two different subsets is empty; the union of all the subsets equals the original set. An equivalence relation follows three properties: (1) reflexive property, (2) symmetric property, and (3) transitive property.

(1) Reflexive Property

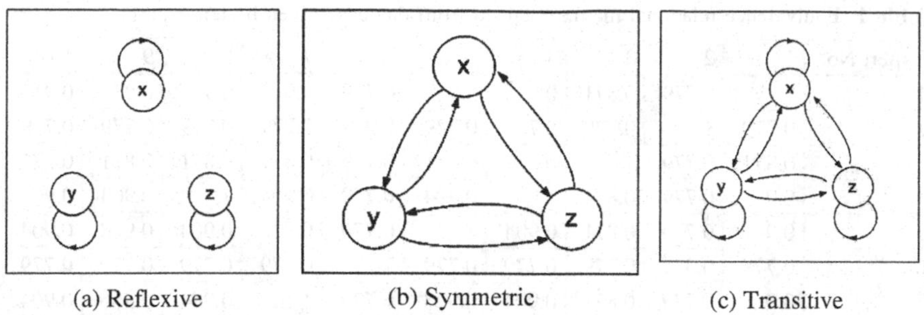

(a) Reflexive (b) Symmetric (c) Transitive

Fig. 2 Properties of equivalence relation

A relation R for all $x, y, z \in S$ is said to be reflexive, if $x \approx x$ as shown in Fig. 2a.

(2) Symmetric Property

A relation R for all $x, y, z \in S$ is said to be symmetric, if $x \approx y$ and $y \approx x$ as shown in Fig. 2b.

(3) Transitive Property

A relation R for all $x, y, z \in S$ is said to be transitive if $x \approx y$ and $y \approx z$, then it is true for $x \approx z$ as shown in Fig. 2c.

(c) Tolerance to Equivalence Relation

Tolerance relation is a relation who follows symmetry and reflexivity but not transitivity. For comparison, relation should be equivalence. Tolerance relation is converted into equivalence relation by self-composition of the relation. Two most familiar methods are max–min composition and max-product composition. For relation R, max–min composition is defined as

$$T = R^{\circ} R,$$

$$\chi_T = \vee(\chi_R(\text{Col}, \text{row}) \wedge \chi_R(\text{row}, \text{col})).$$

Max-product composition is defined as

$$\chi_T = \vee(\chi_R(\text{Col}, \text{row}) \cdot \chi_R(\text{row}, \text{col})).$$

In this paper, max–min composition is used.

4 Results and Discussion

Gower's coefficient between each expert is calculated which shows similarity between the pair of experts. Similarity matrix is obtained by calculating similarity index between each pair of experts. This matrix is then converted into equivalence

Table 1 Equivalence relational matrix obtained from database given by ten experts

Expert No.	1	2	3	4	5	6	7	8	9	10
1	1	0.779	0.871	0.9	0.9	0.779	0.9	0.9	0.9	0.9
2	0.779	1	0.779	0.779	0.779	0.9	0.779	0.779	0.779	0.779
3	0.871	0.779	1	0.871	0.871	0.779	0.871	0.871	0.871	0.871
4	0.9	0.779	0.871	1	0.904	0.779	0.904	1	0.904	0.96
5	0.9	0.779	0.871	0.904	1	0.779	0.928	0.904	0.928	0.904
6	0.779	0.9	0.779	0.779	0.779	1	0.779	0.779	0.779	0.779
7	0.9	0.779	0.871	0.904	0.928	0.779	1	0.904	0.966	0.904
8	0.9	0.779	0.871	1	0.904	0.779	0.904	1	0.904	0.96
9	0.9	0.779	0.871	0.904	0.928	0.779	0.966	0.904	1	0.904
10	0.9	0.779	0.871	0.96	0.904	0.779	0.904	0.96	0.904	1

matrix for comparison. Equivalent relational matrix of all experts is as shown in Table 1. We can observe that this matrix follows reflexivity, symmetry, and transitivity, hence satisfying conditions for classification. For expert-based system, if 80% level of confidence is considered, then Table 2 shows relational matrix after applying α-cut at 0.8. From this table, we can observe that expert No. 2 and 6 are not satisfying the 80% level of confidence. Hence, while considering the database for expert-based system, these two experts can be neglected, in view of other 8 experts. Database of these 8 experts is used to design fuzzy logic-based decision support system for intravenous anesthesia [11, 12], and it is implemented using National Instruments LabVIEW software. Screenshot of the same is as shown in Fig. 3. A set of 270 results noted from decision support system is validated from the expert. Out of 270, expert declines 3 decisions accepting 98.88% result.

Table 2 Relational matrix after applying α-cut at 0.8

Expert No.	1	2	3	4	5	6	7	8	9	10
1	1	0	1	1	1	0	1	1	1	1
2	0	1	0	0	0	1	0	0	0	0
3	1	0	1	1	1	0	1	1	1	1
4	1	0	1	1	1	0	1	1	1	1
5	1	0	1	1	1	0	1	1	1	1
6	0	1	0	0	0	1	0	0	0	0
7	1	0	1	1	1	0	1	1	1	1
8	1	0	1	1	1	0	1	1	1	1
9	1	0	1	1	1	0	1	1	1	1
10	1	0	1	1	1	0	1	1	1	1

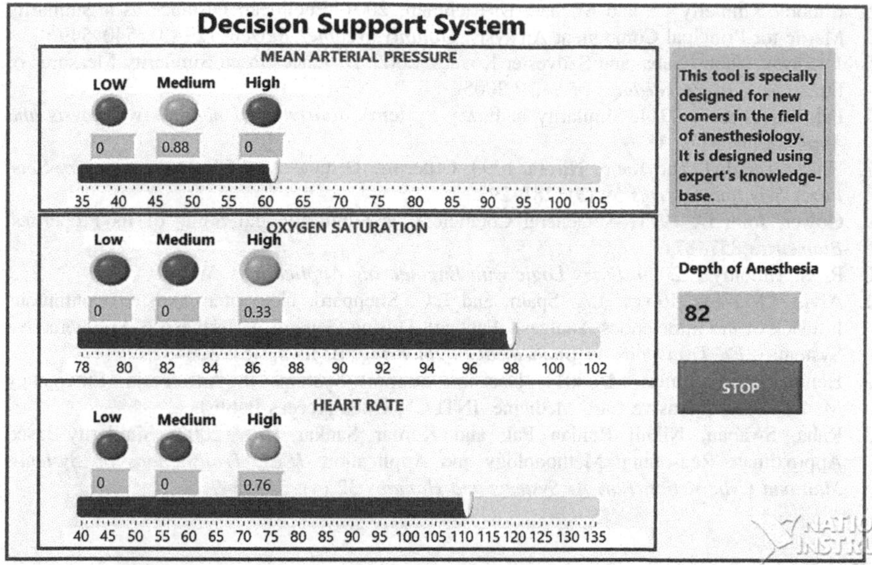

Fig. 3 Decision support system for intravenous anesthesia

5 Conclusion

For applications such as anesthesia, quality of control is based on expert's skill and experience. Being biological control, accuracy is most important here. For such applications, to design decision support system, uncertainty from the experts' knowledge base can be reduced by classification of experts' knowledge base. For fuzzy logic-based experts' system, where knowledge base is in terms of quantitative as well as qualitative measures, Gower's coefficient satisfies the need of comparison. Selecting strong experts out of comparison matrix gives strong knowledge base for decision support system for intravenous anesthesia.

References

1. Zadeh, Lotfi A. 1965. Fuzzy Sets. *Information and Control* 8 (3): 338–353.
2. Bhole, K.A., S.D. Agashe, D.N. Sonawane, Vinayak Desurkar, and Ashok Deshpande. 2011. FPGA Implementation of type I Fuzzy Inference System for Intravenous Anesthetic Drug Delivery. In *World Conference on Soft Computing,* May 23–26.
3. Kalyani Bhole, Sudhir Agashe, and Ashok Deshpande. 2013. FPGA Implementation of Type 1 Fuzzy Inference System for Intravenous Anesthesia. In *IEEE International Conference on Fuzzy Systems (FUZZ)*. IEEE.
4. Bhole, Kalyani, and Sudhir Agashe. 2015. Automating Intravenous Anesthesia with a Fuzzy Inference System Coupled with a Proportional Integral Derivative (PID) Controller. *American Journal of Biomedical Science and Engineering* 1 (6): 93–99.

5. Elmore, Kimberly L., and Michael B. Richman. 2001. Euclidean Distance as a Similarity Metric for Principal Component Analysis. *Monthly Weather Review* 129 (3): 540–549.
6. Johanyák, Zsolt Csaba, and Szilvester Kovács. 2005. Distance Based Similarity Measures of Fuzzy Sets. In *Proceedings of SAMI* 2005.
7. Pal, Asim, et al. 2014. Similarity in Fuzzy Systems. *Journal of Uncertainty Analysis and Applications* 2 (1): 18.
8. Kóczy, LászlóT, and Kaoru Hirota. 1993. Ordering, Distance and Closeness of Fuzzy Sets. *Fuzzy Sets and Systems* 59 (3): 281–293.
9. Gower, John C. 1971. A General Coefficient of Similarity and Some of Its Properties. *Biometrics* 857–871.
10. Ross, Timothy J. 2009. *Fuzzy Logic with Engineering Applications*. Wiley.
11. Alvis, J.M., J.G. Reves, J.A. Spain, and L.C. Sheppard. Computer Assisted Continuous Infusion of the Intravenous Analgesic Fentanyl During General Anaesthesia—An Interactive System. *IEEE Transtions on Biomedical Engineering*, BME-32 (5): 323–329.
12. Hemmerling, Thomas M. 2011. Decision Support Systems in Anesthesia, Emergency Medicine and Intensive Care Medicine. INTECH Open Access Publisher.
13. Raha, Swapan, Nikhil Ranjan Pal, and Kumar Sankar Ray. 2002. Similarity-Based Approximate Reasoning: Methodology and Application. *IEEE Transactions on Systems, Man and Cybernetics, Part A: Systems and Humans* 32 (4): 541–547.

Modeling the Resistance Spot Welding of Galvanized Steel Sheets Using Neuro-Fuzzy Method

L. Boriwal, R.M. Sarviya and M.M. Mahapatra

Abstract The present paper discusses the application of Neuro-fuzzy method (ANFIS) to model resistance spot welding of galvanized steel sheets. A direct search algorithm was implemented in the proposed model for assigning appropriate membership function to each input that saves the time as the manual assigning of membership functions could be avoided. The Neuro-fuzzy model proposed in the present investigation is capable enough to predict process responses based on the input process variables with minimum computational time. Resistance spot welding experiments were conducted as per a design matrix to collect data for the training of the proposed network. A set of test case data, with which the network was not trained, was also used for checking the prediction capability of the network. Further, the comparison between predicted results by model and experimental data has been done that finds good agreement.

Keywords Resistance spot welding · Galvanized steel sheet · Neuro-fuzzy method · Direct search algorithm

1 Introduction

Resistance spot welding (RSW) is one of the widely used joining processes in sheet metal fabrication and automotive industries. Filler material is not required in RSW process, and it can be automated. The steel sheets were held together by the electrode pressure during the welding process. Electrode pressure also prevents the splash of molten metal such as zinc coating and breaking down the surface asperities. The nugget is then formed by the heat generated due to the electrical resistance offered by the metal sheets. The heat generated in the welding process is

L. Boriwal (✉) · R.M. Sarviya
Mechanical Engineering Department, MANIT, Bhopal, India
e-mail: lokeshborimal@gmail.com

M.M. Mahapatra
Mechanical & Industrial Engineering Department, IIT Roorkee, Roorkee, India

M.S. Reddy et al. (eds.), *International Proceedings on Advances in Soft Computing, Intelligent Systems and Applications*, Advances in Intelligent Systems and Computing 628, https://doi.org/10.1007/978-981-10-5272-9_4

sufficient to melt and fuse the faying surfaces between steel sheets. The mechanical strength and nugget size greatly influence the integrity of RSW structures. In recent days, many applications of spot-welded structure are in use for load bearing condition, and their mechanical strength has a strong effect on the integrity of the whole structure. Destructive tests (tensile shear and peel strength) have been a common practice for determining the quality of each spot weld joint.

The effect of electrical resistance between the faying surfaces to nugget formation with respect to the input process variables has been studied by Kaiser et al. [1]. The coating of steels poses problems during the spot welding because the coating material might be of lower melting temperature and thus can melt and vaporize much before the welding of the steels. Hence, coated steels like galvanized ones are needed to be welded with cautions and suitable input process variables to prevent the excessive degradation of the electrodes and defective welds. Gedeon and Eager [2] has investigated the material characteristics and process modification details to achieve sound welds of galvanized steel sheets. William et al. [3] have investigated electrode deformation during spot welding process of coated steel sheets. The thermal cycle of RSW is completed in four stages: squeeze time, weld time, hold time, and off time. During the weld time, temperature is rise rapidly beyond the melting point of the material due to the resistance heating. Na and Park [4] have studied the thermal response of the work sheet material with respect to the input process parameters. In RSW, the contact resistance between the faying surfaces of the sheets significantly affects the power drawn by the electrical circuit and subsequent resistance heating. The contact résistance between the faying surfaces is dependent on the surface nature and the applied electrode pressure. This phenomenon was studied in detail by Davies et al. [5]. Parker et al. [6] investigated electrode degradation mechanism during the spot welding of coated steels. Usually, the input process variables of fusion welding processes are interdependent and often, the deterministic numerical modeling of the process is time consuming and costly. Moreover, numerical methods might not always be suitable for predicting and optimizing the behavior of fusion welds over a widely varying range of process parameters. Hence, many researchers have resorted to experimental modeling to predict the fusion welding characteristics [7–11].

The RSW setup and schematic of the process is shown in Fig. 1. The nugget is mainly dependent on variables like welding current, electrode pressure, and number of welding cycles applied. A further look into the process also indicates some other aspects which affect the RSW nugget quality like the preparation and surface cleanness of the sheets, condition of the electrodes, cooling of electrodes, uniformity of application of pressure, rigidness of clamping, and pressure application mechanisms. Hence, the RSW process is also associated with some degree of uncertainty. Even though widely used, reproducing the desired weld quality in the resistance spot welding process has remained a challenge. Due to these uncertainties, problems, the present investigation is aimed at exploring the use of intelligent modeling techniques like neural network and fuzzy logic to model the process which will help in online monitoring and control of the process.

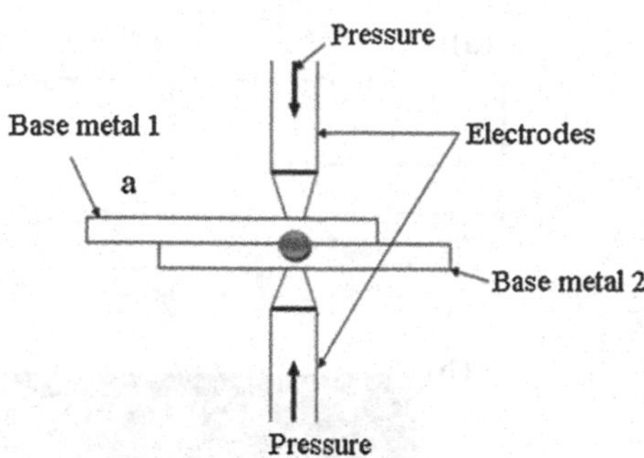

Fig. 1 Resistance welding process

2 Process Parameters for RSW of Galvanized Sheets

The RSW process requires three main input process parameters like welding current, electrode pressure, and welding cycles to form the weld nugget. The above three input process parameters were used for modeling after limit setting of the values. The upper and lower limits of the input process parameter were set based on the trial test runs. The trial test runs were made such that acceptable RSW nugget can be achieved by using the process variables within the limits. The low, medium, and high values of process parameters are given in Table 1.

3 Experimental Procedure and Data Collection

The resistance spot welding was carried out using a constant-alternating current resistance welder (AK-54) having 150 kVA capacity, with full digital setup parameter (F.D.S.P.) controlled by microcomputer and pneumatic application mechanism as shown in Fig. 1. Electrode diameter was checked each time before the commencement of welding because it is observed that electrode wearing occurred during the resistance welding of galvanized steel sheets, might be due to the reason that zinc has a low melting temperature and evaporates early and affects the surface of the electrode. Therefore, electrode was cleaned of scales, ground, and comprehensive to the desired size of electrode tip diameter before each spot weld. After each spot weld cycle, the electrode tip was checked to confirm for constant electrode tip diameter. If it is found under desired tip diameter, it was replaced by new one. The electrode was cooled by flowing water during the spot welding.

Table 1 Range and level of input process parameters

Levels	Welding current (kA)	Weld cycle (s)	Electrode pressure (kg/cm^2)
Low	6	4	2
Medium	7.11	5	3
High	7.9	6	4

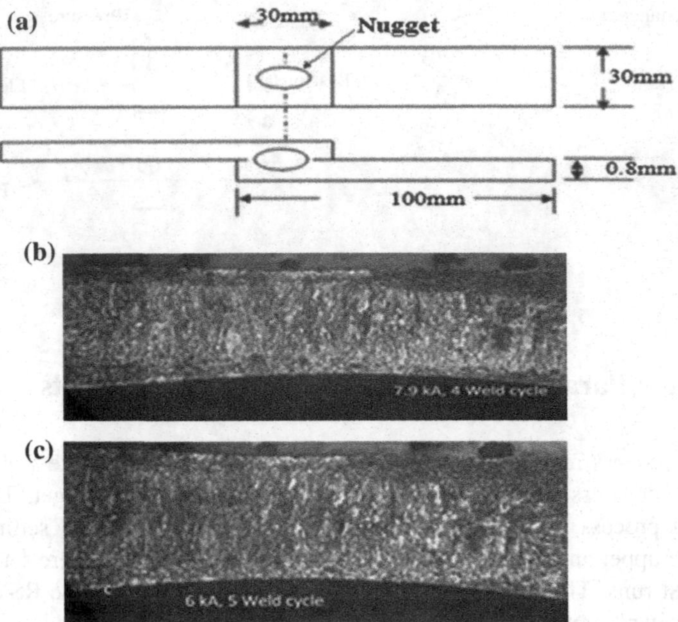

Fig. 2 **a** Schematic of RSW joint, **b** macrostructure of some RSW welds

Galvanized steel sheet was used for the experiments having 0.8 mm thickness with 23 μm galvanized layer. The size and dimension of specimen were 30 mm × 100 mm × 0.8 mm used for the destructive testing are shown in Fig. 2. Copper alloy was used as the electrode material in the experimental process. Steel sheets interfaces were cleaned with acetone and dry air jet before each spot welding. Pilot experiments were performed to observe the range of input process parameters within which welds could be achieved. The lower and upper range of the process parameters were fixed, based on a number of observations. The welding current, weld cycle, and electrode pressure were selected as the variable input process parameters. Squeeze time and hold time were selected as constant parameter during the process. All selected process parameters were set in the microcomputer for spot welding. Prepared specimen was placed between the two electrodes face with the help of center marked on the specimens for producing spot welds at the center. A foot pedal was used for controlling the welding process. The spot weld joints were developed using the galvanized steel sheet specimen. Based on Table 1, build up full factorial design matrix for main experiments with full replication. Main experiments were conducted for all 27 combinations of input process parameters. To avoid any systematical error, each test sample size was five which was required for shear tensile strength, peel strength tests, and for determining the microstructure, nugget size, and micro hardness. Some of the macrostructures prepared after

Table 2 Process parameters for the test cases

S. No.	Welding current (kA)	Welding cycle	Electrode pressure (kg/cm^2)
1	6.66	4	2.6
2	6.66	5	3.2
3	6.66	5	3.4
4	6.66	5	3.6
5	7.02	5	2.8
6	7.02	5	3.2
7	7.02	6	3.6
8	7.8	5	3.6
9	7.8	6	3.6

metallographic examination is shown in Fig. 2. Micro hardness (Mini load 2) tester was used for micro hardness analysis. Both Vickers and spherical indenter were used with the applied load 100 g. Test case study of galvanized steel sheet was done, expect the values of input process parameters used in the experiments. Table 2 shows the input process parameters for the test cases. Tensile shear strength and peel strength test of spot-welded specimen were performed on a universal testing machine with a loading rate 1 mm/min under 50 kN load cell.

4 RSW Process Modeling Using Neuro-Fuzzy Method

Fusion welding processes are essentially nonlinear and multivariate. Often, numerical deterministic modeling has been time consuming and requires much simplification of the physics. Hence, many researchers have resorted to soft computing techniques for modeling the welding processes [10, 11]. Cook et al. [11] had noted that the ANN model prediction capability over a large domain of inputs and outputs is adequate, and an occasional individual case prediction error of even 19% for the unseen test case data is acceptable as the overall percentage error is very less. Excessive training leads to memorization of ANN, and its prediction capability is greatly reduced. Hence, the training schedule of ANN is to be done by trial and error methods for setting parameters like learning rate, number of hidden layers, and neurons. However, as compared to other statistical techniques the soft computing techniques are more preferred for the data handling and relation pattern recognition. The soft computing methodologies are in general complementary in nature. The various soft computing methodologies are tried to obtain better modeling of a system often in combination. It is advocated that the soft computing methodologies, when applied in combination might provide better results rather than a standalone method. In materials processing technologies, the inherent uncertainties and wide range of varying process variables further justify the use of combined soft computing methodologies for better modeling of the systems [12–15]. A very promising combination in this regard is neuro-fuzzy systems. Adaptive neuro-fuzzy interface systems are hybrid intelligent systems that combine features of artificial neural networks and fuzzy logic systems [12]. They are powerful tools which also considers the uncertainties of pattern of relations that exist between the input and output relations of process variables. The neuro-fuzzy method has been successfully applied to modeling and control of several systems [13–15]. Alimardani and Toyserkani [15] developed neuro-fuzzy system for laser solid free form fabrication [15, 16]. The neuro-fuzzy systems can be classified as neural nctwork-driven fuzzy system and hybrid neural network-based systems. In the present investigation, network-based fuzzy inference systems (ANFIS) are used to model the RSW process. The ANFIS is explained in detail in the Appendix. ANFIS is used in various fields of engineering for its prediction and modeling capability [17]. Moreover, the ANFIS computational time is less and can be further improved to make it almost automated for adapting itself to a set of data. For example, in the present investigation, a direct search algorithm is used such that the membership functions which are assigned to the inputs are optimally selected. This decreases the execution time for the ANFIS and increases its usability. For the identification of RSW process, the welding current, electrode pressure, and welding cycle were considered as input to the ANFIS in order to predict the nugget size, tensile strength, and peel strength. The membership functions types and number might be different for different set of input and outputs. The assigning of the membership functions can be done individually manually in the program. However, the ANFIS program can be further improved by enabling it to find the optimum number and type of membership function by following suitable algorithm. To achieve this objective, a direct search algorithm has been used in the present investigation. The direct search algorithm is explained in the appendix in detail, and a flow chart is shown in Fig. 3. Figure 4 shows the ANFIS engine for prediction of RSW output.

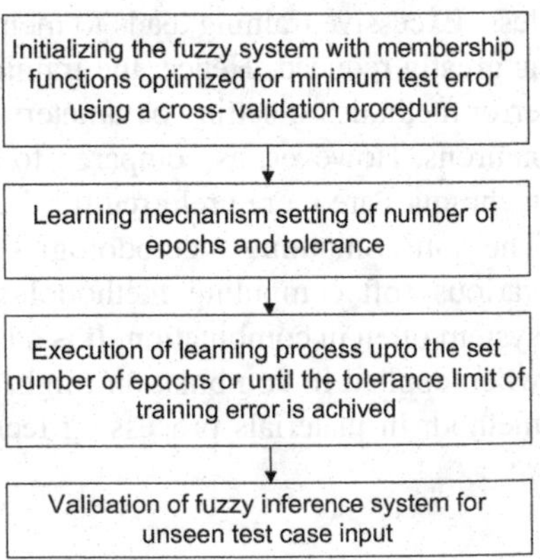
Initializing the fuzzy system with membership functions optimized for minimum test error using a cross- validation procedure
Learning mechanism setting of number of epochs and tolerance
Execution of learning process upto the set number of epochs or until the tolerance limit of training error is achived
Validation of fuzzy inference system for unseen test case input

Fig. 3 Direct search algorithm (discrete optimization)

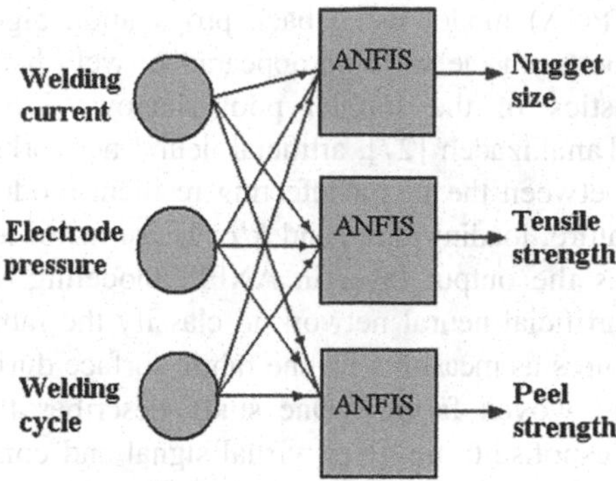

Fig. 4 ANFIS engine for prediction of RSW outputs

5 Soft Computing Modeling of RSW Process

In the recent past, researchers have also used soft computing techniques like particle swarm optimization and genetic algorithms to optimize the process parameters of resistance spot welds [10, 11, 18–21]. The application of soft computing technique like artificial neural network (ANN) to model arc welding problems was studied [10, 11] apart from statistical methods. They have emphasized the nonlinearity and multivariate nature of welding input and output process variables and suggested ANN as a viable technique for modeling purposes. Researchers also used techniques like ANN for modeling, welding, and plate-bending operations [19–21]. The optimization of the fusion welding processes was also done by investigators using ANN and optimization techniques [21, 22]. Although ANN is highly popular for modeling the material processing techniques, it is not without any limitations. For ANN to perform suitably, it is to be trained properly with a set of data through which a pattern of relationship can be learnt. However, the uncertainties of the process parameters and relations are difficult to handle using ANN. Hence, the appropriate choice of soft computing technique is important to model a process [17, 23, 24]. The choice of soft computing technique to be used for modeling a process primarily depends upon the availability of data for training, uncertainties involved in the relations of process variables. Moreover, computational time is also a factor in the choice of soft computing technique. Fuzzy logic in this regard helps to model the process data involving nonlinearity and uncertainties with comparatively less computational time [12]. The ability of fuzzy logic system to handle uncertainties, relation patterns, and low computational time makes it suitable for modeling complex problem like welding and material processing [13].

A new approach for classified the weld defects such as slag inclusion, porosity, longitudinal crack, and transversal crack in radiography film, based on adaptive neuro-fuzzy interference system is discussed here. Each object in the radiography image is identified and described with the combination of geometric features [25]. Principal component analysis (PCA) model to analyze the characteristics of the molten pool shadows has been developed. Principal components analysis

(PCA) model using back propagation algorithm was used for defining relations between the welding appearance (weld height and weld width) and the characteristics of the molten pool shadows [26]. In the study of Shafabakhsh and Tanakizadeh [27], artificial neural networks were employed to predict the relation between the factor affecting resilient modulus of asphalt mixture such as temperature, loading time, and *R/L*. These factor used as input layer and resilient modulus as the output layer in ANFIS modeling. Eldessouki and Hassan [28] employed artificial neural network to classify the fabric sample. Fabric's image textural features as measures for the fabric surface during the quantitative evaluation of pilling in woven fabrics. One study describes the characteristic of the human welder response to be given visual signal and control arm movement in gas tungsten arc welding process using an artificial neural network model [29]. Zhou et al. [30] study the fatigue life of welded joints in the structure using a hybrid genetic algorithm (GA) with the Back propagation neural network (BPNN). Developed artificial neural network (ANN) model to predict the mechanical properties of the A356 matrix reinforced with B4C such as hardness, yield stress, ultimate tensile strength, and elongation percentage. Author found the good agreement between the predicted and experimental data [31].

6 Results and Discussion

To train the ANFIS predictive models, a set of 27 input/output data was used. The data for training covered the complete range of input process variables with which acceptable welds could be achieved. Training was carried on till 3 epochs which were found to be optimum empirically; below that, the validation error was higher, and above that over fitting took place. On the input side of the ANFIS engine, bell-shaped and trapezoidal functions were used based on the results of the direct search algorithm which was employed to choose the best membership function along with the best number of membership functions. The membership function assigned by the direct search algorithm for the three inputs (welding current, cycle, and electrode pressure) are shown in Fig. 5, 6 and 7. There was only a slight change in the membership functions (shown in Fig. 5, 6 and 7) after training showing that our initial model did not require much training; the same fact is reflected by the small number of epochs required to train the model. Coefficients of membership function associated with outputs are shown in Tables 3, 4 and 5. The root mean square error considered as the performance of the model for the prediction of nugget size, tensile Strength prediction, and peel Strength through the training steps are shown in Fig. 8. These errors indicate the difference between the model output and triaging data. It can be observed that for predicting the nugget size (output 1),

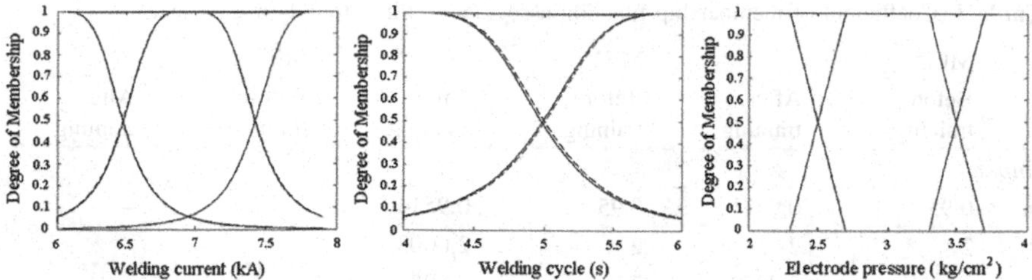

Fig. 5 Assigned membership functions for each input before training to predict nugget size

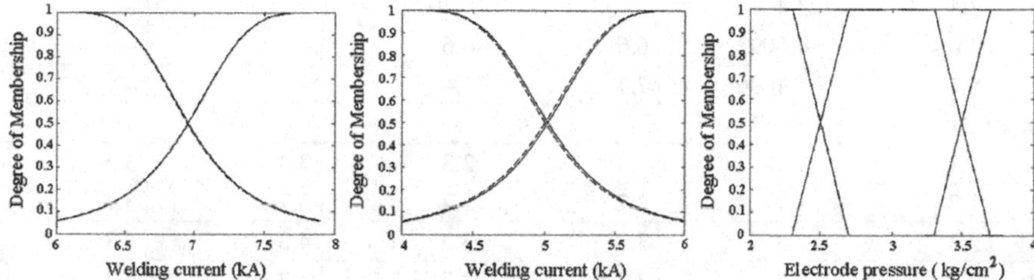

Fig. 6 Assigned membership functions for each input before training to predict tensile strength

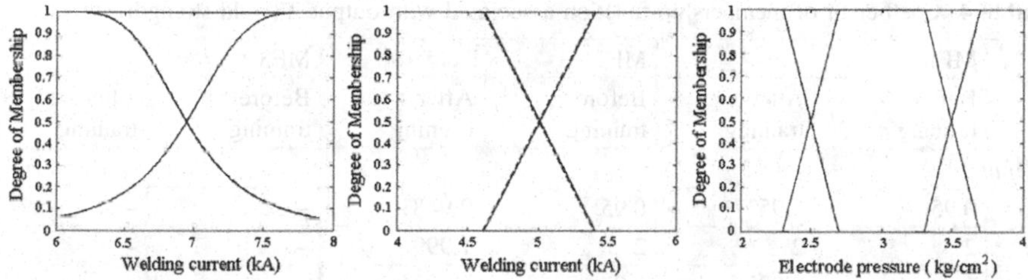

Fig. 7 Assigned membership functions for each input before training to predict peel strength

two membership functions were assigned to inputs 1 and 2 (Tables 3, 4 and 5). Similar assignment was made for predicting tensile strength and peel strength (Tables 3, 4 and 5). Input 3 was assigned with three membership functions for the prediction of all outputs (Tables 3, 4 and 5). The developed model was verified with set of 9 input process parameter data for observing the predicted output. The predicted output of the model with test case unseen data was further compared with the experimental output given in Fig. 9. A good agreement was found between the model output and the actual data. It can be seen that most of the data lie within the 0–10% within the 450 line in Fig. 9.

Table 3 Coefficient of membership function associated with output 1 (nugget size)

	MF1		MF2		MF3	
	Before training	After training	Before training	After training	Before training	After training
Input 1						
a	0.95	0.9471	0.95	0.9514	–	–
b	2	2	2	2.0003	–	–
c	6	5.9972	7.9	7.898	–	–
Input 2						
a	2.6	2.6	4.6	4.6098	–	–
b	3.4	3.4	5.4	5.4097	–	–
c	4.6	4.6097	6.6	6.6	–	–
d	5.4	5.4096	7.4	7.4	–	–
Input 3						
a	1.3	1.3	2.3	2.3	3.3	3.3
b	1.7	1.7	2.7	2.7	3.7	3.7
c	2.3	2.3	3.3	3.3	4.3	4.3
d	2.7	2.7	3.7	3.7	4.7	4.7

Table 4 Coefficient of membership function associated with output 2 (weld strength)

	MF1		MF2		MF3	
	Before training	After training	Before training	After training	Before training	After training
Input 1						
a	0.95	0.9537	0.95	0.9481	–	–
b	2	2	2	1.9996	–	–
c	6	6.0036	7.9	7.9026	–	–
Input 2						
a	1	1.0088	1	0.9909	–	–
b	2	2.0008	2	1.9993	–	–
c	4	4.01	6	6.0101	–	–
Input 3						
a	1.3	1.3	2.3	2.3	3.3	3.3
b	1.7	1.7	2.7	2.7	3.7	3.7
c	2.3	2.3	3.3	3.3	4.3	4.3
d	2.7	2.7	3.7	3.7	4.7	4.7

Table 5 Coefficient of membership function associated with output 3 (peel strength)

	MF1		MF2		MF3	
	Before training	After training	Before training	After training	Before training	After training
Input 1						
a	0.95	0.9471	0.95	0.9514	–	–
b	2	2	2	2.0003	–	–
c	6	5.9972	7.9	7.898	–	–
Input 2						
a	2.6	2.6	4.6	4.6098	–	–
b	3.4	3.4	5.4	5.4097	–	–
c	4.6	4.6097	6.6	6.6	–	–
d	5.4	5.4096	7.4	7.4	–	–
Input 3						
a	1.3	1.3	2.3	2.3	3.3	3.3
b	1.7	1.7	2.7	2.7	3.7	3.7
c	2.3	2.3	3.3	3.3	4.3	4.3
d	2.7	2.7	3.7	3.7	4.7	4.7

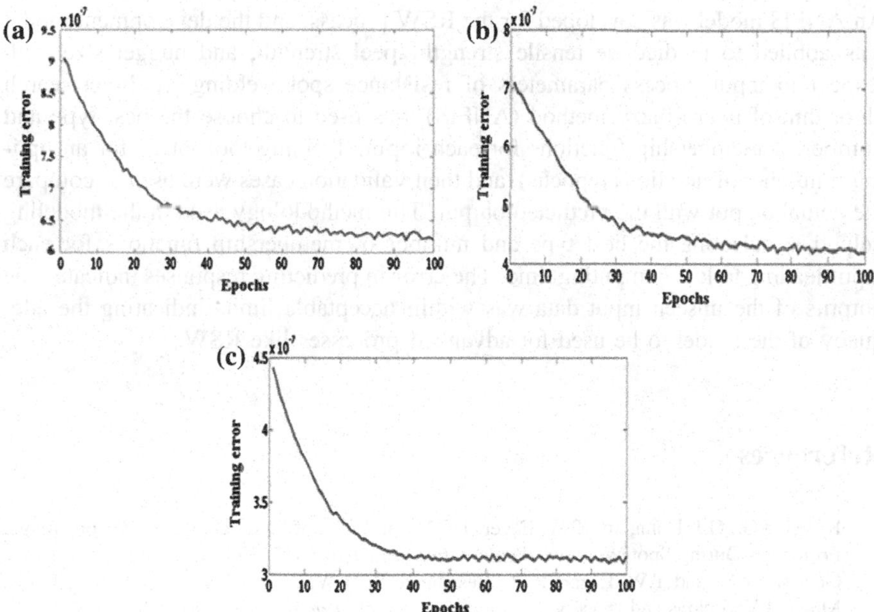

Fig. 8 Root mean square error for prediction of **a** nugget size, **b** tensile strength prediction, **c** peel strength

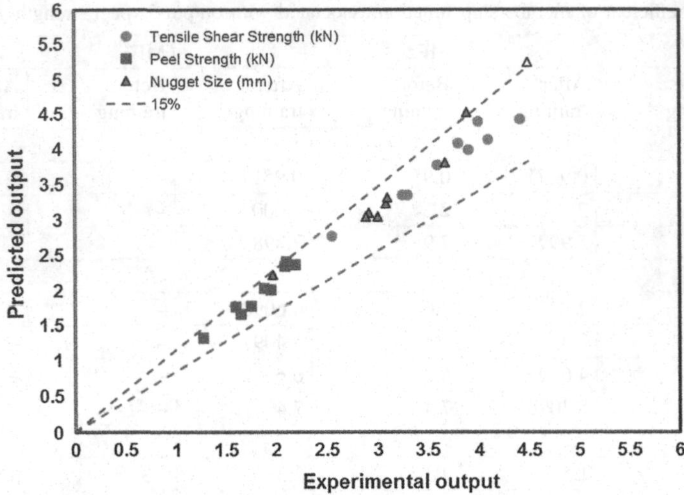

Fig. 9 Prediction error of test cases for **a** nugget size, **b** tensile strength and **c** peel strength

7 Conclusions

An ANFIS model was developed for the RSW process, and the development model was applied to predict the tensile strength, peel strength, and nugget size with respect to input process parameters of resistance spot welding. A direct search algorithm of neuro-fuzzy method (ANFIS) was used to choose the best type and number of membership functions for each input. Training took place for an optimum number of iterations (epochs), and then validation cases were used to compare the actual output with the predicted output. The methodology used in the modeling helped in selecting the best type and number of membership functions for each input leading to less computing time. The error in predicting responses indicated the outputs of the unseen input data was within acceptable limits indicating the adequacy of the model to be used for advanced processes like RSW.

References

1. Kaiser, J.G., G.J. Dunn, and T.W. Eager. 1982. The Effect of Electrical Resistance on Nugget Formation During Spot Welding. *Welding Journal* 61 (6): 167–174.
2. Gedeon, S.A., and T.W. Eagar. 1986. Resistance Spot Welding of Galvanized Steel: Part I. Material Variations and Process Modifications. *Metallurgical Transactions B* 17B: 879–885.
3. Holliday, R., J.D. Parker, and N.T. Williams. 1995. Electrode Deformation When Spot Welding Coated Steels. *Welding in the World* 35 (3): 160–164.
4. Na, S.J., and S.W. Park. 1996. A Theoretical Study on Electrical and Thermal Response in Resistance Spot Welding. *Welding Journal* 8: 233–241.

5. Thornton, P.H., A.R. Krause, and R.G. Davies. 1996. Contact Resistances in Spot Welding. *Welding Journal* 75 (12): 402–412.
6. Parker, J.D., N.T. Williams, and R.J. Holliday. 1998. Mechanisms of Electrode Degradation When Spot Welding Coated Steels. *Science and Technology of Welding and Joining* 3 (2): 65–74.
7. Lee, J.I., and S. Rhee. 2000. Prediction of Process Parameters by Gas Metal Arc Welding. *Proceedings of the institution of Mechanical Engineers, Part B: Journal of Engineering Manufacture* 214: 443–449.
8. Li, W., S. Cheng, S.J. Hu, and J. Shriver. 2000. Statistical Investigation of Resistance Spot Welding Quality Using a Two-Stage, Sliding-Level Experiment. *Journal of Manufacturing Science and Engineering* 123: 513–520.
9. Cho, Y., and S. Rhee. 2003. Experimental Study of Nugget Formation in Resistance Spot Welding. *Welding Journal.* 82 (8): 195–200.
10. Cook, G.E., K. Andersen, G. Karsai, and K. Ramaswamy. 1990. Artificial Neural Networks Applied to Arc Welding Process Modeling and Control. *IEEE Transactions on Industry Applications* 26 (5): 824–830.
11. Cook, G.E., R.J. Barnett, K. Andersen, and A.M. Strauss. 1995. Weld Modeling and Control Using Artificial Neural Networks. *IEEE Transactions on Industry Applications* 31 (6): 1484–1491.
12. Kandel, A. 1988. *Fuzzy Expert Systems*. Addison Wesley.
13. Arafeh, L., H. Singh, and S.K. Putatunda. 1999. A Neuro Fuzzy Logic Approach to Material Processing. *IEEE Transactions on Systems, Man, and Cybernetics.–Part C: Applications and Reviews* 29 (3): 362–370.
14. Duraisamy, V., N. Devarajan, D. Somasundareswari, A. Antony Maria Vasanth, and S.N. Sivanandam. 2007. Neuro Fuzzy Schemes for Fault Detection in Power Transformer. *Journal of Applied Soft Computing* 7 (2): 534–539.
15. Alimardani, M., and E. Toyserkani. 2008. Prediction of Laser Solid Freeform Fabrication Using Neuro-Fuzzy Method. *Applied Soft Computing* 8: 316–323.
16. Sugeno, M. 1985. *Industrial Applications of Fuzzy Control*. New York: Elsevier.
17. Jang, J.S.R. 1993. ANFIS: Adaptive-Network-Based Fuzzy Inference Systems. *IEEE Transactions on Systems, Man and Cybernetics* 23 (3): 665–685.
18. Kim, I.S., J.S. Son, C.E. Park, I.J. Kim, and H.H. Kim. 2005. An investigation into an Intelligent System for Predicting Bead Geometry in GMA Welding Process. *Journal of Material Processing Technology* 159: 113–118.
19. Kanti, M.K., and S.P. Rao. 2008. Prediction of Bead Geometry in Pulsed GMA Welding Using Back Propagation Neural Network. *Journal of Material Processing Technology* 200: 300–305.
20. Dragos, V., V. Dan, and R. Kovacevic. 2000. Prediction of the Laser Sheet Bending Using Neural Network. In *Proceeding of IEEE International Symposium on Circuits and Systems, Geneva, Switzerland*, 686–689.
21. Hamedi, M., M. Shariatpanahi, and A. Mansourzadeh. 2007. Optimizing Spot Welding Parameters in a Sheet Metal Assembly by Neural Networks and Genetic Algorithm. *Proceedings of the institution of Mechanical Engineers, Part B: Journal of Engineering Manufacture* 221: 1175–1184.
22. Lin, H.-L., T. Chou, and C.-P. Chou. 2008. Modelling and Optimization of the Resistance Spot Welding Process Via a Taguchi–Neural Approach in the Automobile Industry. *Proceedings of the institution of Mechanical Engineers, Part D: Journal of Automobile Engineering* 222: 1385–1393.
23. Klir, G.J., and B. Yuan. 2009. *Fuzzy Sets and Fuzzy Logic*. Pearson Singapore.
24. Ross, T.J. 2007. *Fuzzy Logic with Engineering Applications*. New York: Wiley.
25. Zapata, J., R. Vilar, and R. Ruiz. 2010. An Adaptive-Network-Based Fuzzy Inference System for Classification of Welding Fects. *NDT and E International* 43 (3): 191–199.

26. Zhang, Y., X. Gao, and S. Katayama. 2015. Weld Appearance Prediction with BP Neural Network Improved by Genetic Algorithm During Disk Laser Welding. *Journal of Manufacturing Systems* 34: 53–59.
27. Shafabakhsh, G., and A. Tanakizadeh. 2015. Investigation of Loading Features Effects on Resilient Modulus of Asphalt Mixtures Using Adaptive Neuro-Fuzzy Inference System. *Construction and Building Materials* 76: 256–263.
28. Eldessouki, M., and M. Hassan. 2014. Adaptive Neuro-Fuzzy System for Quantitative Evaluation of Woven Fabrics' Pilling Resistance. *Expert Systems with Applications*.
29. Liu, Y., and Y. Zhang. 2014. Control of Human Arm Movement in Machine-Human Cooperative Welding Process. *Control Engineering Practice* 32: 161–171.
30. Zhou, N., et al. 2012. Genetic Algorithm Coupled with the Neural Network for Fatigue Properties of Welding Joints Predicting. *Journal of Computers* 7 (8): 1887–1894.
31. Shabani, M.O., et al. 2012. FEM and ANN Investigation of A356 Composites Reinforced with B 4 C Particulates. *Journal of King Saud University-Engineering Sciences* 24 (2): 107–113.
32. Cormen, T.H., et al. 2010. *Introduction to Algorithms*. Prentice Hall India.

Algorithms for Iterative Applications in MapReduce Framework

A. Diwakar Reddy and J. Geetha Reddy

Abstract Hadoop is an open source framework used to store and analyze large-scale applications. MapReduce offers a quality of services to distributed computing communities. MapReduce framework is used for the large-scale data analysis, but it will not efficiently work for iterative applications. While in the process of iteration to iteration, the components of map and reduce tasks produce unnecessary scheduling overhead, the data must be reloaded and reprocessed for every iteration, wasting I/O, network bandwidth, and CPU resources. Iterative computation is important in many areas such as big data; social networks like Facebook, Twitter; PageRank algorithm. The iterative applications contain large number of datasets, i.e., billions of datasets. However, MapReduce lacks built-in support for iterative applications. iMapReduce supports iterative applications partially. We are proposing a new algorithm for iterative applications, namely incremental iterative MapReduce. I^{2}MapReduce builds on the iMapReduce for further processing, as from the last iteration to the next iteration the changes of the data are only very small compared to the previous state. The new converged state is similar to the last state. I^{2}MapReduce takes this observation to save computations by executing the last converged state; after that, it executes only the changed tasks. I^{2}MapReduce produces efficient results by the converged states.

Keywords Hadoop · MapReduce · Iterative processing
Incremental processing

A. Diwakar Reddy (✉) · J. Geetha Reddy
MSRIT, Banglore, India
e-mail: diwakarannarapu@gmail.com

J. Geetha Reddy
e-mail: geetha@msrit.edu

M.S. Reddy et al. (eds.), *International Proceedings on Advances in Soft Computing, Intelligent Systems and Applications*, Advances in Intelligent Systems and Computing 628, https://doi.org/10.1007/978-981-10-5272-9_5

1 Introduction

Hadoop [1] is used for the large-scale data processing efficiently. After the improvements in the Web 2.0, the usage of online sites, e-commerce, social networks, finance, and health care has been increased. The data produced by the companies are becomingly very large. Relational database [2] contains the billions of records or datasets. For analyzing such datasets or records manually, it takes months to complete, which becomes too expensive and time-consuming. We need a framework which can analyze the data very fast and accurately. Hadoop is such a framework used for analyzing the data accurately and very fast.

Hadoop is a combination of the Hadoop Distributed File System (HDFS) and MapReduce. HDFS is used for the storing and MapReduce is used for analyzing the large datasets. It provides the fault tolerance, load balancing, and distributed execution. Hadoop provides a resource management platform and scheduling. HDFS [3] is used for the storing of the data, fault tolerance, and streaming applications, and it provides scalability to hundreds of thousands of nodes with high aggregate bandwidth. HDFS can again divide into single namenode and multiple datanodes. Namenode is used for storing the metadata and information of datanodes located. Datanode is used for storing the entire data submitted by the user. For fault tolerance, datanodes have to be replicated more than one. The default replication is three, and the default block size is 64 Mb. The secondary namenode is not a fault tolerance to datanode, but it is used to get information of the editLogs. MapReduce [4] framework is used for writing parallel data processing applications. It is a combination of map and reduce tasks. MapReduce is based on the key, value pairs, for both input and output should be key, value pair. The mapper method is used for generating the intermediate (key, value) pairs from the input (key, value) pairs. The reducer which takes input as intermediate values sends back the results to mapper. MapReduce has a job tracker and task tracker. Job tracker acts as master/slave, whereas task tracker acts as a slave only. Job tracker schedules, monitors, and executes failed tasks. Task tracker executes tasks as assigned by the job tracker and sends back results to the job tracker.

2 Limitations of MapReduce Implementation

We have observed several limitations [5] while implementing the MapReduce iterative algorithms. These are as follows.

2.1 Scheduling Overhead

Each MapReduce job has to load the input data from DFS before the map operation. For each and every iteration, MapReduce has to start the new job. The new job should load data from the DFS. In the next iteration, the map function loads the iterated data from DFS again and repeats the process. Through these unnecessary iterations, it produces synchronization overhead.

2.2 Redundant DFS Loading

Sometimes, the data could not be changed from the last iteration to the previous iteration though it will load data from DFS. Through the reloading of the data, the resources of the system can be wasted.

The mapper has to wait for the reducer to complete its operation, and it produces synchronization overhead.

2.3 Termination

Any proper termination has not been indicated. User has to write driver program to stop iterations. The conditions to iterate terminations are as follows: Stop the execution for fixed number of iterations or else stop the execution for two similar successive iterations. MapReduce is not guaranteed to be fast, and it is not is used for streaming of the data.

The above limitations can be addressed by several iterative frameworks that are designed for supporting the iterative applications. These are iMapReduce, iHadoop [6], HaLoop, Twister, Apache Spark and Apache Tej. Each of these frameworks is designed for extending the MapReduce framework to support more complex iterative algorithms.

3 Iterative Frameworks

Iterative algorithms are used for improving the results compared to original MapReduce because it has to execute several numbers of iterations on the same record. The examples of iterative applications are PageRank algorithm, gradient descent algorithm, social network analysis, graph processing, data mining, e-commerce, and finance.

3.1 HaLoop

HaLoop [7] is a modified version of Hadoop that supports iterative large-scale application. To support iterative applications, several modifications have been made in architecture and scheduler. The improvements are given in Fig. 1.

Firstly, HaLoop provides a new programming interface for supporting iterative processing. Secondly, in HaLoop architecture, master node has a module called loop control; to generate loop body, it has to start new map and reduce tasks unless the termination condition is set by the user. HaLoop [8] uses the traditional HDFS itself with slight tweaks. The system is divided into the master and many slaves. If a job is submitted, the master schedules a number of parallel running tasks on all slaves. In the above HaLoop architecture the Demon called task tracker in slave node, it has to communicate with the master for the further processing of the starting the job tasks. Figure 1 shows the major improvements in the HaLoop architecture compared to Hadoop architecture.

The features of the HaLoop [9] are inter-iteration locality, which is used to store the data and use it for total iterations while running the map and reduce tasks. Cache is also used in the HaLoop called reducer input cache and output cache. Using caching technique, we reduce the unnecessary overhead.

3.1.1 Advantages

1. It can reuse the Hadoop [10] MapReduce code.
2. HaLoop has inbuilt controls, so it will be used for the iterative data processing.
3. Caching is accepted, and it supported the terminate execution after fixed number of iterations.

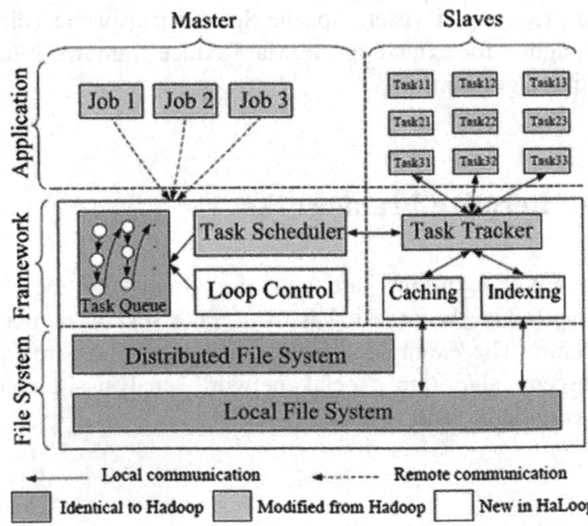

Fig. 1 HaLoop architecture

3.1.2 Drawbacks

1. HaLoop only works with Hadoop 0.20.0.2
2. It has issues while recovering the cached data.
3. Termination condition depends only on the last two recent iterations.
4. Its termination condition is greater than two, and it should fail to work.

3.2 Twister

Twister [11] is a framework used to extend the support for the iterative applications. Twister uses long running jobs for execution of map and reduces tasks, which do not terminate in the entire iteration process. By this, it is not necessary to start new tasks in the each iteration, which reduces the overhead in the system. Twister contains three modules:

(i) Client-side driver program ensures the MapReduce computation.
(ii) Broker network.
(iii) In every working node, there should be one Twister Demon running.

For accessing the input data, Twister directly takes from either worker nodes or local disk. The accessed data should be read by the broker network. Twister's main functionality of the block diagram 2 is Twister Demon, and it is responsible for managing the map and reduce tasks and status and controlling the tasks. The Demon will be running at the every worker node, and it will connect to the broker network to access the read data from the local disk. It uses the distributed architecture. The key features in Twister are distinguishing the static data and variable data.

3.2.1 Advantages

1. The architecture uses the long running jobs, which should be configured once and used for every iteration.
2. Caching concept is used, so unnecessary overhead is reduced.
3. It uses command line arguments to load data, but it is possible in traditional HDFS.

3.2.2 Drawbacks

1. Twister assumes that it uses distributed memory architecture, and it is not possible for every time.
2. Partition is the manual process (Fig. 2).

3.3 *iMapReduce*

In iMapReduce [12], the tasks are alive for total iterative process, so-called persistent tasks. iMapReduce eliminates the basic disadvantage of the normal Hadoop execution, i.e., redundant DFS loading. The normal execution in default Hadoop task tracker involves dumping of data immaterial of whether it has changed or not into the DFS as shown in Fig. 3.

The model is built upon the basic concept of not letting static data get dumped repeatedly into the DFS. There are 3 distinct points in its design. First, persistent tasks avoid repeated task scheduling. Second, the input data from the DFS system is loaded only once. Third, it allows asynchronous execution of map tasks, i.e., mapper does not wait for receiver to complete its operation.

3.3.1 Features

1. **Persistent tasks**: In the default Hadoop MapReduce, the tasks are destroyed as soon as the job is completed. The tasks are active in the whole iteration process. So, instead of adding and retrieving the data from the DFS, we just add it once in the beginning and extract it once at the end.
2. **Data management**: From iteration to iteration, some of the data may not change. To avoid the unchanged data, iMapReduce distinguishes between the state data and static data. The static data should be same for all iterations and state data updated to the DFS system for every iteration.

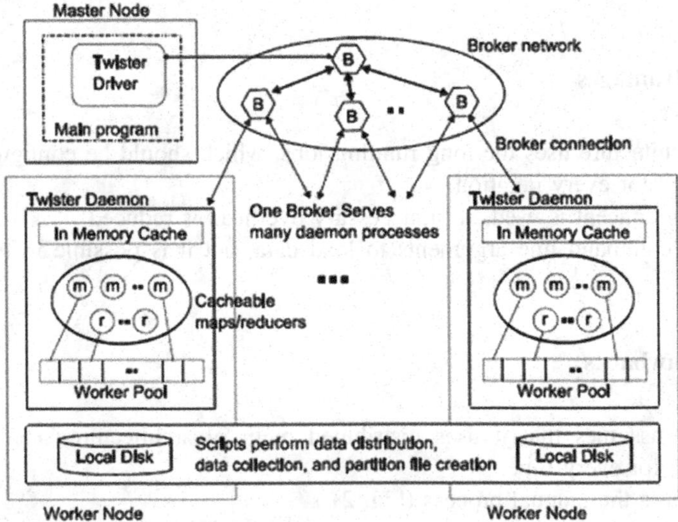

Fig. 2 Architecture of Twister

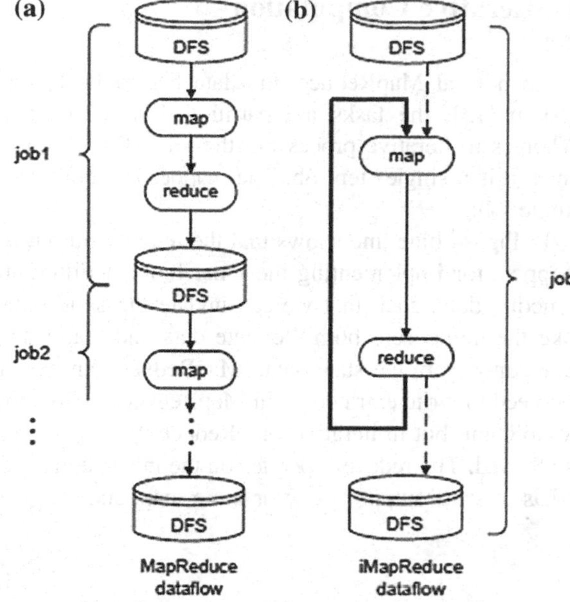

Fig. 3 **a** Data flow of MapReduce and **b** data flow of iMapReduce

3. **Combination of state data and static data**: The static data in the map tasks are combined with the iterated state data for the starting of map operation.

3.3.2 Advantages

1. It removes unnecessary DFS dumping. Scheduling overhead is eliminated.
2. The join operation of static and state data removes communication overhead.
3. The map tasks start as soon as data is available. It does not wait for the reduce task to finish. This allows asynchronous execution of map tasks.

3.3.3 Drawbacks

1. **Single-step jobs**: iMapReduce is assumed as a single-step jobs, and we cannot assure perfect results.
2. **Different keys**: The asynchronous execution does not work if the map and reduce functions have different keys.
3. **Adjacency information**: The map and reduce tasks are same for all iterations, because the tasks are repeated for every iteration. We need to write another program to differentiate between static data and dynamic data before starting the new iteration, and this produces unnecessary communication overhead.

4 Iterative Computation

As in normal MapReduce, the data has to be loaded every time from the DFS system [13], the tasks are partitioned into many smaller jobs for processing. Whereas in iterative processing the jobs should load data from DFS system only once, it is a single-step job. The mapper has combined with reducer to process as single job.

In Fig. 4, blue line shows that the reducer output is sending back the respective mappers for implementing the inner loop. It differentiates the state data from the structure data, such that we can update the state data. The map operation should take the input from both the state data and static data, whereas reduce operation takes only from the state data. iMapReduce can perform the iterative jobs without the need to write extra code. In MapReduce, the mapper has to wait for the process completion, but in iterative MapReduce, the asynchronous execution of map tasks is allowed. The reducer operates on the intermediate results, and for fault tolerance, it has to send output to one or more mappers.

5 Implementation of Incremental Processing

Incremental processing [14] is a very elegant approach to analyzing the datasets in the era of big data. Starting from the scratch, it is very difficult and too expensive, as we are considering base result as iMapReduce. I^2MapReduce builds up the iMapReduce for further processing, as from the last iteration to the next iteration the changes of the data are only very small part compared to the previous state. I^2MapReduce takes this observation to save computations by executing from the last converged state; after that, it has to execute only the changed tasks. For example, input of the two successive computations is *X* and *Y*, respectively, and the both states are similar. The changes are very minor compared to the last state; then, observation is to implement incremental computation from the previous state, say *X*'s state (Fig. 5).

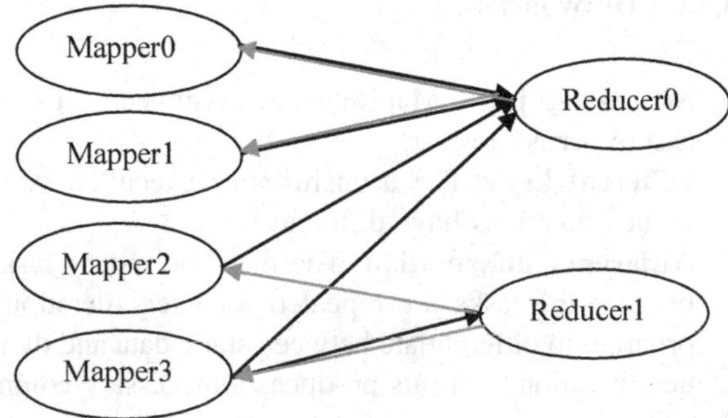

Fig. 4 Iterative processing

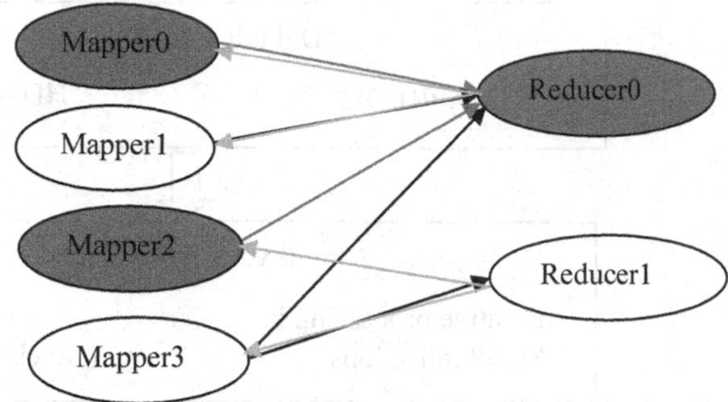

Fig. 5 Incremental processing

Incremental processing is based on the iterative computation. The red color oval represents the changed data compared to the previous state. Mapper 0 and mapper 2 represent the changed states, and the remaining mapper1 and mapper3 are ideal for the next iteration. The changed states only send their tasks to the respective reducer 0. The red line indicates that changed states have sent to the reducer. The reducer [15] receives its intermediate results, performs necessary operations on the results, and sends back to the receiver. The other mappers and reducer are ideal in the incremental computation by that we are reducing the unnecessary overhead.

In MapReduce, the default scheduler is FIFO scheduler. For large-scale applications, we have not considered FIFO scheduler. The schedulers available in the market are capacity scheduler, developed by Yahoo, and fair scheduler, developed by Facebook. Scheduler provides capacity to the queues. For iterative and incremental processing, we are going to change the default job tracker and task tracker configuration files in fair scheduler. Fair scheduler [16] contains pools, and the jobs are positioned into the pools. Fair scheduler provides the minimum guaranteed capacity to the submitted job, and smaller jobs will finish very fast, whereas large jobs are guaranteed that it does not get starved. Fair scheduler also allocates resources to jobs, and the jobs get equal share of all resources. We are changing the requirements based on Hadoop 0.20.2. In default mapred, the job tracker should follow the normal MapReduce data flow. The job tracker status, task attempts, job configuration, assign tasks, termination conditions (start and stop) should be modified as per the incremental processing, as shown in Fig. 6. Finally for building apache ANT, MAVEN software is required. Once we changed the requirements, it should have to rebuild.

6 Related Work

Apache Spark: Apache Spark [17] is an open source platform developed by apache software foundation. Spark uses the in-memory caching of the data specified by the user. The speed of the spark is 100 times more compared to Hadoop MapReduce and 10 times faster compared to disks. To access a datasets, resilient distributed

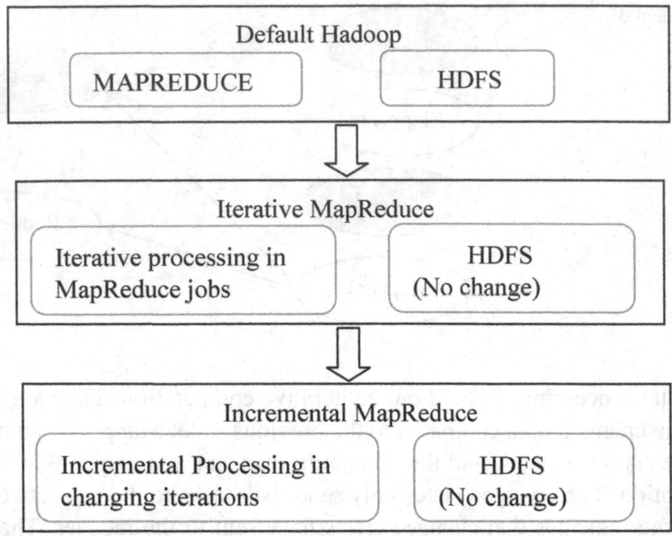

Fig. 6 Data flow diagram of incremental MapReduce

datasets are used. Spark uses distributed system and cluster manager for executing the tasks. Co-grouping, joins, unions, and Cartesian products are supported by spark. It provides some checkpointing services to the user to terminate the execution. One possible disadvantage of spark is control of the distribution of data between several machines.

7 Conclusion

Hadoop is used for processing and analyzing large-scale applications efficiently. Iterative applications are constantly changing data from iteration to iteration, and traditional Hadoop does not support iterative applications efficiently. It produces scheduling overhead, synchronization overhead, and wastage of the resources of the system. The solutions are available to run iterative applications, such as iMapReduce, HaLoop, Twister. These frameworks achieve good performance but produce some compatibility issues like assuming distributed memory architecture, single-step jobs, only works with basic versions of Hadoop. By the incremental processing, we are achieving a greater performance compared to all frameworks available. We are building incremental processing on top of the iterative processing such that saving total cost of the system. Finally, the proposed method reduces the scheduling overhead and synchronization overhead and increases the performance of the system.

References

1. Dwivedi, K., and S.K. Dubey. 2014. Analytical Review on Hadoop Distributed File System. In *2014 5th International Conference on Confluence the Next Generation Information Technology Summit (Confluence)*, 174–181, September 25–26, 2014. doi:10.1109/CONFLUENCE.2014.6949336.
2. Ousterhout, J., P. Agrawal, et al. 2011. The Case for RAMCloud. *Communications of the ACM* 54 (7): 121–130.
3. Shafer, J., and S. Rixner. 2010. The Hadoop Distributed File System: Balancing Portability and Performance. In *2010 IEEE International Symposium on Performance Analysis of Systems & Software (ISPASS2010)*, White Plains, NY, 122–133, March 2010.
4. Dean, J., and S. Ghemawat. Mapreduce: Simplified Data Processing on Large Clusters. In *Proceedings of the 6th Conference on Symposium on Opearting Systems Design and Implementation* (OSDI'04).
5. Zhang, Y., Q. Gao, L. Gao, and C. Wang. 2011. Imapreduce: A Distributed Computing Framework for Iterative Computation. In *Proceedings of the 1st International Workshop on Data Intensive Computing in the Clouds (DataCloud'11)*, 1112.
6. Elnikety, E., T. Elsayed, and H.E. Ramadan. 2011. iHadoop: Asynchronous Iterations for MapReduce. In *2011 IEEE Third International Conference on Cloud Computing Technology and Science (CloudCom)*, 81–90, 29 November, 2011–December 1, 2011. doi:10.1109/CloudCom.2011.21.
7. Bu, Y., B. Howe, M. Balazinska, and M.D. Ernst. 2010. Haloop: Efficient Iterative Data Processing on Large Clusters. In *Proceedings of VLDB Endow*, vol 3, 285–296.
8. Bu, Y., B. Howe, M. Balazinska, and M.D. Ernst. 2012. The Haloop Approach to Large-Scale Iterative Data Analysis. *The VLDB Journal* 21 (2): 169–190.
9. Iterative Mapreduce. http://iterativemapreduce.weebly.com/haloop.html.
10. Jakovits, P., and S.N. Srirama. 2014. Evaluating MapReduce Frameworks for Iterative Scientific Computing Applications. In *2014 International Conference on High Performance Computing & Simulation (HPCS)* 226–233, July 21–25, 2014 doi:10.1109/HPCSim.2014.6903690.
11. Ekanayake, J., H. Li, B. Zhang, T. Gunarathne, S.-H. Bae, J. Qiu, and G. Fox. 2010. Twister: A Runtime for Iterative MapReduce. In *Proceedings of the 19th ACM International Symposium on High Performance Distributed Computing (HPDC '10)*, 810–818. ACM.
12. imapreduce on Google code. http://code.google.com/p/i-mapreduce/. Accessed 2012.
13. Zhang, Y., Q. Gao, L. Gao, and C. Wang. 2012. Accelerate Large-Scale Iterative Computation Through Asynchronous Accumulative Updates. In *Proceedings of Science Cloud'12*.
14. Yan, C., X. Yang, Z. Yu, M. Li, and X. Li. 2012. Incmr: Incremental Data Processing Based on Mapreduce. In *Proceedings of CLOUD'12*.
15. Peng, D., and F. Dabek. 2010. Large-scale incremental processing using distributed transactions and notifications. In *Proceedings of OSDI'10*, 1–15.
16. Hadoop's Fair Scheduler. http://hadoop.apache.org/common/docs/r0.20.2/fair_scheduler.html.
17. Zaharia, M., M. Chowdhury, M.J. Franklin, S. Shenker, and I. Stoica. 2010. Spark: Cluster Computing with Working Sets. In *2nd USENIX Conference on Hot Topics in Cloud Computing, (HotCloud'10)*, 10.

Discovering Frequent Itemsets Over Event Logs Using ECLAT Algorithm

A.S. Sundeep and G.S. Veena

Abstract Event logs are files that can record significant events that occur on a computing device. For example, when a user logs in or logs out, when the device encounters an error, etc., events are recorded. These events logs can be used to troubleshoot the device when it is down or works inappropriately. Generally, automatic device troubleshoot includes mining interesting patterns inside log events and classifying them as normal patterns or anomalies. In this paper, we are providing a sequential mining technique named ECLAT to discover interesting patterns over event logs.

Keywords Event logs · ECLAT algorithm · Sequence patterns

1 Introduction

Today, devices are getting more and more ubiquitous. It is difficult to troubleshoot these devices when they are affected by an internal error. One simple solution is to use event logs to record each and every event that occur in these devices. These event logs can be of various formats. Some can be in non-structured format like text files. And some can be in semi-structured formats like CDF files, XML files, or even files with log extension (i.e., '.log' formats). These files can be used to analyze the device functionality. For example, any windows system provides event logging. To view these event logs, open event viewer under System and security settings and admin istrative tools inside control panel.

Analysis of log events can be done in different ways. Traditionally, this is used to be a manual process. The analysis team used to get the data from the log files in the format that they require understands the device behavior and generates report

A.S. Sundeep (✉) · G.S. Veena
M S Ramaiah Institute of Technology, Bengaluru, India
e-mail: sundeepas4u@gmail.com

G.S. Veena
e-mail: veenags@msrit.edu

M.S. Reddy et al. (eds.), *International Proceedings on Advances in Soft Computing, Intelligent Systems and Applications*, Advances in Intelligent Systems and Computing 628, https://doi.org/10.1007/978-981-10-5272-9_6

on their findings. Automated techniques include tool-based analysis like SWATCH [1], LogSurfer [2], and SEC [3]. Some other automated process involves clustering techniques [4, 5], which includes calculating distance between log files and classifying them. Some other techniques involve mining internal patterns inside a single log file and classify these patterns. Few such approaches use state machine approach to analyze log files [6]. Generally, variants of apriori [7] algorithms are used in log analysis. We made a study on comparison of apriori over ECLAT [8] and decided to use ECLAT for this technique.

In this paper, we use a sequential mining algorithm named éclat and discover interesting rules over a pattern of events. The major objective of the work was to detect anomalies over event logs. However, scope of this paper is restricted to mining interesting patterns. We have considered an example to illustrate this process of generation of all possible subsets of event pattern over event logs.

2 ECLAT Algorithm Over Event Logs

ECLAT algorithm is used to perform itemset mining [9]. ECLAT stands for Equivalence Class clustering And bottom up Lattice Traversal. The algorithm uses tidsets (generally called as transaction ID sets) to avoid generation of unwanted subsets that does not occur in the current context. ECLAT algorithm is applicable for both sequential as well as normal patterns. For the proposed technique, we apply ECLAT algorithm over sequential patterns which are separated by bounded events. Let us illustrate how ECLAT can be modified over event logs. Let E indicate the set of all events of an event log if E_1, E_2, E_3, E_4 € E (Table 1).

When this input is given, the algorithm has to transform these transactions into the following table format format. The right-hand side of Table 2 shows the TID sets for a specific event E_i € E.

In the next step, the algorithm provides the support for every event and is shown in Table 3.

Here if we consider the minimum support as 2 $E3$ will be eliminated and other Events will be considered as Frequent item set. This is illustrated in Table 4.

Table 1 Initial input to ECLAT

TID (sequence ID)	Event sequences
1	E_1, E_3
2	E_1, E_4
3	E_2, E_3, E_4
4	E_2, $E4$

Table 2 Generating TID lists

Events E_i € E	TID list
$E1$	{1, 2}
$E2$	{3, 4}
$E3$	{3}
$E4$	{2, 3, 4}

Table 3 Events with support

Events $E_i \in E$	Support
*E*1	2
*E*2	2
*E*3	1
*E*4	3

Table 4 ECLAT output

Events $E_i \in E$	Support
*E*1	2
*E*2	2
*E*4	3

Table 5 Generating TID lists for two itemsets

Itemset $\{E_i, E_j\}$	TID list	Support
{*E*1, *E*2}	–	0
{*E*1, *E*4}	{2}	1
{*E*2, *E*4}	{3, 4}	2

Table 6 ECLAT output of two item set

Itemset $\{E_i, E_j\}$	Support
{*E*2, *E*4}	2

The above process shown is for single events over a set of sequences. The next step is to include item sets with more than one event. Table 5 illustrates this scenario.

By observing Table 6, it is not possible to perform this process for three item sets. Final output of the algorithm is $\{E_1\}$, $\{E_2\}$, $\{E_4\}$, and $\{E_2, E_4\}$. In the next section of this paper, we have discussed about the implementation process using a sample event log.

3 Implementation of ECLAT Algorithm

In this paper, we are considering a sample device log which is shown in Fig. 1. Our initial step is to generate pattern boundaries, to separate out different sequential patterns. This can be done in different ways. One such technique is to use temporal relationship between patterns. Another approach is to manually identify boundary patterns and generate pattern sequences. In our case, we are using the second approach to generate pattern sequences. Figure 2 shows set of bounded patterns indicating start and end of events.

Implementation of the proposed technique requires three steps mentioned as follows:

1. Creating the event transaction matrix;
2. Deciding minimum support;

	V1	V2	V3	V4	V5	V6
1	236533	2015-07-20T15:34:30	581	Command Requested. CommandName : System log fil...	Developer	None
2	236532	2015-07-20T15:34:19	965	Command Completed. CommandName : Log viewer.	Developer	None
3	236531	2015-07-20T15:33:02	457	*List count: logContainerList[3998] : staticWorker[39...	Developer	None
4	236530	2015-07-20T15:33:02	123	*ListCount:3998 Start:0 End:3997*	Developer	None
5	236529	2015-07-20T15:33:01	817	*Inital Loading done event received*	Developer	None

Fig. 1 Sample event log

	V1	V2	V3	V4	V5	V6	V7	V8
1	236342	2015-07-20T15:31:53	133	Sound on: *TubeOverload.mid*	Service	DebugLowLevel	Information	50010700
2	236321	2015-07-20T15:31:53	31	Sound on: *EndOfExposure.mid*	Service	DebugLowLevel	Information	50010700
3	236316	2015-07-20T15:31:52	474	Sound on: *RadiationOn.mid*	Service	DebugLowLevel	Information	50010700
4	236281	2015-07-20T15:31:47	449	Sound on: *TubeOverload.mid*	Service	DebugLowLevel	Information	50010700
5	236256	2015-07-20T15:31:47	85	Sound on: *EndOfExposure.mid*	Service	DebugLowLevel	Information	50010700
6	236250	2015-07-20T15:31:46	510	Sound on: *RadiationOn.mid*	Service	DebugLowLevel	Information	50010700

Fig. 2 Boundary events

3. Applying ECLAT;

A. *Creating the event transaction matrix*

From the above data, every event is identified by an identifier under column V8. These identifiers categorize events into a number of categories. For example, two events indicating user log in and log out can have the same identifier. These two events can be considered as events under user session category. Similarly, if user enters a set of commands to input variables into memory, increment a memory pointer, etc., such events can be categorized as memory events and can have a same identifier. Using the help of these identifiers, we can generate boundary events as shown in Fig. 2.

Now by considering this bounded events, sequence of events can be generated from the above data and can be included inside a single table as shown below. Each sequence is assigned with a sequence ID or pattern ID; in the order, they are generated indicated by column V1. Figure 3 illustrates the generation of these sequences. Each sequence has a different length. Hence, they are called variant patterns.

The above patterns are generated by considering the identifiers present in the original data, which can indicate the event flow. After obtaining these event sequences, they should be converted into transaction matrix to avoid "NA" characters in the item sets. Figure 4 shows the transaction matrix that can be obtained from the above log sequence.

B. *Deciding support*

Before applying the ECLAT over transaction matrix, we need to decide the support value. The support value can be decided by an item frequency plot. If the plot indicates a specific support value to be used by the algorithm. This plot is

1	500107001025	500107001030	500610003532	500610003534	500610003587	501200033120	50120000000
2	500107001025	500100002041	502100000259	501000000230	502100000002	500107001025	NA
3	500107001025	501401304510	500100002036	501200000002	501200000001	501400002010	50230000053
4	500107001025	500610003532	500610003534	501000000230	501000000230	501200000002	50120000000
5	500107001025	500610003360	500100002035	500100002041	501000000230	502100000002	50010700102
6	500107001025	501401304510	501200000002	501200000001	500100002036	501400002010	50010000204
7	500107001025	500610003360	500100002035	500100002041	502100000002	501000000230	50010700102
8	500107001025	501200000002	501200000001	500100002036	500100002046	501401304510	50140000201
9	500107001025	502100000259	502100000002	501000000230	500107001025	NA	NA

Fig. 3 Sequence of events

```
 500100002035    500100002041    500107001025    500107001030    500610003111
"500100002041" "500107001025" "501000000230" "502100000002" "502100000259"
"500100002034" "500100002036" "500100002046" "500100002047" "500107001025"
"500107001025" "500610003111" "500610003112" "500610003532" "500610003534"
"500100002035" "500100002041" "500107001025" "500610003360" "501000000230"
"500100002034" "500100002035" "500100002036" "500100002046" "500100002047"
"500100002035" "500100002041" "500107001025" "500610003360" "501000000230"
"500100002034" "500100002035" "500100002036" "500100002040" "500100002041"
"500107001025" "501000000230" "502100000002" "502100000259"
"500100002034" "500100002036" "500100002041" "500100002046" "500100002047"
"500107001025"
..   ..          ..   ..          ..   ..          ..   ..          ..
```

Fig. 4 Transactions sets

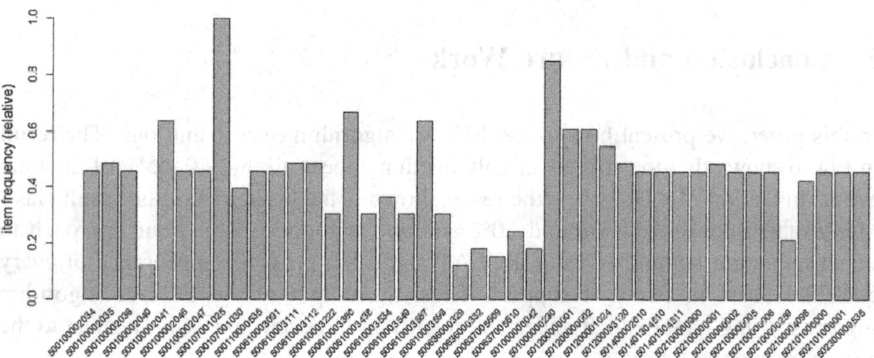

Fig. 5 Item frequency plot

shown in Fig. 5 under result section. Here, all the events appear in about 50% of the sequences. Hence, support can be kept nearer to 0.5. Hence in our paper, we have selected a support equal to 0.4.

C. *Applying ECLAT*

ECLAT is applied over the transaction matrix with a minimum support from the previous step, and the process follows the steps mentioned in previous section. The result of the algorithm also is shown in Fig. 5 in the result section.

```
items                                                                  support
{500107001025,501000000230,502100000002}                               0.4545455
{500107001025,502100000002}                                            0.4545455
{501000000230,502100000002}                                            0.4545455
{500100002035,500100002041,500107001025,500610003360,501000000230}     0.4242424
{500100002035,500100002041,500107001025,500610003360}                  0.4545455
{500100002035,500100002041,500610003360,501000000230}                  0.4242424
```

Fig. 6 Frequent itemsets with minimum support

4 Results

In the previous section, we discussed about transaction matrix and an item frequency plot was generated from the transactions to decide the minimum support. This plot is shown in Fig. 5.

From Fig. 5, it can be noticed that most of the events appear in 30–50% of the mined patterns. Hence, minimum support can be selected anywhere between 0.3 and 0.5. In our case, we have selected 0.4.

After selecting the support, the transaction matrix and minimum support can be fed to the algorithm and the output result would be itemsets of various lengths. The result of our implementation is shown in Fig. 6.

5 Conclusion and Future Work

In this paper, we present how to use ECLAT algorithm over event logs. The result in Fig. 6 shows that the subsets of patterns that appear around 40–45% of the total events in the log file. However, the result shown is the head part of the result class. Many other subsets have around 50% and above support. Thus, our approach to select minimum support is valid. ECLAT algorithm provides appearance of every possible subset in terms of support. Thus, the output of the ECLAT algorithm suggests strong patterns that appear frequently in the event logs with respect to the minimum support selected.

After obtaining frequent patterns from the event logs, our further job would be to classify the patterns as anomalies or normal events. One way this can be done is by using Naïve Bayes filter and classify the events into anomalous or normal events which is a supervised learning technique, wherein we feed the system with known error patterns and normal patterns and classify the results. Another way is to consider time series analysis for anomaly detection. In our approach by observing the data anomaly detection over time series is difficulty as the difference between time is very less. Thus, we are using Naïve Bayes approach.

References

1. Hansen, Stephen E., and E. Todd Atkins. 1993. Automated System Monitoring and Notification With Swatch. In *Proceedings of the USENIX 7th System Administration Conference*.
2. Prewett, J.E. 2003. Analyzing Cluster Log Files Using Logsurfer. In *Proceedings of Annual Conference on Linu x Clusters*.
3. Rouillard, John P. 2004. Real-Time Log File Analysis Using the Simple Event Correlator (SEC). In *Proceedings of LISA XVIII*, 133–149 (Print).
4. Dickinson, W., D. Leon, and A. Podgurski. 2001. Finding Failures by Cluster Analysis of Execution Profiles. In Proceedings of ICSE.
5. Vaarandi, Risto. 2003. A Data Clustering Algorithm for Mining Patterns from Event Logs. In *Proceedings of the 2003 IEEE Workshop on IP Operations and Management*, 119–126.
6. Tan, J., X. Pan, S. Kavulya, R. Gandhi, and P. Narasimhan. 2008. SALSA: Analyzing Logs as State Machines. In *Proceedings of WASL*.
7. Pei, Jian, Jiawei Han, Behzad Mortazaviasl, and Hua Zhu. 2000. Mining Access Patterns Efficiently from Web Logs. In *Proceedings of the 4th Pacific-Asia Conference on Knowledge Discovery and Data Mining*, 396–407.
8. Kotiyal, Bina, Ankit Kumar, Bhaskar Pant, R.H. Goudar, Shivali Chauhan, Sonam Junee. 2013. User Behavior Analysis in Web Log Through Comparative Study of Eclat and Apriori. In *2013 7th International Conference on Intelligent Systems and Control (ISCO)*, 421, 426, January 4–5, 2013.
9. Kaur, Manjitkaur, Urvashi Grag. 2014. ECLAT Algorithm for Frequent Itemsets Generation. *International Journal of Computer Systems* 01 (03). ISSN: 2394-1065.

CRSA-Enriched Mining Model for Efficient Service Innovation Discovery and Redesign in BI Applications

K.M. Rajasekharaiah, Chhaya S. Dule and P.K. Srimanmi

Abstract In present business competition, organizations are emphasizing on optimal decision support systems (DSS) to enhance its growth-oriented decision support system process. To meet these requirements, the enterprise solutions facilitate certain robust and productive business intelligence (BI) applications. On the other hand, to retain market, organizations are emphasizing on service innovation discovery and its redesign (SIDRD) paradigm using BI applications. To accomplish SIDRD, DSS-oriented BI applications require huge feedback datasets to process where the data could be retrieved from certain feedback channel or data warehouses (DWs). The data warehouses encompass datasets from varied resources, and to ensure optimal data extraction and retrieval, these DWs need robust data mining schemes compatible with multidimensional data model (MDDM). On the other hand, to provide most precise and accurate DSS applications, the data security and its privacy preservation are inevitable. Considering all these requirements, in this paper, a robust privacy-preserved mining model called Commutative RSA (CRSA) has been developed and implemented with C5.0 decision tree algorithm to achieve SIDRD objectives with BI utilities. The developed paradigm has exhibited optimal performance for optimized accuracy, overheads, and secure rule set

K.M. Rajasekharaiah (✉)
Department of Computer Science and Engineering, Dr. Sri Sri Sri Shivakumara Mahaswamy College of Engineering, Bangalore, India
e-mail: raju2007_bng@rediff.com

K.M. Rajasekharaiah · C.S. Dule
Visvesvaraya Technological University (VTU), Belgaum, Karnataka, India

K.M. Rajasekharaiah
SCSVMV University, Kanchipuram, Tamil Nadu, India

C.S. Dule
Department of Computer Science and Engineering, Jyothy Institute of Technology, Bangalore, India
e-mail: chhaya0671@gmail.com

P.K. Srimanmi
Department of Computer Science & Maths, Bangalore University, Bangalore, India
e-mail: profsrimanipk@gmail.com

M.S. Reddy et al. (eds.), *International Proceedings on Advances in Soft Computing, Intelligent Systems and Applications*, Advances in Intelligent Systems and Computing 628, https://doi.org/10.1007/978-981-10-5272-9_7

generation for BI. The performance for the developed SIDRD has been analyzed in terms of coverage accuracy and F1 scores, and the overall system has performed optimal.

Keywords Business intelligence · Data warehouse · Service innovation Service redesign · C5.0 · CRSA

1 Introduction

The decision support system (DSS) plays a very vital role in ascertaining optimal organizational decision process, and on the other hand, the intricacy of modern business scenario needs a highly robust and effective automation software support to optimize growth-oriented decision support [1]. This certain advanced technologies such as business intelligence (BI) are employed [2]. BI application encompasses numerous significances like dispersion of information, also provides facilities for growth-oriented features to interact with customers, and makes associated decision making to retain markets and flexible scheme for data access, versatility, and litheness in adapting BI in organizational decision making [3]. BI makes understand that optimal and well-calibrated sales information and guidance for service innovation and discovery to meet market demands and to attract new customers along with retaining old customers with value-added products [4]. BI paradigm, on the other hand, facilitates optimal solution by facilitating the integration of information flows from clients and suppliers, service innovation, redesign and formalization of business processes. It also might be significantly assisted due to its appropriate and relevant data assortment, processing, and objective-oriented information retrieval for decision support. The data retrieved or collected only from local-based resources could not be effective to ensure optimum results for certain DSS utilities. Therefore, the data are required to be collected from varied heterogeneous data sources. In such circumstances, the datasets are required to be stored (DW) at certain well-structured medium or infrastructure where it could be effectively mined and processed to ensure ultimate DSS.

The data storage infrastructure or framework called data warehouse states a "goal-oriented, data-centric and integrated, time-variant, nonvolatile collection of gigantic datasets which are implemented for effective decision support for BI utilities." It assists for online analytical processing (OLAP) resulting into superior performance needs as compared to the online transaction processing (OLTP). Data warehousing is a collection of decision support technologies, aimed at enabling the knowledge workers (executive, manager, and analyst) to make better and faster decisions. It serves as a physical implementation of a decision support data model and stores the information on which an enterprise needs to make strategic decisions [5, 6]. In such situations, the consideration of data warehouses and OLAP becomes inevitable. To ensure efficient QoS and effective data processing, DWs or OLAPs are required to be structured in certain multidimensional data model (MDDM) [7].

Hence, the data security and its redundancy play a very significant role to ensure most precise BI functions for DSS utilities.

In this paper, a highly robust and effective BI system has been developed employing C5.0 decision tree algorithm-based mining module [8] that ensures optimal classification of data for MDDM and DWs. To have a privacy-preserved mining model for DWs, a hybrid system model based on the Commutative RSA (CRSA) cryptosystem and C5.0 decision tree algorithm has been developed which has been further employed for retrieving various feedback data for service innovation discovery and redesign for BI applications. It not only strengthens optimal secure data mining or processing but also reduces huge computational overheads.

Thus, the BI application for SIDRD can be employed for service prototyping, blue printing, innovation, and its redesign to meet customers' expectations so as to sustain in competitive market. The system performance has been evaluated by increasing user-based and reducing risk factors, and results have been obtained in terms of accuracy, coverage, and F1 score. The developed system ensures optimal. The other sections of the presented manuscript discusses like: Sect. 2 presents related work and in Sect. 3 the proposed system model or our contributions or results and analysis, in Sect. 4. In Sect. 5 the conclusions drawn. The references considered for research work have been provided at the last of the manuscript.

2 Related Work

A hybrid mining model for DSS was advocated in [9] where a unified BI architecture was developed using stream mining concept. For E-commerce utilities in [10], a BI-driven data mining (BidDM) framework was developed using four-layered architecture based on mining processing. Chang et al. in [11] proposed an integrated mining model with Web information which was further enhanced in [12] where a bankruptcy prediction approach was developed using a qualitative as well as quantitative optimization with GA [13]. For DSS utilities, the data derived from UCI machine learning repository for data clustering was advocated in [14]. To ensure data security in mining, [15] advocated a k-anonymity-based privacy preservation approach that was further optimized in [16] using classification and weighted attributes. A privacy-preserving model for decision tree mining algorithm was developed in [17] based on the homomorphism encryption approach.

3 Our Contributions

In this paper, a BI model has been developed that emphasizes on service innovation discovery and service redesign (SIDRD). The system encompasses varied uniqueness in terms of classification efficiency and accuracy as well as data security under multidimensional data model (MDDM). The system has multiple data warehouses in the form of MDDM that is considered where the individual data cube behaves like a data source. To implement SIDRD, the feedback from customers is collected from these multiple data warehouses where the datasets under consideration represent a heterogeneous environment. In such scenario, the security of data plays a very significant role to ensure optimal accuracy and DSS support. Hence, in order to facilitate optimal security of the multiple data sources, we have employed a novel cryptosystem called *Commutative RSA* (*CRSA*). Unlike traditional RSA cryptosystem, CRSA possesses numerous advantages are as follows: It does not introduce huge computational overheads for key computation, distribution, and management, and it also reduces computational complexities. Since the organization might have huge datasets and even with multiple DWs, in such circumstances, the data mining and resulting classification accuracy could have concluding significances. Therefore, taking into account these requirements, in this paper C5.0 decision tree algorithm has been used for mining data. Thus, the combination of MDDM, C5.0 DTA, and CRSA cryptosystem makes the overall system much robust and efficient for DSS utilities and provides solution for business intelligence (BI) application. A brief description of the proposed research model has been discussed in the following sections.

The overall functional procedure of the system is illustrated in Fig. 1.

A. CRSA-Enriched C5.0 Mining Model for SIDRD Business Intelligence

In this paper, a novel scheme for BI application has been developed using Commutative RSA (CRSA) cryptosystem amalgamated with C5.0 decision tree algorithm for data mining in MDDM-based BI environment. The implementation of CRSA for mining process is accomplished in three consecutive steps. These are as follows:

- **Step 1**: Here, the data security is facilitated for locally generated rule sets which are followed by combined secure rule set generation which depicts the collection of all generated rules by encompassing MDDMs in form of encrypted data elements.

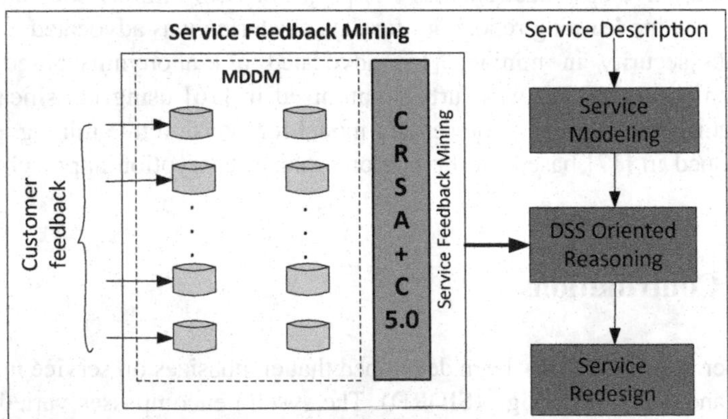

Fig. 1 Proposed system model for DSS-oriented BI solution for SIDRD

- **Step 2**: Once the combined secure rule set has been generated for C5.0 DTA to perform data mining, the individual data sources are initiated to propagate secure rule sets throughout the encompassing data elements or MDDM components. Retrieving the secure rule sets, the MDDM components exhibit decryption of data elements. Here, it must be mentioned that the initiating data element or MDDM component performs decryption of secure rule set that is supposed to accomplish combined rule sets.
- **Step 3**: This phase the data classification for unified BI representation takes place where the employed C5.0 data mining algorithm is implemented for data mining in MDDM elements and to classify data elements for BI utilities and DSS functions.

B. System Model

The overall system modeling for C5.0 DTA implementation with CRSA for SIDRD has been discussed in the following sections.

a. System Initialization

The pseudoalgorithm for initialization is given in Fig. 2.

In expression, the combined rule set (YC_{RSet}) can be stated as $YC_{RSet} = \{Nj_1 \cup Nj_2 \cup Nh_3 \cdots \cup Nj_y\} \forall y$ which is employed for achieving data mining results. Here, the data analysis can be done for entity MDDM elements.

Initialization

Key In: Prime Numbers$A$ and B (with $A > B$)
Key Out: Encryption key $=(y_{Cub_y}, e_{Cub_y})$,
Decryption key $= (y_{Cub_y}, d_{Cub_y})$
Local classification rules, Nj_y *for entity feedback datasets or MDDM component*
Consider 2 variables A and B
For Each $Cub_y \in Cub$
Initialize for data security

$$A_{Cub_y} = A$$
$$B_{Cub_y} = B$$

Evaluate $y_{Cub_y} = A_{Cub_y} \times B_{Cub_y}$
Evaluate greatest common divisor (G.C.D)

$$\varphi_{Cub_y} = \varphi\left(p_{Cub_y}\right) \times \varphi\left(q_{Cub_y}\right) = \left(A_{Cub_y} - 1\right) \times (B_{Cub_y} - 1)$$

Estimate $E_{Cub_y} \mid GCD\left(E_{Cub_y}, Q_{Cub_y}\right) = 1$
Estimate $d_{Cub_y} = e_{Cub_y}{}^{-1} \mathcal{M}od(\emptyset_{Cub_y})$
Party A_{Cub_y} *encryption key parameter*$= (n_{Cub_y}, E_{Cub_y})$
Feedback data sets of MDDM component A_{Cub_y} *decryption key pair* $= (n_{Cub_y}, d_{Cub_y})$
End
Initialize local classification rule $C^{5.0}_{\mathcal{R}jgen}(U_d, ij_{\mathcal{R}})$
Estimate Pre classified data set of $Cub_y U_{d_y}$
Estimate set of data classification $ij_{\mathcal{R}}$
Estimate local classification rules $C^{5.0}_{\mathcal{R}jgen}(U_d, ij_{\mathcal{R}}) = Nj_y$
End
*End (*Here*$ij_{\mathcal{R}}$ *states classification set)*

Fig. 2 System initialization for CRSA

b. **Secure Rule Set Generation for Classification**

The pseudocode for securing rule sets generated is given in Fig. 3.

c. **Combined Rule Set Generation**

The pseudoalgorithms for combined rule set generation is given in Fig. 4

The developed system ensures the optimal performance in terms of mining efficiency, execution time, accuracy, and data security which are the key requirements for service innovation and redesign (SIDRD)-oriented BI utilities.

d. **Amalgamation of C5.0 with CRSA for SIDRD-Oriented BI Application**

In our previous work [8], the comparative analysis of CRSA with C4.5 and C5.0 had been done where C5.0 exhibited much better as compared to C4.5 algorithm.

Algorithm: Generation of Secure rule set ZC_{RSet}

Key In: Encryption key pairs = $\left(n_{Cub_y}, E_{Cub_y}\right)$, Nj_y

Key Out: ZC_{RSet}

Initialize ZC_{RSet}

For each $Cub_y \in Cub$

Estimate $ZNj_y = \zeta_y(Nj_{yn}) = Nj_y^{E_{Cub_y}} Mod(n)$

Estimate $Cub_{RSet} = Cub \cap Cub_y = \{Cub_1, Cub_2, Cub_3, \ldots\ldots Cub_x\}$ *where* $x \neq y$

For Each $Cub_x \in Cub_{Rset}$

Estimate $ZNj_{xy} = \zeta_{xy}(ZNj_y) = ZNj_y^{e_{Cub_x}} Mod(n)$

End

$$ZC_{RSet} = ZC_{RSet} \cup ZNj_{123\ldots\ldots xy}$$

End

Fig. 3 Pseudocode for rule set generation

Algorithm: Combined Rule Set Generation C_{RSet}

Key In: Key pair for decryption(n_{Cub_y}, d_{Cub_y})

Key Out: C_{RSet}

Initiate rule sets

Initiate Cub_{Init}

Estimate $Cub_{rem} = Cub \cap Cub_{Init} = \{Cub_1, Cub_2, Cub_3, \ldots\ldots Cub_x\}$

For each $Cub_x \in Cub_{rem}$

Initiate a variable $T_x C_{RSet}$

For each $ZNj_y \in ZC_{RSet}$

Compute $V_x(ZNj_y) = ZRb_y^{d_{Cub_x}} Mod(n)$

$$T_x C_{RSet} = T_x C_{RSet} \cup V_x(ZNj_y)$$

End

$$ZC_{RSet} = T_x C_{RSet}$$

End

Initiate $T_{Init} C_{RSet} = \emptyset$

For each $ZNj_y \in ZC_{RSet}$

Compute $V_{Init}(ZNj_y) = ZNj_y^{d_{Cub_{Init}}} Mod(n)$

$$T_{Init} C_{RSet} = T_{Init} C_{RSet} \cup U_{init}(SNh_n)$$

End

$C_{RSet} = Temp_{Init} N_{Set}$

Fig. 4 CRSA cryptosystem combined rule set generation

Thus, the robustness of C5.0 has been incorporated in our research, and SIDRD objective has been achieved with hybridization or amalgamation of CRSA with C5.0 decision tree algorithm. The combined rule set generated for C5.0 data mining for y feedback datasets is given by $C^{C5.0}_{RSety} = C^{c5.0}_{\mathcal{R}jx\mathcal{RG}}(C_{RSet}, y)$. In other words, $C^{C5.0}_{RSety} = C^{c5.0}_{\mathcal{R}jx\mathcal{RG}}(\{Nj_1 \cup Nj_2 \cup Nj_3 \cdots \cup Nj_y\}, y)$

$$C^{C5.0}_{RSet\,y} = C^{c5.0}_{\mathcal{R}jx\mathcal{RG}}(\{\{\mathcal{R}j_1, \mathcal{R}j_{21} \ldots, \mathcal{R}j_{1xCub1}\} \cup \{\mathcal{R}j_{12}, \mathcal{R}j_{22} \ldots, \mathcal{R}j_{1xCub2}\} \cup \cdots \cup \{\mathcal{R}j_{1y}, \mathcal{R}j_{2y} \ldots, rl_{1xCuby}\}\}, y)$$

Here, $\mathcal{R}j_{1xCuby}$ is stated for the highest rules generated by certain feedback sources. Similarly, $C^{C5.0}_{\mathcal{R}jx\mathcal{RG}}$ depicts C5.0 mining model-based function for combined rule generation. The derived algorithm for ultimate model with C5.0 algorithm and CRSA for BI application in multidimensional data model (MDDM) scenario is given in Fig. 5.

e. **Service Innovation Discovery and Service Redesign (SIDRD) Implementation**

Considering the ultimate requirement of a robust and highly effective decision support-based business intelligence (BI) utility, the system model has been implemented for service innovation discovery and service redesign (SIDRD). The overall system implementation has been illustrated in Fig. 1. There are predominantly 3 processes for accomplishing SIDRD objective. The implementation details are as follows:

- **Phase 1: Service Feedback Mining**

In this phase, the feedback data from various data heterogeneous data sources are processed for robust classification. In this research phase, we have implemented our developed C5.0 data mining and CRSA-based mining model for service feedback mining to ensure optimal classification accuracy and genuine classified data. In fact,

<u>Algorithm: C5.0 Algorithm based CRSA approach for SIDRD utilities in MDDM scenario</u>

Key In: C_{RSet}
Key Out: Analyzed and mined data results $fc^{C5.0}_{d}$

Initiate $C_{RSet} = \emptyset$
Initiate $C_{pt} = 0$
For each $Nj_y \in N_{pool}$
For each $\mathcal{R}j_{1xCuby} \in Nj_y$
$C_{pt} = C_{pt} + 1$
$C^{C5.0}_{ypt} = C^{C5.0}_{\mathcal{R}jx\mathcal{RG}}(\mathcal{R}j_{1xCuby}, y)$
End
Update $C_{RSet}{}^{C5.0} = C_{RSet} + C^{c5.0}_{ypt}$
End
Calculate $fc^{C5.0}_{d} = C^{C5.0}_{clss}(C_{RSet}{}^{C5.0}, U_d)$

Fig. 5 C5.0 algorithm-based CRSA approach for SIDRD utilities in MDDM scenario

this model exhibits the crawling of service feedback data retrieved from various data sources, and then, the mining is accomplished in MDDM warehouse infrastructure. The contribution of this research phase is to facilitate most precise and accurate classified data which can be employed for decision support system (DSS).

- **Phase 2: Service Modeling**

Since the SADT flow-based service descriptions are not optimal and cannot be employed for DSS-oriented reasoning, therefore, the retrieved and processed data are converted into an optimal structured and classified paradigm to accomplish reasoning. In this paper, we have implemented strategic rationale model (SRM) for presenting the service flow architecture. In fact, strategic rationale model represents a graph-based presentation paradigm where the nodes could be goal, feedback data resources, or tasks. In our work, the node elements have been linked together with the help of means–end relationship and the robust task decomposition relationship.

- **Phase-3 Reasoning**

It might be considered as the justifying element of the SIDRD model as it executes reasoning for service modeling and service feedback mining. In fact, the significances of reasoning are refinement of service optimization. In our system model, the reasoning component performs learning of C5.0 decision trees from the strategic rationale model as considered earlier. And in DSS-oriented BI utilities, managers or business analysts define the organizational goals or objectives. On the basis of objectives defined, we have constructed decision tree using SRM. The data elements in the considered privacy-presented decision tree (CRSA+C5.0) are analyzed against the data mined, and thus, the validity of decision is verified against mined and classified data.

4 Results and Discussion

In this paper, a robust and hybrid system for business intelligence (BI) application has been developed. To accomplish the ultimate objective of the research work, we have exploited and employed the contributions of our previous researches [1, 2, 31, 32]. The proposed and hence developed system has considered the effectiveness of C5.0 decision tree algorithm [2] and Commutative RSA (CRSA)-based privacy preservation to ensure data genuinity in multiple dimensional data model [1]. Further to ensure effective performance for service innovation discovery and service redesign (SIDRD), we have implemented the developed hybrid system with multiple datasets where data are in heterogeneous form as feedback from customers. The refactor services are accomplished in two consecutive terms: by increasing user base and then increasing user base. The system has also been analyzed while reducing user risk. The overall system performance has been analyzed by means of varying parameters such as overall accuracy which is nothing else, but the

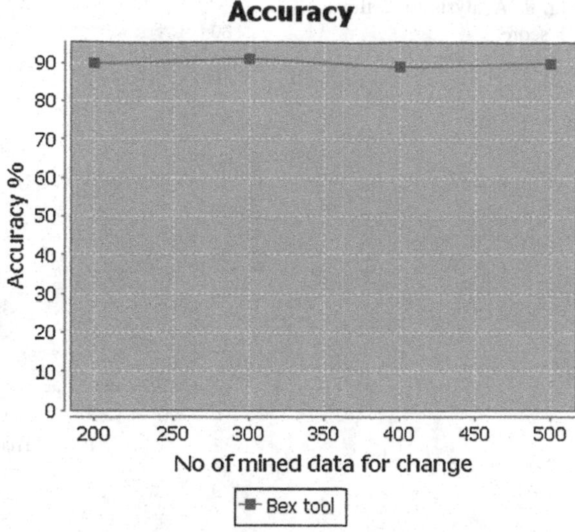

Fig. 6 Analysis for accuracy with respect to number of mined data

BI-oriented classification accuracy, coverage, and ultimately the developed system are analyzed for F1 score. The simulation framework has been developed on Java platform (Figs. 6 and 7).

Figure 8 (Fig. 4) represents the F1 score stating the weighted average of precision and recall for the system under evaluation. Here, considering the results obtained, it can be found that the F1 score is approaching toward unity (1) in relation to the number of mined data. It illustrates the system under consideration is much robust for higher data samples and it performs better even with higher data counts. The higher F1 score depicts the effectiveness of the developed system for

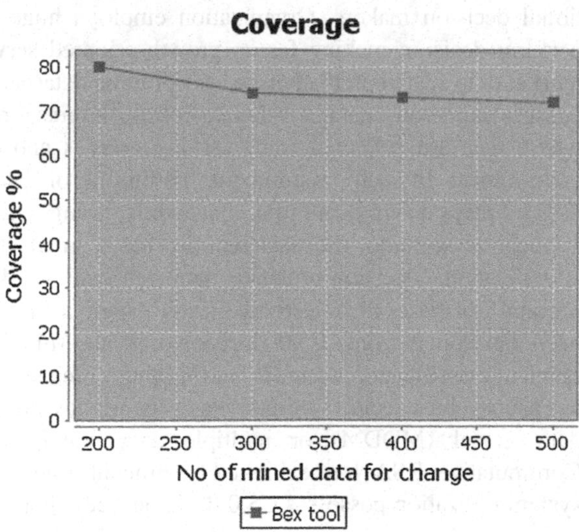

Fig. 7 Analysis of coverage with respect to number of mined data for change

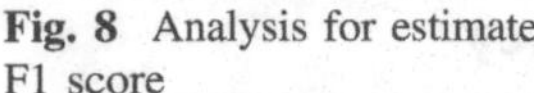

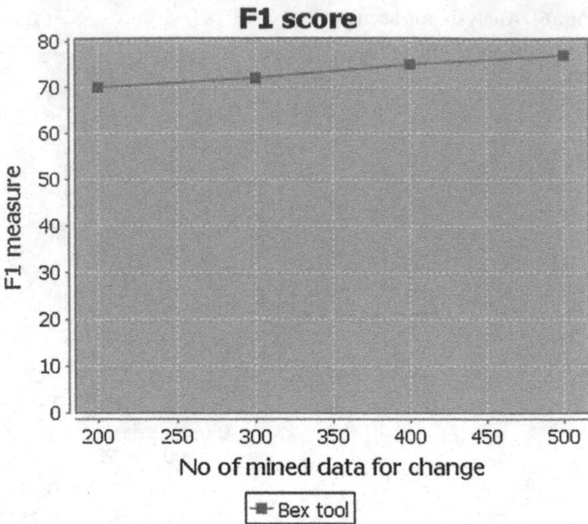

Fig. 8 Analysis for estimate F1 score

ensuring optimal information retrieval and query classification or classification. Thus, the results exhibit that the proposed system can be a potential solution for BI utilities where every aspect of DSS-oriented BI can be fulfilled while ensuring optimal classification, security, accuracy, and minimal computational complexity.

5 Conclusion

In present-day competitive business scenario, there is a great significance of business intelligence (BI) to provide decision support to managers and organizational decision makers. Organization employs huge datasets and information to perform decision making for its growth-oriented service redesign so as to retain market or to sustain in market. In case of huge datasets for analysis, business houses possess numerous data warehouse infrastructures or online analytical processes where data are collected from various sources and are stored in heterogeneous environment. In such environment, the mining of data and its classification for BI utilities plays a significant role; meanwhile, being a heterogeneous multiuser-based application scenario, the data security has a great significance to ensure precise classification, decision-oriented data retrieval, and presentation. In order to accomplish these all objectives in this paper, a highly efficient BI paradigm has been developed using C5.0 decision tree algorithm which is recognized for its efficient classification accuracy and mining efficiency. On the other hand, in order to ensure robust security of datasets or its privacy preservation in multidimensional data model (MDDM) or multiple data warehouses, the implementation of Commutative RSA has exhibited significant role for data security. The hybrid system utilization possesses C5.0 decision tree algorithm with CRSA cryptosystem for MDDM or multiple data warehouse and its implementation for service innovation discovery and service redesign (SIDRD). The ultimate system is enriched with the effectiveness of C5.0 data mining algorithm and robust privacy preservation approach that ensures optimal data processing and classification. Thus, the final implemented system has exhibited optimal performance for accuracy, coverage, and F1 score. Unlike other existing paradigms, the developed system has emphasized on overall system optimization and decision support-oriented BI efficiency.

References

1. Bocij, P., A. Greasley, and S. Hickie. 2009. *Business Information Systems: Technology, Development and Management.* Ft Press.
2. Marjanovic, O. 2007. The Next Stage of Operational Business Intelligence: Creating New Challenges for Business Process Management. In *Proceedings of 40th Annual Hawaii International Conference on System Sciences.*
3. Lönnqvist, A., and V. Pirttimäki. 2006. The Measurement of Business Intelligence. *Information Systems Management* 32–40.
4. Green, A. 2007. Business Information—A Natural Path to Business Intelligence: Knowing What to Capture. *VINE* 37: 18–23.
5. Eckerson, W.W. 2010. *Performance Dashboards: Measuring, Monitoring and Managing Your Business.*
6. Ranjan. 2008. Business Justification with Business Intelligence. *Vine* 461–475.
7. Dr. Srimani, P.K., and K.M. Rajasekharaiah. 2011. The Advantages of Multi Dimensional Data Model—A Case Study. *International Journal of Current Research* 3 (11): 110–115.
8. Dr. Srimani, P.K., and K.M. Rajasekharaiah. 2016. CRSA Cryptosystem Based Secure Data Mining Model for Business Intelligence Applications. In *International on Electrical, Electronics and Optimization Technique—ICEEOT 2016*. IEEE Explorer Digital Library, March 3–5, 2016 (paper code ORO328).
9. Srimani, P.K., and K.M. Rajasekharaiah. 2014. A Data Centric Privacy Preserved Mining Model for Business Intelligence Applications. *American International Journal of Research in Science, Technology, Engineering & Mathematics* 6 (3): 245–252.
10. Hang, Yang, and S. Fong. 2010. Real-time business intelligence system architecture with stream mining. In *2010 Fifth International Conference on Digital Information Management (ICDIM)*, 29–34, July 5–8, 2010.
11. Hang, Yang, and S. Fong. 2009. A Framework of Business Intelligence-Driven Data Mining for E-Business. In *Fifth International Joint Conference on INC, IMS and IDC*, 1964–1970, August 25–27, 2009.
12. Chung, Ping-Tsai, and S.H. Chung. 2013. On Data Integration and Data Mining for Developing Business Intelligence. In *Systems, Applications and Technology Conference (LISAT)*, 1–6, May 3, 2013.
13. Martin, A., T.M. Lakshmi, and V. Prasanna Venkatesan. 2012. An Analysis on Business Intelligence Models to Improve Business Performance. In *International Conference on Advances in Engineering, Science and Management*, 503–508, March 30–31, 2012.
14. Martin, A., T. Miranda Lakshmi, and V. Prasanna Venkatesan. 2012. A Business Intelligence Framework for Business Performance Using Data Mining Techniques. In *2012 International Conference on Emerging Trends in Science, Engineering and Technology (INCOSET)*, 373–380, December 13–14, 2012.

15. Usman, M., and R. Pears. 2010. A methodology for integrating and exploiting data mining techniques in the design of data warehouses. In *2010 6th International Conference on Advanced Information Management and Service (IMS)*, 361–367 November 30–December 2, 2010.
16. Zare-Mirakabad, M.-R., F. Kaveh-Yazdy, and M. Tahmasebi. 2013. Privacy Preservation by k-Anonymizing Ngrams of Time Series. In *2013 10th International ISC Conference on Information Security and Cryptology (ISCISC)*, 1–6, August 29–30, 2013.
17. Wu, Jiandang, Jiyi Wang, Jianmin Han, Hao Peng, and Jianfeng Lu. 2013. An Anonymized Method for Classification with Weighted Attributes. In *2013 IEEE International Conference on Signal Processing, Communication and Computing (ICSPCC)*, 1–5, August 5–8, 2013.

Market Production Assessment Using Optimized MapReduce Programming

Mohit Dayal and Nanhay Singh

Abstract Big data, an easily catching phrase is given to huge and complex amount of data where data size exceeds by terabytes and zetabytes and more. Twitter, Facebook, and other sites are generating enormous amount of data on regular basis and performing analysis on it using various tools and technologies. There is a surge in development of analyzing tools as this much amount of data cannot be processed by the traditional tools. Big data analytics refer to the techniques that can be used for converting raw data into meaningful information which helps in business analysis and forms a decision support system for the executives in the organization. In this paper, we analyzed and evaluated the market of 14 countries through optimized MapReduce program to find out the total percentage of products produced by different market of different countries. Based on the probability, the products are categorized on the basis of high, medium, and low and then for each country we found out the total count of the market that is producing widely. The objective of this analysis was to evaluate the market of each country on the basis of various goods produced by different market. Production of some goods is highly concentrated while others are widely produced. Market experts have to come up with new and innovative idea and technologies in order to increase the production of the concentrated products. So that even distribution of production can be seen in the country.

Keywords Big data · Unstructured data · Hadoop distributed file system MapReduce

M. Dayal (✉) · N. Singh
Ambedkar Institute of Advanced Communication Technologies and Research, New Delhi, India
e-mail: mohitdayal.md@gmail.com

N. Singh
e-mail: nsingh1973@gmail.com

M.S. Reddy et al. (eds.), *International Proceedings on Advances in Soft Computing, Intelligent Systems and Applications*, Advances in Intelligent Systems and Computing 628, https://doi.org/10.1007/978-981-10-5272-9_8

1 Introduction

Amount of data being generated and produced are perpetually snowballing over the years and will continue to do so in the coming future. A survey has indicated that there will be 50% growth in data from 2010 to 2020 only [1]. It is not easy to look out for information from the silos of data from our naked eyes. An upward step is imperative and essential to harness large data sets. Storing of data is not restricted by any constraint like data type, data format, scope of data, and age of data, and so on. Many companies focus on gathering old files, archive files, and current data to predict results and to gain better insights [2]. Big data is everywhere. Any standardized definition has not been given by anyone for Big data. Many companies define it according to their requirements and which fits for the best. There is a surge in development of analyzing tools as this much amount of data cannot be processed by the available database tools [3]. Gartner has defined Big data by giving three characteristics prevalently known as 3 V's of Big data. Later on, IBM defined one more characteristic and provided the theory of 4 V's. It also includes the 3 V's given by Gartner. Analytic souk is leaving no stone unturned to tame the available data that yield interesting results and better insights of industry or organization [4].

- Volume: Size of data is what responsible for this characteristic. Big data means not only data which are in large in size, but data which outstrip a particular limit and cannot be analyzed using on-hand database tools. It corresponds to data at rest. Many firms like eBay, Twitter use Big data analytics to salvage important information.
- Velocity: It refers to the rate at which data arrives and action is performed on it. It is related to data in motion. It is mandatory to process and respond data that arrives in few milliseconds to seconds. This is an essential condition for real-time applications which require abrupt response [5].
- Variety: This characteristic defines diversity in data that can be considered as a part of Big data. Big data can handle various data type formats. Structured data, being the conventional data form, is a part of data format. Other types of data formats are semi-structured and unstructured forms. Audio, video, images formats all are admissible data types which can be stored and harnessed [5].
- Veracity: With large amount of data, uncertainties and ambiguities in data comes as well. It corresponds to data in doubt [6]. It is not an acknowledgment feature and must be removed before proceeding to further stages.

In this paper, we analyzed and evaluated the market of 14 countries through optimized MapReduce program. The objective of this analysis was to evaluate the market of each country on the basis of various goods produced by different market. Production of some goods is highly concentrated while others are widely produced. Market experts have to come up with new and innovative idea and technologies in order to increase the production of the concentrated products. So that even distribution of production can be seen in the country.

2 Research Methodology

Figure 1 describes the research methodology adopted. The first step involves identification of problem and then adopting appropriate approach and strategy to solve the problem.

Data cleansing is one of the most important component required in data analytics. It includes data cleansing, data extraction, removing duplicates, and converting it into standardized schemas. Hadoop MapReduce offers a perfect framework to perform these tasks in parallel with large data sets. Data can be loaded through: (i) Apache Sqoop: It has been designed to move voluminous data from structured data stores. It specializes in connecting with RDBMS and moves the structured data into the HDFS; (ii) Apache Flume: Flume has a simple and flexible architecture which is used for injecting unstructured and streaming flow of data into Hadoop distributed file system (HDFS). The Hadoop distributed file system (HDFS) is a highly scalable and distributed file system for Hadoop. HDFS is able to store file big in size across multiple nodes in a cluster [7]. It is reliable as it replicates the data across the nodes, and hence, theoretically does not require redundant array of integrated devices (RAID) storage. MapReduce is a programming paradigm which can do parallel processing on nodes in a cluster. It takes input and gives the output in form of key–value pairs [8]. MapReduce is able to achieve all this by simply dividing each problem into two broad phases: the map phase and the reduce phase. After collecting the data, it has to be processed so that meaningful information can be extracted out of it which can serve as decision support system. Therefore, the analysts need to come up with a good technique for the same. One way of achieving this is MapReduce; it permits filtering and aggregation of data stored in the HDFS so as to gain knowledge from the data. However, writing MapReduce requires basic knowledge of Java along with sound programming skills

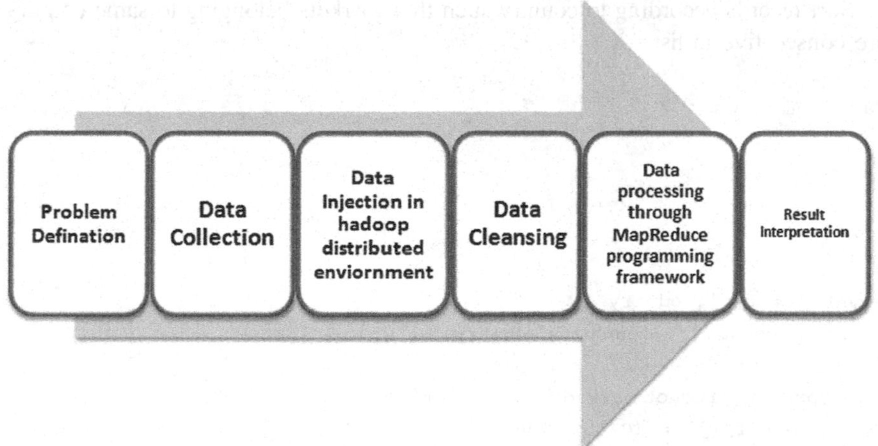

Fig. 1 Research methodology

[9]. Assuming one does possess these skills, even after writing the code, which is itself a labor-intensive task, an additional time is required for the review of code and its quality assessment. But now, analysts have additional options of using the Pig Latin or Hive QL. These are the respective scripting languages to construct MapReduce programs for two Apache projects which run on Hadoop, Pig and Hive [10]. The benefit of using these is that there is a need to write much fewer lines of code which reduces overall development and testing time. These scripts take just about 5% time compared to writing MR programs.

3 Experimental Analysis

In this paper, we analyzed and evaluated the market of 14 countries through optimized MapReduce program to find out the total percentage of products produced by different market of different countries. Based on the probability, the products are categorized on the basis of high, medium, and low and then for each country we found out the total count of the market that is producing widely. The objective of this analysis was to evaluate the market of each country on the basis of various goods produced by different market. Production of some goods is highly concentrated while others are widely produced. Market experts have to come up with new and innovative idea and technologies in order to increase the production of the concentrated products. So that even distribution of production can be seen in the country.

The data set was in .csv format which was first cleaned by removing the missing values and other inconsistent values. Fully distributed Hadoop cluster was used to store and process the voluminous data

N = number of records

Percentage [*i*][*j*] = percentage of products produced by *j*th market of *i*th country

Sort records according to country such that markets belonging to same country are consecutive in list

```
Mapper Algo
{
i = 1
while( i <=N)
{
j = 1;
Country = r[i].country
while(j <=N && r[j].country = country)
{
C = count of product marked as 'y' by market
percentage[i][j] = (C/15) *100
j ++
}
```

```
i = j;
}
c[24] : count of markets producing ith product
for ( i = 1 to N)
{
for all products p produced by ith market, c[p] ++
}
category[i] = category of ith product (high/med/low)
for ( i = 1 to 15)
{
if (c[i]/24) > 0.6
{
category[i] = high
}
elseif( (c[i]/24) < 0.6 && (c[i]/24) > 0.4)
{
category[i] = med
}
else
(
category[i] = low
}
count_high = count of market that comes under high category for all prod-
ucts 'p' that are marked as high
{
for( i = 1 to N)
{
if ith market produce product p,
count.high ++;
}
}
Reducer Algo
{
percentage[][] : percentage of products produced by different mar-
ket by different countries
category[] = category of all 15 products [high/med/low]
count_ = count of markets that come under category high
}
```

4 Results Interpretations

(1) Through the first graph, we can conclude that some of the goods are widely produced by the markets of different countries while production of some goods is highly concentrated. Market experts have to come up with new and innovative idea and technologies in order to increase the production of the concentrated products. So that evenly distribution of production can be seen in the country (Fig. 2).
(2) Through second table, we conclude that production of some products is high in some country while low in others. Market experts have to come up with powerful ways to increase the output in order to meet the growing demands of the customers (Fig. 3).
(3) Through third graph, it could be outlined that Germany is the major product-producing country. Experts may take the ideas by analyzing the production trends of Germany in order to uniformly distribute the production of each product in their own countries (Fig. 4).

5 Limitation

The various limitations of this analysis include:

1. There could be more techniques through which aadhaar data set could be analyzed.

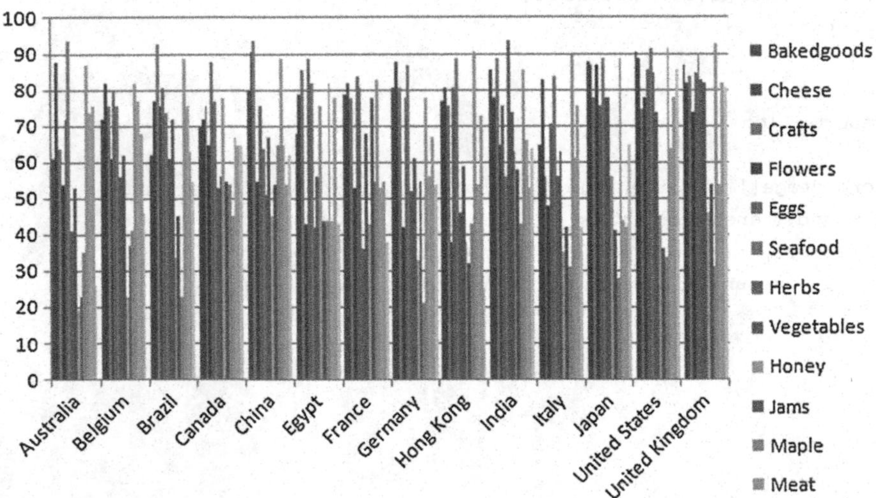

Fig. 2 Total percentage of products produced by different countries

	Bakedgoods	Cheese	Crafts	Flowers	Eggs	Seafood	Herbs	Vegetable	Honey	Jams	Maple	Meat	Nuts	Wine	Soap
Australia	high	high	high	med	high	high	med	med	low	low	low	high	high	high	low
Belgium	high	high	high	high	high	high	med	high	low	low	med	high	high	high	med
Brazil	high	high	high	high	high	high	high	high	low	med	low	high	high	high	med
Canada	high	high	high	high	high	high	med	med	high	med	med	high	high	high	high
China	high	high	high	med	high	high	med	high	med	med	high	high	high	med	high
Egypt	high	high	high	med	high	high	med	med	high	med	med	high	med	high	med
France	high	high	high	med	high	high	low	high	med	med	med	high	med	med	low
Germany	high	high	high	med	high	high	med	high	low	high	low	high	med	high	med
Hong Kong	high	high	high	low	high	high	med	med	low	med	med	high	med	high	low
India	high	high	high	high	high	high	high	high	high	low	med	high	high	med	high
Italy	high	high	high	med	high	high	med	high	low	med	low	high	high	high	med
Japan	high	high	high	high	high	high	high	high	med	med	low	high	med	med	high
United States	high	high	high	high	high	high	high	high	med	low	low	high	high	high	high
United Kingdom	high	high	high	high	high	high	high	high	med	med	low	high	med	high	high

Fig. 3 Category of all 15 products

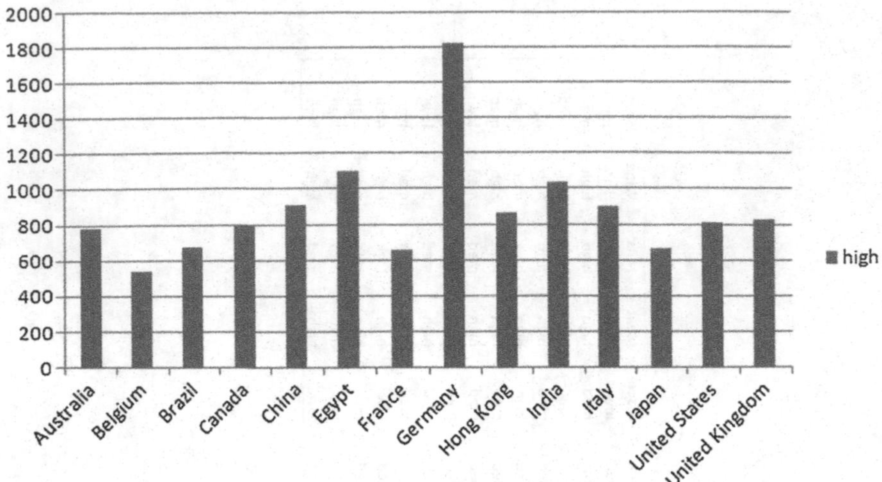

Fig. 4 Count of market coming under high category

2. Fully distributed Hadoop cluster mode is used.
3. The main limitation includes few publications and non-English publications which were excluded; therefore as authors, we could not claim that the work has not been printed in other languages.

6 Conclusion

In this paper, we analyzed and evaluated the market of 14 countries through optimized MapReduce program to find out the total percentage of products produced by different market of different countries. Based on the probability, the products are categorized on the basis of high, medium, and low and then for each country we found out the total count of the market that is producing widely. The objective of this analysis was to evaluate the market of each country on the basis of various goods produced by different market. Production of some goods is highly concentrated while others are widely produced. Market experts have to come up with new and innovative idea and technologies in order to increase the production of the concentrated products. So that even distribution of production can be seen in the country.

Acknowledgements We would like to thank Ambedkar Institute of Advanced Communication Technologies and Research, New Delhi, for successfully accomplishing the research work.

References

1. Howe, Doug, Maria Costanzo, Petra Fey, Takashi Gojobori, Linda Hannick, Winston Hide, David P. Hill, Renate Kania, Mary Schaeffer, Susan St Pierre, Simon Twigger, Owen White, and Seung Yon Rhee. 2008. Big Data: The Future of Biocuration. *Nature International Weekly Journal of Science* 455: 47–50.
2. Lynch, Clifford. Big Data: How Do Your Data Grow? *Nature International Weekly Journal of Science* 455: 28–29.
3. Jacobs, Adam. 2009. The Pathologies of Big Data. *Communications of the ACM—A Blind Person's Interaction with Technology* 52 (8).
4. Chen, Min, Shiwen Mao, and Yunhao Liu. 2014. Big Data: A Survey. *Springer-Mobile Networks and Applications* 19 (2): 171–209.
5. Lee, Yeonhee, and Youngseok Lee. 2013. Toward Scalable Internet Traffic Measurement and Analysis with Hadoop. *ACM SIGCOMM Computer Communication* 43 (1).
6. Labrinidis, Alexandros, and H.V. Jagadish. 2012. Challenges and Opportunities with Big Data. *ACM- Proceedings of the VLDB Endowment* 5 (12).
7. Bhandarkar, M. 2010. MapReduce Programming with Apache Hadoop. In *IEEE International Symposium on Parallel & Distributed Processing (IPDPS)*, April 19–23, 2010.
8. Ibrahim, Shadi, Hai Jin, Lu Lu, Li Qi, Song Wu, and Xuanhua Shi. 2009. Evaluating MapReduce on Virtual Machines: The Hadoop Case. In *Springer: Cloud Computing Lecture Notes in Computer Science*, vol 5931, 519–528.
9. Fedoryszak, Mateusz, Dominika Tkaczyk, and Łukasz Bolikowski. 2013. *Large Scale Citation Matching Using Apache Hadoop*. Springer-Research and Advanced Technology for Digital Libraries. Lecture Notes in Computer Science, vol. 8092, 362–365.
10. Verma, J.P., B. Patel, and A. Patel. 2015. Big Data Analysis: Recommendation System with Hadoop Framework. In *IEEE International Conference on Computational Intelligence & Communication Technology (CICT)*, February 13–14, 2015.

Ultra-Wideband Slotted Microstrip Patch Antenna for Cognitive Radio

M. Abirami and K.R. Kashwan

Abstract In this chapter, an ultra-wideband (UWB) slotted E-shaped microstrip patch antenna is designed, and simulation tests are carried out for the applications of cognitive radio and navigation system. The antenna is designed in such a way that it suits for cognitive radio networks. Normally, a bandwidth remains underutilized in cognitive radio network. To utilize the bandwidth efficiently, two different types of antennas, ultra-wide bandwidth antenna and reconfigurable antenna, are used. FR4 substrate is used with inset feeding technique employment. The simulation tests are carried out on ADS toolbox. It has reduced both the fabrication complexity and cost. E-shaped microstrip patch antenna is tested for functional performance with the frequency range 2–4 GHz. A return loss is observed in the range from −12 to −35 dB. Gain is about 9 dB, and efficiency varies from 72 to 100%.

Keywords Cognitive radio networks · UWB slotted microstrip patch antenna
ADS tool · FR4 substrate · Inset feed

1 Introduction

Cognitive radio is an intelligent radio which allows changing the system parameters according to the environment. It can adapt to a new situation easily to keep up the quality of service. Cognitive radio networks provide solutions for spectrum congestion issues and uncertainty in occupancy of spectrum. The case can be that of continuous use of one spectrum band while other frequency spectrum is unused most of the time. This problem can be minimized by using one frequency band by other user who faces congestion. System and antenna flexibility can improve the cognitive

M. Abirami (✉) · K.R. Kashwan
Department of Electronics and Communication Engineering, Sona College of Technology (Autonomous), Salem 636005, Tamil Nadu, India
e-mail: abiramimanoharanece@gmail.com

K.R. Kashwan
e-mail: kashwan.kr@gmail.com

M.S. Reddy et al. (eds.), *International Proceedings on Advances in Soft Computing, Intelligent Systems and Applications*, Advances in Intelligent Systems and Computing 628, https://doi.org/10.1007/978-981-10-5272-9_9

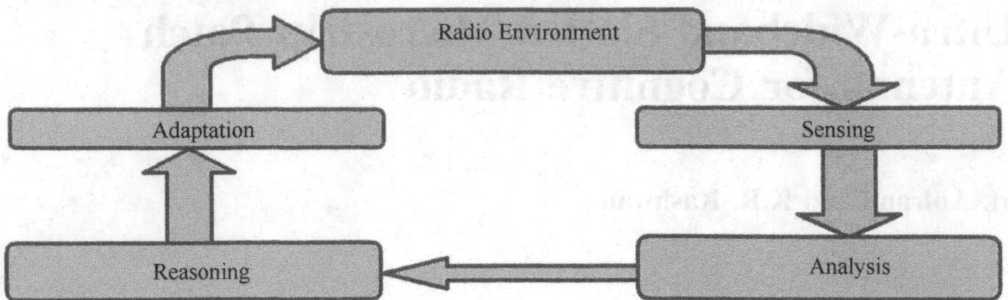

Fig. 1 General block diagram showing operation of cognitive radio system

radio performance. Cognitive radio is operated in two modes, sensing mode and operating mode. UWB antenna is used for sensing mode, and narrowband antenna is used for operating mode. Sensing antenna senses the radio environment. Subsequently, the system can analyze the environment to reach on prediction and decision. Then, the system is reconfigured or adapted, depending upon the environment. The process of sensing and reconfiguration improves the system performance. There are two types of users in cognitive radio network, primary users and secondary users. Primary users are licensed user, and secondary users are unlicensed user. Operation of cognitive radio system helps the secondary user to use the spectrum of licensed band while the primary user is not using spectrum. If the primary user needs spectrum, then secondary user has to vacate the licensed band. Operation of cognitive radio system is shown in Fig. 1, as a flow diagram.

Microstrip patch antenna is a low profile and a low volume antenna. The microstrip patch antenna is also called as a printed or patch antenna. It is used in spacecraft, aircraft, satellite, and missile applications. It is of small size, light weight, and inexpensive. It has high performance, can be easily installed, and has low profile. These requirements are suitable for commercial applications such as mobile radio and wireless applications that have related specifications.

For design and implementation, an UWB antenna is used for cognitive radio networks. E-shaped antenna is integrated with slots to improve the performance for cognitive radio communication. Inset feed is also selected based on suitability and better performance. It provides superior gain and is easy to implement the architecture. The FR4 substrate is selected, since it is inexpensive and easily available. The substrate has reasonably high dielectric consent and performs with high stability even in extreme environment. The substrate sheets are also easily available with various thicknesses.

2 Literature Review

E-shaped antenna with frequency reconfigurable property is used to solve the challenging cognitive radio problems. Reconfigurable antenna has the ability to switch over the frequency, polarization, and patterns. Reconfiguration is achieved with help of RF MEMS switches and two algorithms such as nature-inspired optimization and particle swarm optimization technique. Operating frequency range of this design is from 2 to 3.2 GHz. Slot dimensions are varied to achieve wide bandwidth which is double the times of original E-shaped patch bandwidth [1].

Wideband E-shaped microstrip patch antenna with two slots is introduced in order to increase bandwidth. The slot length, width, positions, etc., are optimized to achieve wide bandwidth, and an operating frequency range varies from 1.9 to 2.4 GHz. This design has features such as small size and simplicity which are applicable to modern time wireless communications standards. An array of antenna is used to optimize and to maximize the performance for WiMAX applications, especially for the transceiver system [2].

Frequency reconfigurable steering wheel-shaped antenna along with four RF MEMS switches was used to improve performance of cognitive radio systems at a frequency range from 6.25 to 8.25 GHz. Rectangular–circular patch antenna design was used by research work reported in [3]. The antenna structure is rectangular patch at the center, and a ring-shaped circular patch antenna is placed around the rectangular patch. The reconfiguration of antenna is obtained by placing four RF MEMS switches in between encircled patch and driven patch. Frequency reconfiguration is done by altering the shape of the antenna. It operates at seven different frequencies. It provides sufficient radiation characteristics, directivity, bandwidth, and minimum return loss without any extra impedance matching device [3].

Frequency reconfigurable wideband E-shaped patch antenna using particle swarm optimization is reported by [4]. Reconfigurability is achieved through ideal switches. This design has frequency range from 2 to 3.25 GHz. Particle swarm optimization technique is proposed in this design. Frequency reconfigurable concept is done by turning ON and OFF the switches to short and open the bridges [4]. The satellite downlink applications have found newer designs to suit the specific requirements [5].

E-shaped microstrip patch antenna is used for wideband application. It provides broad bandwidth in various applications such as remote sensing, biomedical application, satellite communication, and mobile radio applications. Simulations are performed using high-frequency structure simulator (HFSS). Coaxial or probe feed technique is used for this design. Operating frequency range of this design is from 8.80 to 13.49 GHZ. Improved VSWR is obtained in this system [6].

A selective frequency reconfigurable antenna is used for cognitive radio application. Reconfiguration is achieved with help of inserting PIN diode switches into 'T' slot filter. This proposed antenna design operates at 3–10 GHz of wideband and six narrowband of 3–5 GHz. Antenna patch is a combination of two semicircular patches. UWB antenna is used for sensing the entire band and then switch its bandwidth to selective suitable sub-band. Pre-filter is used to communicate without any interference. It has low distortion which increases the gain more than 70% [7].

PSO and FDTD algorithms are combined together and implemented in rectangular patch antenna and E-shaped patch antenna to verify the results from these two antennas. These techniques are mainly used to reduce the computational time. PSO/FDTD algorithm is like bees searching for food. This algorithm is used to find the best space for users in wireless communication. PSO kernel is used for this algorithm. Particle swarm, iterations, and particle position are allocated randomly. Personal best is defined as the particle achieves its best fitness value. Global best is perfect fitness value conquered by any particle. RPC scheme regenerates the record for fitness value. Non-repeated position is updated each time. Repeated position is assigned directly from the previous iteration value. Processing time is further reduced by employing parallel processor. This technique is implemented in rectangular and E-shaped patch antenna to improve the performances of the antenna with these algorithms [8]. A design for cognitive radio networks consists of two structures. First, structure is an ultra-wideband antenna used for channel sensing. Second, structure is triangular-shaped antenna for frequency reconfiguration purpose. Frequency reconfiguration is done by rotating triangular-shaped antenna. Both structures are implanted into the same substrate. A coupling loss is less than -10 dB. FPGA can be used in future to control the rotating motion of antenna and to cover wider bandwidth from 700 MHz to 11 GHz [9].

An UWB sensing antenna and a frequency reconfigurable antenna are integrated into the same antenna substrate. The UWB antenna searches for vacant spectra, whereas the reconfigurable antenna concurrently tunes its operating frequency to the equivalent frequency found out by the sensing antenna. Frequency reconfigurability is implemented via a rotational motion of antenna patches [9]. This eliminates the use of any bias lines as with the case of RF MEMS, PIN diodes, and lumped elements. A sample patch antenna is fabricated to check the proposed method. In the fabricated model, a stepper motor is incorporated which is controlled by computer [10].

A wideband E-shaped patch antenna is designed for high speed wireless communication systems. This design is operated at 5.15–5.825 GHz frequency. In a rectangular patch, two parallel slots are introduced which forms the E-shape antenna. Low dielectric constant is used for the substrate to get a compacted radiating structure which meets the demanding bandwidth necessity. It achieves the return loss less than −10 dB. The other general reference on patch antenna theory and specific applications can be referred from [11–15].

3 Methodology

3.1 Antenna Design Layout

E-shaped UWB antenna is designed with the ADS tool. FR4 substrate with dielectric constant of 3.5 and inset feed techniques is introduced with this design. Ground plane is designed with the dimensions of 100 mm × 120 mm. Layout design of UWB patch antenna is done with the dimensions as shown in Fig. 2a, b with architectural view as illustrated therein.

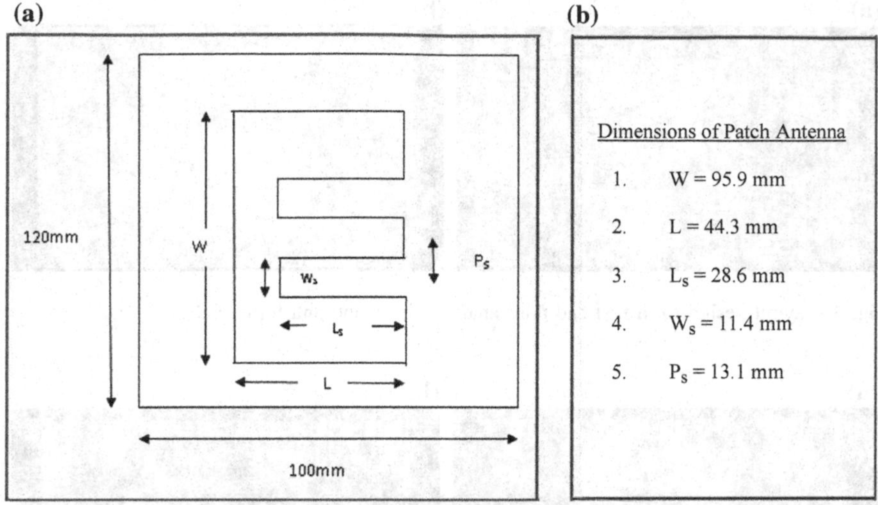

Fig. 2 **a** Design architecture and **b** dimensions of E-shaped patch antenna

Here, L is the length of patch antenna, W is width of the antenna, W_s is width of the slot, L_s is length of the slot, and P_S is slot position. Figure 2b shows the dimensions of UWB antenna design.

3.2 *Analysis for Different Designs*

Different substrate and different feeding techniques are used with the E-shaped patch antenna, and the results are compared to analyze that which type of technique is suitable for the proposed system of antenna. Four different analyses are summarized as follows.

(i) E-shaped patch antenna with Roger substrate and inset feed
(ii) E-shaped patch antenna with Roger substrate and coaxial feed
(iii) E-shaped patch antenna with FR4 substrate and inset feed
(iv) E-shaped patch antenna with FR4 substrate and coaxial feed

3.3 *Layout Work*

Layout design of (i) and (iii) is similar. Likewise, layout (ii) and (iv) are similar. These are illustrated, respectively in Fig. 3a, b.

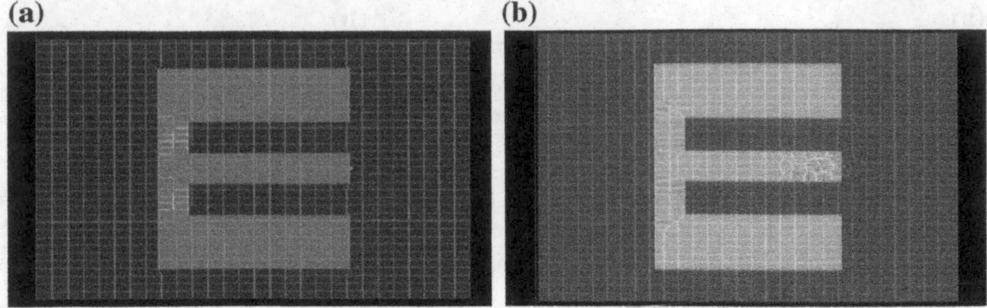

Fig. 3 Layout design for **a** first and third analyses **b** second and forth analyses

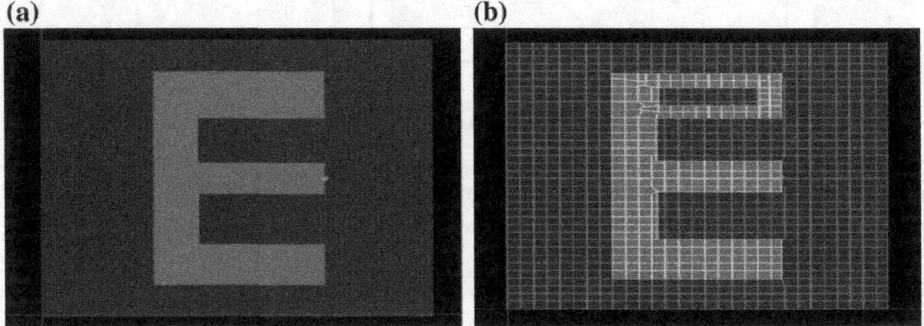

Fig. 4 Layout design of E-shaped patch antenna **a** without slot and **b** with slot, respectively

For UWB slotted patch antenna, inset feed design is chosen. Layout design of E-shaped patch antenna. Rectangular slot is introduced in the patch antenna. Return loss is improved which in turn improves overall performance of the antenna. Figure 4a shows general view of design of antenna. The slot is illustrated in Fig. 4b.

3.4 *Different Substrates in Design*

Substrate is designed with Roger with dielectric constant of 2.2, and FR4 which has 3.5 as a dielectric constant. Both are designed with thickness of 1.6 mm. These are shown in Fig. 5a–d, respectively.

3.5 *Substrate Layout Design*

FR4 substrate is used for this antenna design with the thickness of 1.6 mm. Ground plane is designed with Roger_RO4350 material with 10 mm thickness. Frequency is set as 2–4 GHz. Simulation is done with this design. These are shown in Fig. 6.

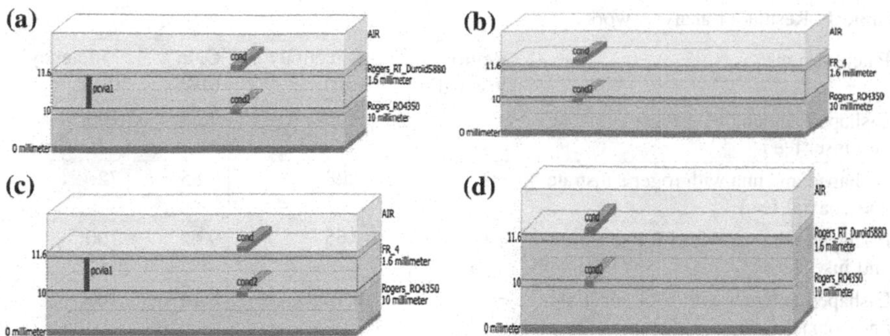

Fig. 5 Substrate layers layout with feedback types and conductor planes layout **a** roger substrate with coaxial type feed **b** FR4 substrate with inset type feed **c** FR4 substrate with coaxial feed **d** roger substrate with inset feed

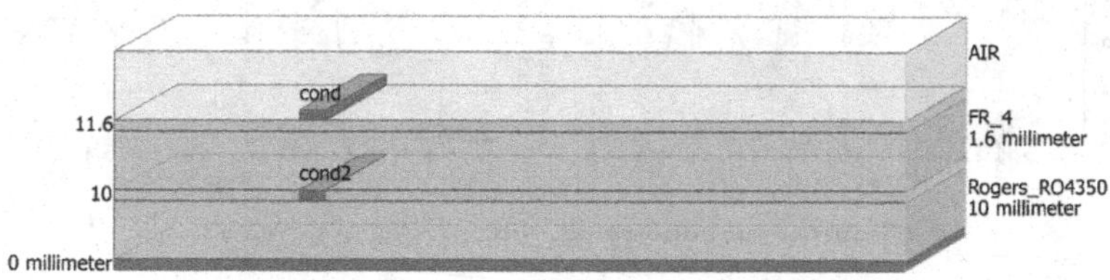

Fig. 6 FR4 substrate design of UWB antenna for both slotted and unslotted

4 Results and Discussions

Coaxial feed provides better return loss and radiation efficiency, but it is difficult to fabricate. Inset feed provides superior gain with ease of fabrication. FR4 substrate is less expensive compared to Roger substrate. Based on the analysis, antenna is designed with FR4 substrate and inset feed. Then, finally, the slots are set up in patch antenna. Return loss is obtained with less than −10 dB. While the slots increased, better return loss is obtained. Feed and substrate design are chosen based on the results analyses. Table 1 shows the antenna parameters.

4.1 Results of Patch Antenna

Results obtained for UWB patch antennas without slot are shown in Fig. 7a, b. The return loss is at about −20 dB. Gain is about 9 dB. These are −30 and 9 dB, respectively, for slotted design, as shown in Fig. 7c, d. Performances of the antenna are improved for the cognitive radio application. FR4 substrate and inset feed

Table 1 Results of analysis work

Patch designs	Return loss (dB)	Directivity (dB)	Gain (dB)	Efficiency (%)
E-shaped antenna with roger substrate and inset feed	−24	9.55	9.55	100
E-shaped antenna with roger substrate and coaxial feed	−23	8.28	6.85	72
E-shaped antenna with FR4 substrate and inset feed	−12	9.65	9.65	100
E-shaped antenna with FR4 substrate and coaxial feed	−35	8.14	8.14	100

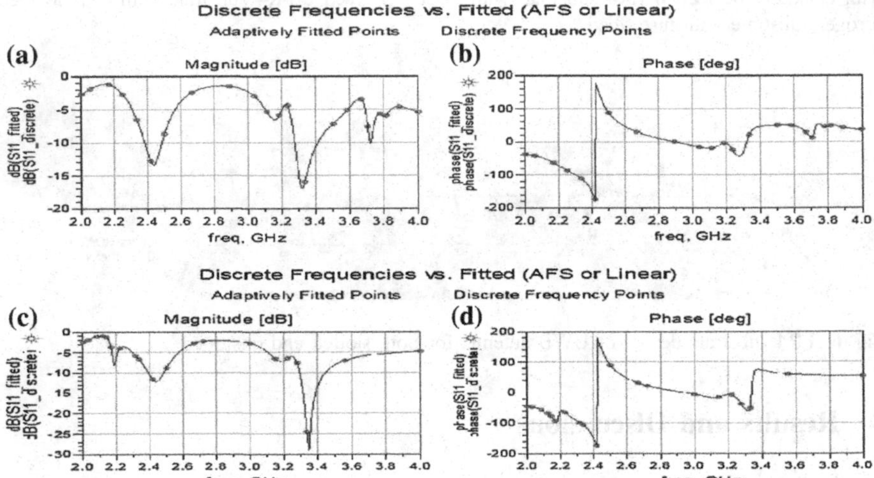

Fig. 7 FR4 substrate design of UWB antenna with inset feed, **a** return loss and **b** phase margins without slots, **c** return loss and **d** phase margin with slots, respectively

reduce the cost and complexity. Antenna parameters are shown for both UWB antennas with and without slots in Table 2.

4.2 *Results for Slotted Patch Antenna*

Figure 7 shows the results of UWB patch antenna design with and without slots. It provides better return loss at −30 dB. Antenna parameters are analyzed which also provide better performance with the UWB antenna which is designed without slots. Figure 7c, d shows antenna parameters of slotted UWB antenna.

Table 2 Antenna parameters

S.No.	Parameters	UWB antenna	Slotted UWB antenna
1	Frequency (GHz)	2	2
2	Input power (mW)	1.19	1.08
3	Radiated power (mW)	1.19	1.08
4	Directivity	9.12	9.64
5	Gain (dB)	9.12	9.64
6	Radiation efficiency (%)	100	100
7	Maximum intensity (mW/Sr)	0.775	0.796
8	Effective angle (sr)	1.538	1.365
9	Angle of U max (Theta, Phi)	35, 118	35, 121
10	E(theta) max(mag, phase)	0.654, −141.52	0.563, −149.05
11	E(phi) max(mag, phase)	0.394, 78.36	0.532, 65.274
12	E(x) max(mag, phase)	0.224, −55.51	0.292, −87.49
13	E(y) max(mag, phase)	0.626, −130.6	0.640, −135.08
14	E(z) max(mag, phase)	0.375, 38.47	0.322, 30.95

5 Conclusion

UWB slotted microstrip patch antenna is used for solving cognitive radio network problems. It can also be used as a navigation system for navigating and guiding visually challenged persons. In this case, two antennas are required to communicate from bus to blind people and vice versa. Slots reduce return loss which in turn improve overall antenna performance. UWB slotted microstrip patch antenna is designed with frequency range from 2 to 4 GHz. The antenna design is simulated and validated using ADS software. The performance of an antenna is analyzed by varying the substrate and properly selecting the dimensions of the slots. Finally, if the appropriate characteristics of an antenna are obtained, then it is fabricated. In future, slots and patch dimensions are varied to obtain better results. The wider applications can be applied for the design of patch antennas.

References

1. Rajagopalan, Harish Member, IEEE, Joshua M Kovitz Student Member, IEEE, and Yahya Rahmat-Samii Fellow, IEEE. 2014. MEMS Reconfigurable Optimized E-Shaped Patch Antenna Design for Cognitive Radio. *IEEE Transactions Antennas Propagation* 62 (3).
2. Kashwan K.R, and V. Rajeshkumar. 2011. Microstrip Array Antenna for Wi-MAX Based Transceiver System. VLSI, Communication and Instrumentation (IVCI). In *International Conference on, Department of Applied Electronics & Instrumentation and the Department of Electronics and Communication Engineering*, 33–36. Kottayam, India: Saintgits College of Engineering.

3. Singh, Harbinder, Ranju Kanwar, and Malika Singh. 2014. Steering Wheel Shaped Frequency Reconfigurable Antenna for Cognitive Radio. *International Journal of Engineering Research and Application*, 4 (5) (Version 1):138–142 (ISSN: 2248-9622).
4. Ang, B.-K., and B.-K. Chung. 2007. A Wideband E-Shaped Microstrip Patch Antenna for 5–6 GHz Wireless Communications. *Progress in Electromagnetic Research, PIER* 75: 397–407.
5. Jin, Nanbo, and Yahya Rahmat-Samii. 2008. Particle Swarm Optimization for Antenna Designs in Engineering Electromagnetics. Hindawi Publishing Corporation Journal of Artificial Evolution and Applications.
6. Amsavalli A., and K.R. Kashwan. 2013. Design and Characterization of Downlink Patch Antenna for DS-CDMA Based Mobile Communication Applications. *International Journal of Engineering and Technology, Engineering Journals Publications*, 5 (6): 4536-4542 (ISSN: 0975-4024).
7. Pauria, Indu Bala. Sachin Kumar, and Sandhya Sharma. 2012. Design and Simulation of E-Shape Microstrip Patch Antenna for Wideband Applications. *International Journal of Soft Computing and Engineering* (*IJSCE*) 2 (3) (ISSN: 2231-2307).
8. Rajagopalan, Harish, Joshua Kovitz, and Yahya Rahmat-Samii. 2012. Frequency Reconfigurable Wideband E-shaped Patch Antenna: Design, Optimization, and Measurements. In *Proceedings IEEE Antennas Propagation Society International Symposium* (*APSURSI*), 1–2.
9. Kumar, Nishant., P. Ananda Raju, and Santanu Kumar Behera, Senior Member IEEE. 2015. Frequency Reconfigurable Microstrip Antenna for Cognitive Radio Applications.
10. Thirumalai, T., and K.R. Kashwan. 2015. Split Ring Patch Antenna for FPGA Configurable RFID Applications. *JNIT: Journal of Next Generation Information Technology* 6 (3): 47–58.
11. Tawk, Y., and C.G. Christodoulou. 2009. A New Reconfigurable Antenna Design For Cognitive Radio. *IEEE Antennas Wireless Propagation Letter* 8: 1378–1381.
12. Thirumalai T, and Kashwan K. R. 2015. Improved Ultra Frequency Design of RFID Patch Antenna. *Advances in Natural and Applied Science* 9 (6): 381–384 (ISSN:1995-0722).
13. Tawk, Y., J. Costantine, and C.G. Christodoulou. 2006. A Rotatable Reconfigurable Antenna for Cognitive Radio Applications.
14. Kashwan K.R., V. Rajeshkumar, Gunasekaran T, and K.R. Shankar Kumar. 2011. Design and Characterization of Pin Fed Microstrip Patch Antennae, Fuzzy Systems and Knowledge Discovery (FSKD). In *Eighth IEEE International Conference on, INSPEC*: 12244584, vol. 4, 2258–2262 (Print ISBN: 978-1-61284-180-9).
15. Fan, Yang, Student Member, IEEE, Xue-Xia Zhang, Xiaoning Ye, and Yahya Rahmat-Samii, Fellow, IEEE. 2001. Wide-Band E-Shaped Patch Antennas for Wireless Communications. IEEE Transactions on Antennas and Propagate 49 (7).

A Robust Framework for the Recognition of Human Action and Activity Using Spatial Distribution Gradients and Gabor Wavelet

Dinesh Kumar Vishwakarma, Jaya Gautam and Kuldeep Singh

Abstract The objective of this paper is to propose a new approach for video-based human action and activity recognition using effective feature extraction and classification methodology. Initially, the video sequence containing human action and activity is segmented to extract the human silhouette. The extraction of the human silhouette is done using texture-based segmentation approach, and subsequently, the average energy images (AEI) of human activities are formed. To represent these images, the shape-based spatial distribution of gradients and view independent features are computed. The robustness of the spatial distribution of gradients feature is strengthened by incorporating the additional features at various views and scale which is computed using Gabor wavelet. Finally, these features are fused and result a robust descriptor. The performance of the descriptor is evaluated on publicly available datasets. The highest recognition accuracy achieved using SVM classifier is compared with similar state-of-the-art and demonstrated the superior performance.

Keywords Human action and activity recognition · Spatial distribution of gradients · Gabor wavelet · Support vector machine

D.K. Vishwakarma (✉) · J. Gautam
Department of Electronics and Communication Engineering,
Delhi Technological University, Delhi, India
e-mail: dvishwakarma@gmail.com

J. Gautam
e-mail: jaya2791@gmail.com

K. Singh
Central Research Laboratory, Bharat Electronics Limited, Ghaziabad, UP, India
e-mail: kuldeep.er@gmail.com

M.S. Reddy et al. (eds.), *International Proceedings on Advances in Soft Computing, Intelligent Systems and Applications*, Advances in Intelligent Systems and Computing 628, https://doi.org/10.1007/978-981-10-5272-9_10

1 Introduction

In recent years, human action recognition (HAR) has been of great interest in computer vision due to its various applications and need in surveillance, assistive health care, ambient intelligence and human–computer Interaction systems [1–4]. Since HAR system has become a recent field of research, human action classification has come out with much advancement [5, 6]. The changing shape and size of a person moving in different directions with respect to a single camera affects HAR system. The other challenging factors include illumination condition, body postures variations, occlusion, cluttered background and performance rate [7]. The system which can adapt to these changes and perform classification accurately is considered as a sober HAR system. The task is challenging due to the environmental conditions and human body taxonomy. In this paper, a novel approach is proposed in the feature extraction part of recognition system for accurate activity recognition like running, jogging, skipping, boxing. There are two types of feature descriptors, namely shape- and motion-based descriptors [8].

The philanthropy given by this paper for efficient representation of human actions is in feature extraction. Firstly, entropy-based texture segmentation is used for foreground detection in which human activity silhouettes are extracted from the video sequences [9]. The various silhouettes from the activity are used to get the average energy image (AEI) features for each activity. The average energy image is a unique feature for an activity in such a way that the actions performed by the same person, i.e. intra-class variations are reduced and the actions performed by different persons are maximized, such that more discriminative information can be extracted for classification. The spatial distribution of gradients (SDGs)-based descriptor is applied on average energy images for computation of feature vector in [9–11]. The SDGs are computed for three levels and instead of going to the fourth level, the previously computed three level parameters are concatenated for reducing the dimension of feature vector. SDGs have a disadvantage of finding features in single orientation. To overcome this problem, Gabor filters are applied to the SDGs feature vector for effective representation of human activity as used in [12]. Gabor filters extract information by changing scale and orientation. These SDGs-Gabor features are exploited for each human activity category of dataset, on which support vector machine is applied for classifying the activity performed in a video.

The rest of the paper is organized as follows: Sect. 2 presents the proposed framework in which the methods used for feature extraction are described. The experimental work, the details of the dataset used and discussion of the result are given in Sect. 2.

2 Proposed Framework

The basic framework of our proposed approach is depicted in Fig. 1 and each block is described in following subsections. Our method is based on the silhouette extraction of the human body by segmentation for each video of various activities. Then, average energy image (AEI) features are computed from segmented silhouettes and region of interest (ROI) is obtained for each activity. Thereafter, SDGs-Gabor feature extraction and classification are used.

3 Extraction of Human Silhouette

In human activity recognition, silhouette extraction by segmentation is considered as a challenging task due to illumination changes, background variation, noise, occlusion, shadows, etc. In silhouette extraction, the fundamental step is to detect the foreground by using textural property differences of human body from the

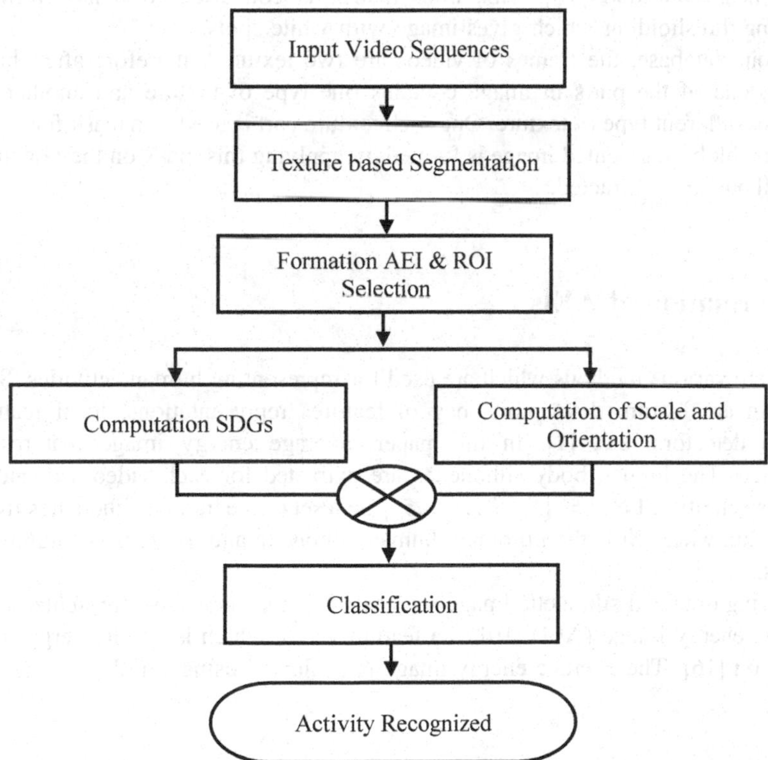

Fig. 1 Flow diagram of proposed framework

whole scene. In the past, Gaussian mixture model (GMM) and local binary pattern (LBP)-based methods have been widely used. A textural-based segmentation technique using gray level co-occurrence matrix (GLCM) is proposed in [13]. After that, various textural feature-based segmentation methods have been proposed [14]. Different techniques describing the texture properties were originally proposed in [13] in which a matrix called gray level co-occurrence matrix is presented that describes the texture features on the basis of intensity variations in different directions.

There are different features for textures in which entropy is the most widely used parameter for describing the textural properties. Entropy measures the randomness of the gray level distribution of an image and can be computed using Eq. 1.

$$\text{Entropy} = -\sum_i \sum_j \rho(i,j) \log \rho(i,j), \tag{1}$$

where $\rho(i,j)$ represents the (i,j)th entry in normalized gray level co-occurrence matrix. A high value of entropy means a random distribution. For representing different textures present in an image, an entropy filter is made in an image. For each pixel, a filter matrix is generated and the entropy value is determined in its 9×9 neighbourhood [15]. This filter matrix is converted to binary form by applying thresholding which gives image with white spots.

In our database, the frames of videos are two textured, therefore after thresholding, one of the parts in image contains one type of texture and another part contains different type of texture. One such texture part is used as a mask for getting a human blob. Segmented image is formed by applying this mask on the raw image and silhouette is extracted.

4 Formation of AEIs

There are various methods which are used for representing human activities. Some of them are feature descriptors, bag of features representations, local features, feature detectors, etc. [1]. In this paper, average energy image features are exploited. The human body silhouettes are extracted for each video sequence of various activities. Let $S = \{S_1, S_2, , \ldots ..S_n\}$ be a set of the human silhouettes from a video clip, where Si is the *i*th binary human silhouette and *n* represents number of frames.

Having obtained silhouette image set, each video sequence is represented by an average energy image (AEI) [10] as a feature vector which is a gait energy image extension [16]. The average energy image is evaluated using Eq. 2.

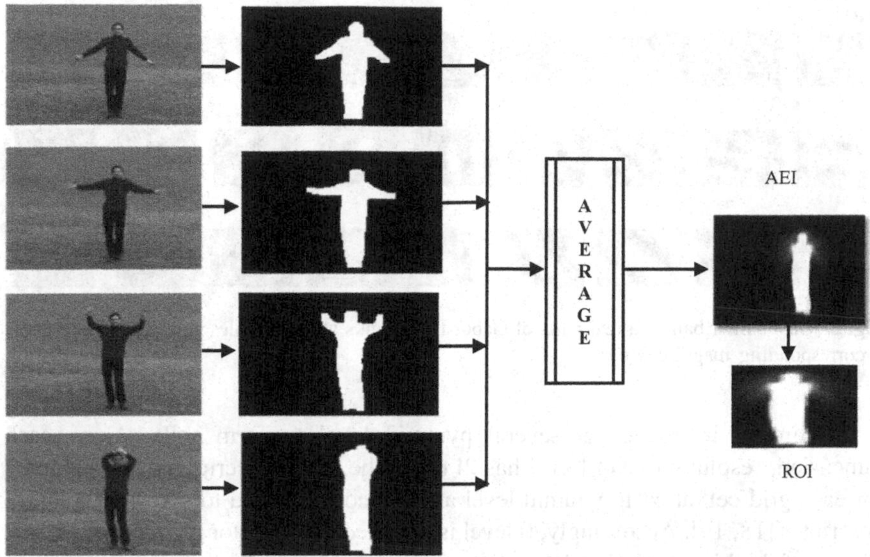

Fig. 2 Process of AEI formation and ROI extraction

$$\text{AEI}(p,q) = \frac{1}{n}\sum_{i=1}^{n} \text{Si}(p,q), \tag{2}$$

where p and q are coordinates of Si. The average energy images are cropped to determine the region of interest (ROI) to reduce the dimension of the image. The ROIs are resized to 64 × 38 for better representation of the average energy images. Average energy features are exploited such that it minimizes the intra-class variations. Figure 2 depicts the ROIs and AEIs of various video clips of an activity.

5 SDGs Computation

In this section, an image is represented by its distribution among edge orientations, i.e. local shape and its spatial arrangement which is dividing the image into blocks at multiple resolutions. SDGs descriptor is a spatial pyramid representation of HOG descriptor The SDGs descriptor is evaluated by computing histogram of orientation gradients in each image sub-region at multiple resolution level and past study shows that it reached good performance [17]. The quantization of edge orientations into K-bins over an image sub-region describes the local shape. Each of the bins represents the number of edges having a particular range of angular orientation. The weight of each edge is calculated by its magnitude and contributes accordingly.

Fig. 3 Gabor filter banks: **a** real parts of Gabor filter banks at three scales and eight orientations, **b** corresponding magnitude part

The image is divided at several pyramid level to form cells. Along each dimension, resolution level l grid has 2l cells The HOG descriptors are computed for each grid cell at each pyramid level and are concatenated to give SDGs image descriptor [18, 19]. Accordingly, 0 level is depicted by *K* vector in analogous to the K-bins of the histogram, level 1 represented by a 4 K-bin histogram, etc., and for the complete image, SDGs descriptor is a vector with dimensionality $K\sum_{l \in L} 4l$. For example, L = 1 level and K = 8 bins has 40 dimension SDGs vector. In our method, the levels are constrained up to L = 3. To assure that images having more edges are not weighted strongly than others, SDGs is normalized to unity. The vectors at each level are concatenated such that the loss of spatial information is minimized. Figure 3 shows the SDGs descriptor at each level.

6 Computation of Scale and View

The Gabor filter is widely used in numerous applications, and it has come out as a robust method under view point and illumination changes [20]. The most prominent characteristic of the Gabor filter is that it captures the spatial location, orientation and frequency characteristic features [21]. In addition to that, Gabor function gives the best time-frequency resolution for signal analysis. Gabor filters extract the edge information by changing the scale and orientation. A 2D Gabor filter defined as a complex sinusoidal plane of certain frequency and orientation modulated by a Gaussian envelope [22]. A two-dimensional Gabor filter in the spatial domain [23] is given as:

$$G(x,y,\theta,u,\sigma) = \frac{1}{2\pi\sigma^2} e^{-\frac{x^2+y^2}{2\sigma^2}} \times e^{\{2\pi i(ux\cos\theta + uy\sin\theta)\}}, \quad (3)$$

where u denotes the frequency of the sinusoidal wave, orientation of the function is controlled by θ and σ is the standard deviation of the Gaussian envelope [24]. In

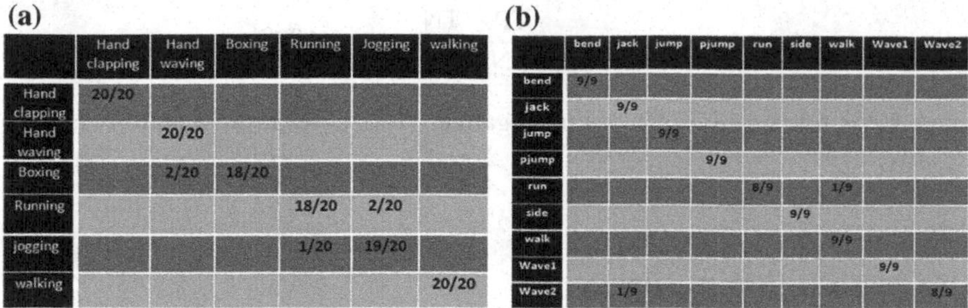

(a)

	Hand clapping	Hand waving	Boxing	Running	Jogging	walking
Hand clapping	20/20					
Hand waving		20/20				
Boxing		2/20	18/20			
Running				18/20	2/20	
jogging				1/20	19/20	
walking						20/20

(b)

	bend	jack	jump	pjump	run	side	walk	Wave1	Wave2
bend	9/9								
jack		9/9							
jump			9/9						
pjump				9/9					
run					8/9		1/9		
side						9/9			
walk							9/9		
Wave1								9/9	
Wave2		1/9							8/9

Fig. 4 Classification results **a** KTH dataset. **b** Weizmann dataset

this work, Gabor Filter banks of five scales and eight orientations are implemented with a total of 40 filters. Gabor filters are applied on the average energy images to give Gabor features. These features are concatenated with SDGs features to give SDGs-Gabor feature vector. Figure 4 shows the Gabor filter banks for three scales and eight orientations. One part is showing the real part of Gabor filter, whereas another is showing the imaginary part of Gabor filter bank.

7 Experimental Result

To demonstrate the effectiveness of the proposed method, we implemented our approach on two public datasets, the KTH dataset [25] and the Weizmann dataset [26]. In this work, for each activity video, the video frames are segmented to give silhouettes which are then averaged to evaluate average energy image features. The region of interest (ROI) containing only the human body is extracted, i.e. frame of size 64×48 is used to represent the average energy image of an activity sequence.

The SDGs features are evaluated at level 0 to give 8 dimension long vector, level 1 has 40 dimension vector, level 2 has 168 long vector. The concatenated SDGs vector of all three levels has a 216 dimension feature vector. In addition to this, Gabor features are also exploited. Gabor filters of five scales and eight orientations are generated and are applied to the average energy image to have 40 images. The mean and variance values of each image are computed and are arranged to give 80 dimensions Gabor feature vector.

The final feature vector in our method comprises of both the SDGs and Gabor features to give 296 dimension feature vector. In training of the classifier, 60 videos are used while testing is done using 20 video sequences for KTH dataset. For Weizmann dataset, six are used for training purpose, whereas three are used for testing of classifier. For each dataset, recognition accuracy is calculated as a parameter to show the efficiency of our method. The average accuracy is evaluated using Eq. 4 and expressed as:

$$\text{Accuracy} = \frac{\text{TP} + \text{TN}}{\text{TP} + \text{TN} + \text{FP} + \text{FN}} \times 100\%, \tag{4}$$

where TP is true positive, TN is true negative, FP is false positive and FN is false negative.

8 Classification Accuracy

The classification results of the proposed approach are depicted in Table 1 and Table 2 on two different datasets. The proposed method is compared with earlier works. The purpose of the comparison is to show the efficiency of the proposed feature descriptor with the existing methods. The comparison is made by analysing the highest ARA achieved on each dataset. The highest ARA is achieved using our method comprises of SDGs-Gabor-based feature description.

The confusion matrix for KTH dataset by giving 20 sequences for validation purpose is shown in Fig. 4a. It can be seen that the classification accuracy of 94.5% is achieved which is a satisfactory result as compared to other methods. Figure 4b shows the cross validation test results of Weizmann dataset without the skip action.

Table 1 Comparison of recognition accuracy with similar state-of-the-art techniques on Weizmann dataset

Method	Parameters		
	Classifiers	Test scheme	ARA (%)
Gorelick et al. [26]	KNN	LOSO	97.5
Chaaraoui et al. [27]	KNN	LOSO	92.8
Wu and Shao [28]	SVM	LOSO	97.78
Goudelis et al. [29]	SVM	LOPO	95.42
Touati and Mignotte [30]	KNN	LOO	92.3
Proposed method	SVM	LOO	97.8

Table 2 Comparison of recognition accuracy with similar state-of-the-art techniques on KTH dataset

Method	Parameters		
	Classifiers	Test scheme	ARA (%)
Sadek et al. [31]	SVM	–	93.30
Saghafi and Rajan [32]	KNN	LOO	92.6
Goudelis et al. [29]	SVM	LOPO	93.14
Rahman et al. [33]	KNN	LOO	94.49
Conde and Olivieri [34]	KNN	–	91.3
Proposed method	SVM	–	94.5

The success rate of 97.8% is achieved in our work which is better in comparison to the state-of-the-art methods.

The highest recognition accuracy achieved on these datasets are compared with the other similar techniques and as given in Tables 1 and 2. Table 1 presents the comparison on Weizmann dataset, and Table 2 presents the comparison on KTH human action datasets. The test methodology used in these approaches are leave-one out (LOO), leave-one-sequence-out (LOSO). Leave-one-person-out (LOPO) and classifier is SVM. Therefore, the comparison is fair enough as the experimental set-up used in our approaches is similar to that in other methods. The high accuracy is achieved due to the use of spatial and multiview information exploited using SDGs-Gabor-based feature description. Hence, it can be said that the proposed framework is robust to handle the view problems.

9 Conclusion

In this work, a human activity recognition system is made robust by exploiting the SDGs-Gabor-based features on average energy images. The recognition rate is increased by employing this technique of feature extraction. Firstly, the silhouettes are extracted accurately using texture segmentation, and then the average energy image features are computed. Secondly, the SDGs vectors are evaluated which are exploiting the spatial properties. The last step is to make the feature descriptor invariant to rotations and scales by computing the Gabor-based features. In regardless to the satisfactory performance, this approach is still less effective under object occlusion and can be further optimized. For future work, one can optimize the method further to achieve good performance even with the challenges of occlusion.

References

1. Agrawal, J.K, and M.S. Ryoo. 2011. Human Activity Analysis: A Review. *ACM Computing Surveys* (*CSUR*) 43 (3): 16.
2. Weinland, D., R. Ronfard, and E. Boyer. 2011. A Survey of Vision-Based Methods for Action Representation, Segmentation, and Recognition. *Computer Vision and Image Understanding* 115: 224–241.
3. Vishwakarma, D.K, and R. Kapoor. 2012. Simple and intelligent system to recognize the expression of speech disabled person. In *4th IEEE International Conference on Intelligent Human Computer Interaction*, Kharagpur, India.
4. Weinland, D., E. Boyer, and R. Ronfard. 2007. Action recognition from arbitrary views Using 3D exemplars. In *IEEE International Conference on Computer Vison.*
5. Gkalelis, N., A. Tefas, and I. Pitas. 2009. Human identification from human movements. In *ICIP*.
6. Iosifidis, A., A. Tefas, and I. Pitas. 2012. Activity based person identification Using fuzzy representation and discriminant learning. *TIFS* 7 (2): 530–542.

7. Brutzer, S., B. Hoferlin, and G. Heidemann. 2011. Evaluation of background subtraction techniques for video surveillance. In *IEEE conference on computer vision and pattern recognition* (*CVPR*).
8. Guha, T., and R.K. Ward. 2012. Learning Sparse Representations for Human Action Recognition. *IEEE Transactions on Pattern Analysis and Machine Intelligence* 34 (8): 1576–1588.
9. Vishwakarma, D.K., A. Dhiman, R. Maheswari, and R. Kapoor. 2015. Human Motion Analysis by Fusion of Silhouette Orientation and Shape Features. *Procedia of Computer Science* 57c: 438–447.
10. Vishwakarma, D.K., and R. Kapoor. 2015. Integrated Approach for Human Action Recognition Using Edge Spatial Distribution, Direction Pixel, and R-Transform. *Advanced Robotics* 29 (23): 1551–1561.
11. Vishwakarma, D.K., R. Kapoor, and A. Dhiman. 2016. Unified Framework For Human Activity Recognition: An Approach Using Spatial Edge Distribution and $\mathfrak{R}$-Transform. *AEU —International Journal of Electronics and Communications* 70 (3): 341–353.
12. Vishwakarma, D.K., R. Kapoor, and R. Rawat. 2015. Human Activity Recognition Using Gabor Wavelet Transform and Ridgelet Transform. *Procedia Computer Science* 57: 630–636.
13. Haralick, R.M., K. Shanmugam, and I. Dinstein. 1973. Textural Features for Image Classification. *IEEE Transactions on Systems Man, and Cybernetics* 6: 610–621.
14. Chua, T.W., Y. Wang, and K. Leman. 2012. Adaptive texture-Color based background subtraction for video surveillance. In *19TH IEEE International conference on image processing* (*ICIP*).
15. Vishwakarma, D.K., and R. Kapoor. 2015. Hybrid Classifier Based Human Activity Recognition Using the Silhouette and Cells. *Expert Systems with Applications* 42 (20): 6957–6965.
16. Han, J., and B. Bhanu. 2006. Individual Recognition Using Gait Energy Image. *Pattern Analysis and Machine Intelligence, IEEE Transactions* 28 (2): 316–322.
17. Bosch, A., A. Zisserman, and X. Munoz. 2007. Representing shape with a spatial pyramid kernel. In *International Conference on Image and Video Retrieval*.
18. Zhang, B., Y. Song, and S. U. Guan. 2010. Historic Chinese Architectures Image Retrieval by SVM and Pyramid Histogram of Oriented Gradients Features. *International Journal of Soft Computing* 5 (2): 19–28.
19. Vishwakarma, D.K., R. Kapoor, and A. Dhiman. 2015. A Proposed Unified Framework for the Recognition of Human Activity by Exploiting the Characteristics of Action Dynamics. *Robotics and Autonomous Systems* 77: 25–38.
20. Long, F., H. Zhang, and D.D. Feng. 2003. *Multimedia Information Retrieval and Management*. Technological Fundamentals and Applications: Springer.
21. Moghadam, P., J.A. Starzyk, and W.S. Wijesoma. 2012. Fast Vanishing-Point Detection in Unstructured Environments. *Image Processing, IEEE Transactions* 21: 425–430.
22. Manjunath, B.S., and W.Y. Ma. 1996. Texture Features for Browsing and Retrieval of Image Data. *IEEE Transactions on Pattern Analysis and Machine Intelligence* 18 (8): 837–842.
23. Weldon, T., and W.E. Higgins. 1991. Designing Multiple Gabor Filters for Multitexture Image Segmentation. *Optical Engineering* 38 (9): 1478–1489.
24. Lim, T.R., and A.T. Guntoro. 2002. Car Recognition Using Gabor Filter Feature Extraction. *Circuits and Systems, APCCAS'02* 2: 451–455.
25. Schuldt, C., I. Laptev, and B. Caputo. 2004. Recognizing human actions: A local SVM approach. In *International Conference on Pattern Recognition*.
26. Gorelick, L., M. Blank, E. Shechtman, and M. Irani. 2007. Actions as Space–Time Shapes. *Pattern Analysis and Machine Intelligence* 29(12): 2247–2253.
27. Chaaraoui, A., P.C. Perez, and F. Revuelta. 2013. Sihouette-Based Human Action Recognition Using Sequences of Key Poses. *Pattern Recognition Letters* 34: 1799–1807.
28. Wu, D., and L. Shao. 2013. Silhouette Analysis-Based Action Recognition Via Exploiting Human Poses. *IEEE Transactions on Circuits and Systems for Video Technology* 23 (2): 236–243.

29. Goudelis, G., K. Karpouzis, and S. Kollias. 2013. Exploring Trace Transform for Robust Human Action Recognition. *Pattern Recognition* 46 (12): 3238–3248.
30. Touati, R., and M. Mignotte. 2014. MDS-based multi-axial dimensionality reduction model for human action recognition. In *Proceedings of IEEE canadian Conference on Compter and Robot Vision*.
31. Sadek, S., A.A. Hamadi, M. Elmezain, B. Michaelis, and U. Sayed. 2012. Human action recognition via affine moment invariants. In *21st International conference on Pattern Recognition*.
32. Saghafi, B., and D. Rajan. 2012. Human Action Recognition Using Pose-based Disriminant Embedding. *Signal Processing: Image Communication* 27: 96–111.
33. Rahman, S., I. Song, M.K.H. Leung, and I. Lee. 2014. Fast Action Recognition Using Negative Space Features. *Expert Systems with Applications* 41: 574–587.
34. Conde, I.G., and D.N. Olivieri. 2015. A KPCA Spatio-Temporal Differential Geometric Trajectory Cloud Classifier for Recognizing Human Actions in a CBVR System. *Expert Systems with Applications* 42 (13): 5472–5490.

Design of a Cloud-Based Framework (SaaS) for Providing Updates on Agricultural Products

Kuldeep Sambrekar, Vijay S. Rajpurohit and S. Bhagyashri Deyannavar

Abstract Nowadays cloud computing is providing a step which brings technology into all the fields to make them convenient and more useful. Here this technology is used in agriculture field to design a cloud-based framework which provides the updates on mobile phone that would be helpful to the user to get required information anytime and anywhere. The cloud computing technology is provided with a knowledge base expert system and automatic updates, which gives an establishment and recognition to the farmers and their products. In this proposed system, the data will be provided to the cloud. The framework is designed, which may vary from pricing, availability, storage, weather, need of various products on the market and receiving some cropping advices. Here a cloud-based framework gives a great recognition to farmers and the agriculture. It helps to know the best prices for products and varies exact demand in the market. This proposal is the initial step to induce a new green revolution in our nation.

Keywords Cloud computing · Agriculture · Knowledge base · SMS updates

1 Introduction

Farmers are the backbone of our nation; our whole world population depends on farmer's cultivation. Agriculture contributes a major role of the economical value in India. Growth and wealth of the Indian economy depends on the path of agricultural production. This meets the complete demands of some individuals. In the field of agriculture, marketing is deciding the value of the agricultural products in terms of

K. Sambrekar (✉) · V.S. Rajpurohit · S. Bhagyashri Deyannavar
KLS Gogte Institute of Technology, Belagavi, Karnataka, India
e-mail: kuldeep.git@gmail.com

V.S. Rajpurohit
e-mail: vijaysr2k@yahoo.com

S. Bhagyashri Deyannavar
e-mail: bhagya.cs10@gmail.com

M.S. Reddy et al. (eds.), *International Proceedings on Advances in Soft Computing, Intelligent Systems and Applications*, Advances in Intelligent Systems and Computing 628, https://doi.org/10.1007/978-981-10-5272-9_11

money. Most of the farmers promote their products through village-level marketers/vendors to the local market without knowing their actual price in Indian market. So this brings an economical loss for the farmers and also to the customers. This proposed work discusses everything about providing updates on various agricultural products as per the user requirements on user's mobile phone. Basically, this will be expected to be helpful for farmers around the country. The Internet and mobile networks have the potential to provide agricultural information services that are at affordable, relevant (timely and customized), searchable and up to date. The mobile phone technology benefits the customers to get information anytime/anywhere, with physical contact-less services. Keeping these factors and needs of Indian farmers in mind, various applications and services are provided for agricultural information. The impact of used technology in this project is on cloud computing which is provided with the knowledge base expert system for automatic updating which matches the requirements of the particular farmers. This work is specifically concerned to our farmer that provides a convenient support to them. Here farmer can get the all available information about the marketing transactions on agricultural products on his mobile phone, by which the farmer owns the direct control of the price. In this way, they would be able to establish a relationship of trust and exchange of information without intermediaries. This relationship would be established between producers and consumers that would induce in marginal development of rural areas.

As today's generation, all are moving towards the technology, where it making things with ease of access. This is really trying to help and build a well-formatted environment by which everyone can afford the services through this technology. So somewhere a thought may arise how if the farmer who really supporting our nation's population can get some accessibility with this upcoming technology by which they can get updates on their field-related information which will provide such service to them. That provides anywhere and anytime access based on cloud-based work.

In this proposed system, the application and the computation are moved into Cloud Server. The cloud technology is chosen, because it is a ready to serve high business segment, which is our area of concern. Cloud computing makes it possible to configure general-purpose, online data by giving command to support any software application quickly. These services are commercially used and viable as the costs, which are related to the data centre that can be shared among many users. The leading forecasting institutions are expecting that India will play a bigger role in world markets in future.

2 Literature Survey

The work aimed [1] is to provide the characteristics of supply chain, here it focussing on the particular food short chain and highlighted what are the weakness and strengths of the agricultural organizations and industries, in which these analyses are proceeded by the management of the agricultural organizations using latest IT innovations by specially focusing on the cloud. It is providing the role of improving processes and also market information. The concept is induced for supply and demand of short chain. The essential advantages are economically benefited by direct sales of food, and reduction of energy consumed in packing and transporting process. So the overall applications and services require only governing them to check the business needs in terms of availability, cost and performance agility. Cloud provides the supportive role by giving flexibility, security and scalability to data management and its applications.

The major challenges in agriculture sector bring certain difficulties in green revolution and evergreen revolution which is introduced in supply chains with overregulated of increased cost, risk and uncertainty, slow progress in implementing land reforms and inefficient finance and marketing services, government policy in which taxes often changes without any notice, no storage facility for food which yields to the wastage of agricultural products which tends to a huge amount of loss to the Indian economy with a lack of knowledge and information. So the easy-to-use interface [2] is provided using mobile phones to both the farmers and agencies concern and updates on market status of various products as per the user demand on daily or weekly bases on farmers mobile phones via SMS-based format, where using the Frontline SMS controller design and maintenance of a Backend Database of all the products whose control is given for the dealers.

Green commerce [3] will open farmer's products to a wide market. From earliest times to the till date, the retailers depend on a third party to get their demands. Green commerce has proposed in such a way that it gives an establishment and recognition to farmers and their products, where it provides a direct marketing phase between farmers and retailers which make a good relationship between them. In the current scenario, the farmer fails to get the exact price value of a product which is completely not entertained in Green commerce.

The development of agriculture is still under development from the past few years due to lack of knowledge and environmental changes. Here, the motto [4] is to reach farmers to provide awareness, usage and perception in e-agriculture. The agriculture field in India is currently facing with problems due to inadequate investments, lack of attention, non-given fair prices for farmers crops. The study used in this statistical survey design technique is to collect data from farmers for their awareness in e-commerce. The results obtained indicated that the level of awareness is less such that there is a need for e-agriculture for their support, so this e-agriculture is a platform for supporting marketing of agricultural products and an emerging field in connection of information, development and entrepreneurs related to agriculture. This system uses data mining method of clustering techniques for group of farmer's data. It verifies in terms of clustering efficiency to compare the parameters such as time and accuracy. Clustering technique will be provided to group the data of farmers to make a comparison with a similar substance.

Nowadays, the latest area of ICT is cloud computing, which is providing IT as a service to the cloud user's on-demand basis with greater flexibility, scalability, reliability and availability with utility computing model. The use of IoT [5] paradigm provides advantages of ubiquitous interconnection of billion embedded devices that can be uniquely identified, localized and communicated. Supply chains are increasingly building as synchronizing supply with demand on competitive infrastructure and measuring the performance. So this concept enhances the virtualization of supply chains in agricultural sector with a propagation technique where farmer can also track up-to-date farming and check the whole process of production and distribution to consumption.

Cloud computing [6] for government in any other organization needs considerations about who the service users will be and who the providers will be, what control and ownership of the services which are made technically possible. Also deals with the implementations across various organizations elaborated on the theoretical framework in which essentially the work has been done by building cloud computing data centres for load balancing and computing virtulization. It brought the great opportunities in China's agriculture [7], presented the framework applications and promotion of cloud technology, which also provides the tracking and monitoring the quality of the products based on supply chain technology. Need of awareness [8] and promotion among the prime stakeholders is required to acquire the correct and massive information about the agro-sector which gives the awareness to the farmers and have a well-established information base for the nation. This will in return lead to a well-connected world.

3 Proposed System

Figure 1, in this proposed system, we are designing a cloud-based framework which can be helpful to the farmer/user to get the information about the monsoonal rainfall, market price, stock availability, and also they can get query processing. Here, we are proposing such system which is merely providing the future prediction

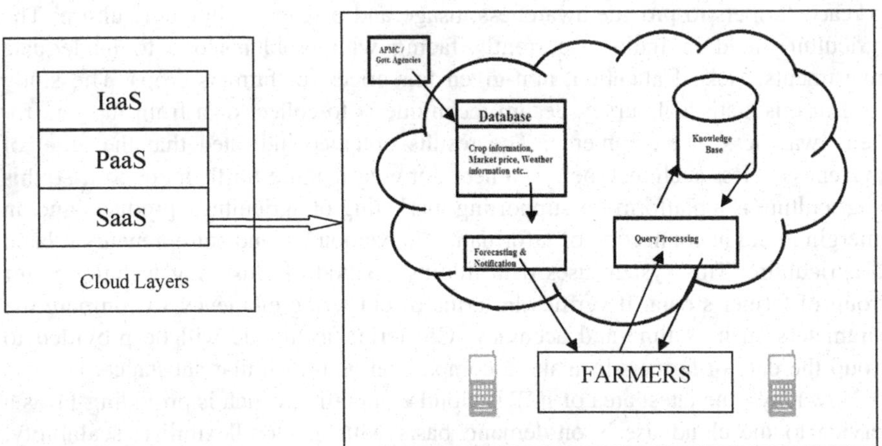

Fig. 1 Proposed architecture

on the particular data by depending on its previous value which are available and already trained as a data set. Forecast model system is used to provide the predicted value based on the data it has been trained and the user requirements. So here, user can login to the system on cloud from anywhere and anytime and get access to it. Where farmer/user can get the values as per there requirements for what they have registered, it may be price, rainfall, so on. And as well user can search the query by giving a query to the system and can get the relevant information by knowledge base system which has been provided with open natural language process (NLP). So with respect to these functions, the proposed system works with aspects of the user requirements.

Following are some of the possible solutions which can be offered by cloud service providers to the farmers.

- Store and maintain a database for the information generated in daily activities
- Database for crop-related information
- Database for market-related information
- Database for production-related data

The following are the implementation of different modules that can be used to implement or propose this project.

Naive Forecasting

Naive forecasting model is a constraint-based algorithm in which dependency value between two variables will not consider the effect of other variable on their relationship. Forecasting based on this naive function is easily based on the history of weather condition data sets after discretization. It can be vision in cloudy, sunny, windy, raining and so on form. So here in this, we have described this naïve function for predicting the future values provided for monsoonal rainfall which will be helpful for the farmers to plant their crops. **This naive algorithm is effective and efficient for machine learning views to solve the problems by predicting the future views on weather.**

For weather, it can be assumed as:

Every″Ti = $a_i + a_{i+1}/2$;″
DataSet observations = new DataSet();
ForecastingModel model = new NaiveForecastingModel();

Polynomial Regression Forecasting

Once the polynomial regression function is initialized with a value, then it can be applied to another data set using the forecasting method to forecast the values of the dependent variable. So once it get trained and initialized with the form of dependent variables, it can be helpful for the polynomial method to predict the number of different period sample values, respectively. Here, this function can give up to 10 coefficients of variables. This polynomial regression model is a highly dynamic method. It implements such interface, which can be used for prediction of the market price values for different crops according to the user choice. **This method is**

effective and efficient, motivated for machine learning views to solve the problems by predicting the future views on market price predictions.

```
DataSetobservations = new DataSet();
ForecastingModelmodel = new PolynomialRegressionModel("powerconsumption", 4);
```

ARIMA Forecasting

Seasonal and Non-Seasonal ARIMA Forecasting

"Seasonal ARIMA" Forecasting model is formed by including the additional seasonal terms in the ARIMA models. **It can be used to predict the future values depending on the seasonal terms.** And similarly for **"Non-Seasonal ARIMA", it will predict the values as per non-seasonal terms depending on the availed data on the particular, respectively**.

```
DataSet observations = new DataSet();
ForecastingModel model = new MovingAverageModel(4);
```

Query Answering/Processing

Query answering or processing is to be done on the knowledge base system which is basically carrying with machine language views. In this, query processing can be done in the format of natural language processing (NLP) which is provided by the word match score with the comparison of trained database.

```
String searchpath;
searchpath = folder;
TestOpenNLP.load();
```

Knowledge Base

In this module, it will specify the knowledge base provided to this framework, where it can be designed with the NLP, neural networks, fuzzy logic, etc., which basically provides understanding between the machine and human beings. So by this, it will be helpful by giving a word like adjective as a search to the machine and as it searches for the particular word with the trained dataset and will find the word score for the input which user has given. With the help of that, it can recognize the word containing with the similarity in the data set, and by this, it will be helpful for the user to find his required information from the particular file with collected data score as per his input to the system. In this way, it will find the values for the queries related to it and process it to the user. Knowledge base can be trained with the regular query information; with this, it can be automatically process the query which has given to it and will proceed.

Advantages and Limitations
Advantages

- Economic Benefits: It provides the direct sales prices of the products which are more profitable for the producers which intercept a greater share of added value.
- Social Benefits: The farmer owns the direct control of the price, in this way; they would be able to establish a relationship of trust and exchange of information without any intermediaries. By this, it will form social bonding between producer and consumers.

Limitations

- With applications moving to the cloud, there is a real risk of the network becoming the single point of failure.
- It should have the existing "ICT" structure to provide such cloud framework on it.
- Because of a large number of connections to the "cloud", the bandwidth may need to be upgraded to provide a continuous network connection.

4 Conclusion

This proposed model, if implemented, reduces the gap between farmers and the retailers, and thereby making a good profit to the farmer's community. With this proposed model, farmers can get all updates related to agriculture and also the crops that can be cultivated in future.

The service provided by the proposed model will be product- and market-specific. A farmer can choose his products and will get the related information of which he is in need of. Also, this model will create a social bonding between the farmers that are associated with the Agriculture Product Market Community.

References

1. Contoa, Francesco, Nicola Faccilongoa, Piermichela La Salaa, and Raffaele Diceccaa. (2013). Cloud approach for short chain administration. In *6th International Conference on Information and Communication Technologies in Agriculture, Food and Environment* (*HAICTA*).
2. Praveen, B., and M. Viswesh. 2013. Agriculture Updates via SMS—A Cloud Computing Approach. *International Journal of Innovation, Management and Technology* 4 (5).
3. Dr. Manimekalai1, S. 2013. A Cognitive Approach to Mobile Application in Green Commerce. *International Journal of Emerging Trends & Technology in Computer Science* (*IJETTCS*) 2 (6).
4. Thankachan, Sumitha, and Dr. S. Kirubakaran. 2014. E-Agriculture Information Management System. *International Journal of Computer Science and Mobile Computing* 3 (5).

5. Satpute, Prashant, Dr. D.Y. Patil, and Omprakash Tembhurne. 2014. A Review of Cloud Centric IoT based Framework for Supply Chain Management in Precision Agriculture. *International journal of advance research in computer science and management studies, research article/survey/case study* 2 (11).
6. Festus onyegbula, maurice dawson and jeffrey stevens. 2011. Understanding the Need and Importance of the Cloud Computing Environment Within the National Institute of Food and Agriculture. *Journal of Information Systems Technology & Planning* 4 (8).
7. Yanxin Zhu, Di Wu and Sujian Li. 2013. Cloud Computing and Agricultural Development of China. *IJCSI International Journal of Computer Science Issues* 10 (1). ISSN (Print): 1694-0784| ISSN (Online): 1694-0814 www.IJCSI.org.
8. Patel, Rakesh, and Mili Patel. 2013. Application of Cloud Computing in Agricultural Development of Rural India. *International Journal of Computer Science and Information Technologies* 4 (6).

A Novel Approach for Pomegranate Image Preprocessing Using Wavelet Denoising

R. Arun Kumar and Vijay S. Rajpurohit

Abstract Image preprocessing is the first and foremost unit in any image processing/machine vision application, because preprocessing inspirits the accuracy of the other imaging processes such as morphological processing, segmentation, feature extraction. The major difficulty in preprocessing is the variation in the levels of intensity values of the image and the presence of noise. Therefore, histogram equalization needs to be performed prior to any further processing of the image. This paper reports on the development of a novel approach for enhancement of the pomegranate images. The major contribution of the present paper is to perform histogram equalization of pomegranate images in combination with wavelet denoising. The proposed method not only enhances contrast through histogram equalization but also has advantage of compensating for the loss of information of the images. The paper compares experimentally the informational entropy of the images. Different alternatives of preprocessing were applied, and informational entropy was computed. To evaluate the robustness and accuracy of the proposed method, tests were conducted for 166 sample images. The results showed an accuracy of 90% success rate in retaining informational entropy of the images when wavelet processing was applied in combination with histogram equalization. The results illustrate that the proposed method can be easily extended to other image processing-based agricultural applications.

Keywords Image preprocessing · Pomegranate · Histogram equalization
Wavelet denoising · Informational entropy

R. Arun Kumar (✉) · V.S. Rajpurohit
KLS Gogte Institute of Technology, Belagavi, India
e-mail: kumararun37@gmail.com

V.S. Rajpurohit
e-mail: vijaysr2k@yahoo.com

M.S. Reddy et al. (eds.), *International Proceedings on Advances in Soft Computing, Intelligent Systems and Applications*, Advances in Intelligent Systems and Computing 628, https://doi.org/10.1007/978-981-10-5272-9_12

1 Introduction

"The Histogram of a digital image with intensity levels in the range [0, L − 1] is a discrete function $h(r_k) = n_k$, where r_k is the kth intensity value and n^k is the number of pixels in the image with intensity r_k" [1]. In general, the image histogram represents the tonal distribution of the gray level values in a digital image [2]. Histogram modeling is a powerful tool for real time image processing with various advantages such as (1) histograms are the foundation for various spatial domain processing techniques (2) used for image enhancement (3) information inherent in histograms is useful for allied image processing tasks such as segmentation, compression. (4) histograms are simple to calculate in software (5) offer themselves for economic hardware implementations.

Improving the quality of the image plays a crucial role in any image processing application. In the present paper, image enhancement is done by applying histogram equalization as the spatial domain technique and wavelet denoising as the frequency domain technique.

Histogram equalization achieves the contrast improvement of the image by uniformly distributing the intensity across the histogram. But, it has the disadvantage that it decreases the informational entropy of the image. Histogram equalization may over-enhance the contrast and may produce enhancements which are unnatural [3]. This drawback can be overcome by further enhancing the image based on wavelet denoising. Hence, goal of the present work is to enhance the image quality through histogram equalization not compensating for loss of informational entropy [4].

Pomegranate has been chosen as the object of interest, and experiments were conducted. The rest of the paper is organized as follow: general information on pomegranate fruit has been outlined in Sect. 2. Section 3 describes the literature review. The materials, methods, and tools used to carry out the experimentation are outlined in Sect. 4. Results are outlined in Sect. 5. Finally, the paper is concluded.

2 Pomegranate

Fruits are the foremost sources of vitamins and minerals. Pomegranate (Punica granatum), the so-called fruit of paradise is one of the major fruit crops of arid region. It is popular in Eastern and Western parts of the world. The fruit is grown for its attractive, juicy, sweet–acidic, and fully succulent grains called "Arils." Magical therapeutic values and increasing demand for table and processed products as well as high export potential have made pomegranate a popular fruit of tropical and subtropical regions in recent times. Important commercial cultivars grown in India are Bhagwa, Ruby Dholka, Ganesh, Kandhari, Jyoti, Arakta, Mridula G-137, Jalore seedless, Jodhpur red. Figure 1 shows few cultivars of pomegranate.

India is the world's leading country in pomegranate production. Estimated global cultivated area under pomegranate is around 3 lakh ha and production

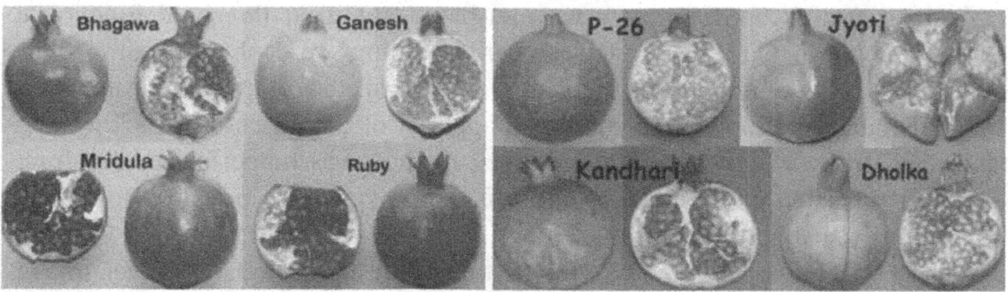

Fig. 1 Few cultivars of pomegranate (courtesy: [5])

3.0 million tonnes. As per recent advance estimates for the year 2014, available at the National Horticulture Board of India, total production in India is 822.80 thousand MT. Export of pomegranate is 31.33 thousand MT (Rs. 2985 million) in 2013–14. There is scope for exporting Indian pomegranates to Bangladesh, Bahrain, Canada, Germany, U.K, Japan, Kuwait, Sri Lanka, Omen, Pakistan, Qatar, Saudi Arabia, Singapore, Switzerland, U.A.E., and U.S.A. Unfortunately, there are no organized marketing systems for pomegranate. Farmers normally dispose their produce to contractors who take the responsibility of transport to far off markets [6, 7].

3 Literature Review

Fu et al. conducted a research [8] where in histogram equalization (HE) was applied on gastric sonogram images and found that there was a reduction in the information in the image after HE. Hence, the authors proposed a new algorithm to enhance the image based on wavelets to recompense for the loss of information. Authors found that drawbacks of an algorithm based solely on wavelets or histogram equalization can be eradicated by wavelet post-processing of the histogram equalization output. The results showed that wavelet-based equalization not only increased the informational entropy but also reduced the variance in informational entropy among gastric wall images for different cases.

Fuzeng et al. used the method of wavelet transform [9] on agricultural image enhancement. Xiankelai flower, sectional view of orange and Chinese date were the objects of interest. Wavelet transform was applied to low contrast images, and results of image enhancement were satisfactory when compared against the traditional histogram equalization, log transform, and LoG filters. The results gave rise to a well-defined detection and classification of dates.

Yin et al. proposed new algorithm [10] is coalescing with histogram equalization and wavelet transforms to enhance the infrared images. The drawbacks of low contrast, low resolution, bad visual effects were overcome using histogram equalization. But, at the same time, image details were lost in the process of enhancement. Hence, the authors proposed a new algorithm that applies a wavelet transform for the histogram equalized images. Results demonstrated that the algorithm was effective for the infrared images.

Yi-qiang et al. proposed a new wavelet-based nonlinear threshold function to denoise the image [11]. Best effects of thresholding were obtained due to the deliberation of two important denoising factors of the optimal threshold and threshold function. Proposed method was applied for the process of picking vegetables and fruits in the field. The proposed method was outperformed the traditional soft threshold and hard threshold with improved peak signal-to-noise ratio (PSNR) and visual appeal.

Zhang et al. proposed a method for texture analysis of ultrasonic liver images [12]. Images were initially applied with wavelet denoising and histogram equalization. The preprocessed images werc conductive to extract quantitative feature parameters from gray level difference statistics (GLDS). The features are fed to a neural network classifier to distinguish between fatty liver and normal liver. Results proved to be satisfactory for clinical diagnosis.

4 Materials and Method

4.1 Overall Methodology

Figure 2 depicts the overall methodology. Altogether, 6 methods are applied for the input image, viz. (1) original image, (2) histogram equalization, (3) wavelet processing, (4) histogram equalization followed by wavelet processing, (5) spatial filtering followed by histogram equalization followed by wavelet processing, and

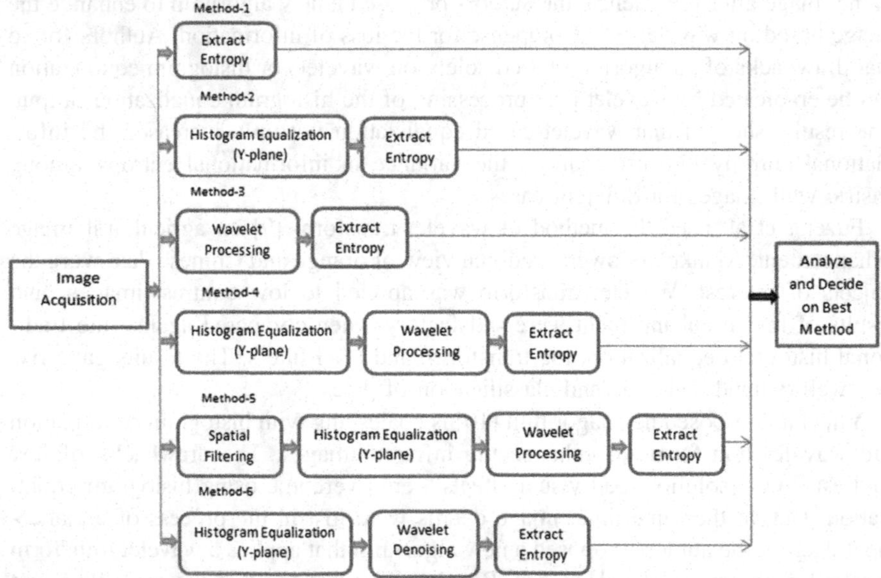

Fig. 2 Overall methodology of the proposed work

(6) histogram equalization followed by wavelet denoising. The value of informational entropy extracted for every method, and finally, an optimal method is decided based on the highest value of the entropy.

4.2 *Sample Collection*

The sample images of pomegranate are collected from a local fruit market at Belagavi, Karnataka, India. The varieties considered are mix of Bhagwa, Ganesh and Arakta. Altogether, 126 samples are collected and used for experimentation.

4.3 *Image Acquisition*

Image acquisition is the first and foremost step before we begin to process the image. In the present work, images of pomegranate fruit samples are captured using a closed metal compartment with provision for light source and cameras placed inside, mimicking the packing lines in industries. Figure 3 shows the image acquisition compartment. The camera used for image acquisition is Logitech Webcam C905 with 2MP sensor and 8MP photos. All the acquired images are stored in jpg format.

4.4 *Histogram Equalization*

The images are captured with a light source. Therefore, the light has effect on results after image processing. In order to neutralize the effect, we need to equalize the spread of intensity values. The equivalent digital image processing method is histogram equalization.

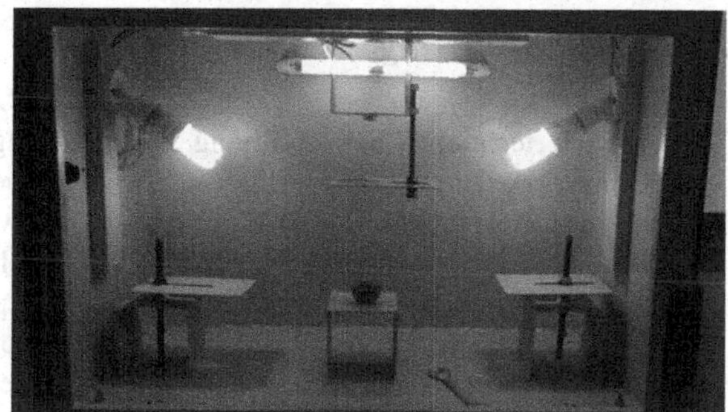

Fig. 3 Image acquisition compartment

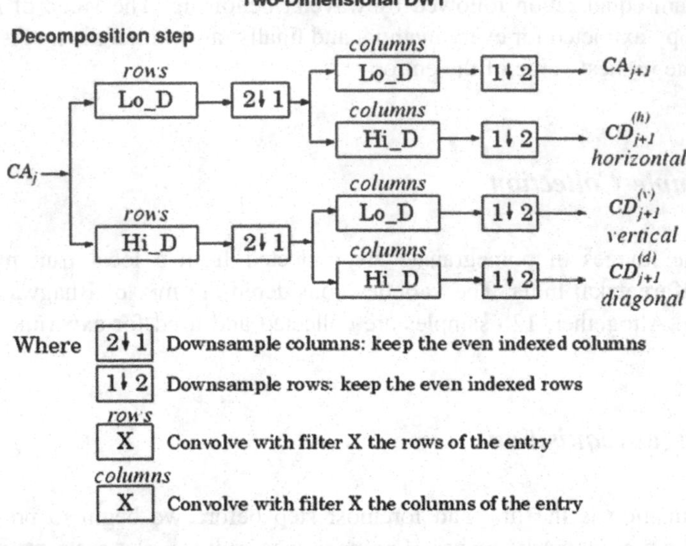

Fig. 4 Basic decomposition steps

Histogram equalization is a technique for adjusting image intensities to enhance contrast. Histogram equalization can be applied for single channel image. Since the original image is a RGB image, we first need to convert to other color model. The suitable model is Y, Cb, and Cr. Then, split the image into 3 images of different channels: Y, Cb, and Cr. Histogram equalizes Y component and finally merge into 3 channel image and revert to RGB image.

But histogram equalization would reduce the information of the image. This can be analyzed by comparing entropy value of original image and histogram equalized image. In order to improve the information of the image, we can apply wavelet processing followed by denoising. This will drastically increase the informational entropy of the image.

4.5 Spatial Filtering

Filtering is a technique for modifying or enhancing an image. Filtering is a neighborhood operation, in which the value of any given pixel in the output image is determined by applying some algorithm to the values of the pixels in the neighborhood of the corresponding input pixel. A pixel's neighborhood is some set of pixels, defined by their locations relative to that pixel. Linear filtering of an image is accomplished through an operation called convolution. Convolution is a neighborhood operation in which each output pixel is the weighted sum of neighboring input pixels. The matrix of weights is called the convolution kernel, also known as the filter.

In the present work, Gaussian low pass filter (GLPF) has been used to perform spatial filtering.

4.6 Wavelet Denoising

To compensate for the loss of information after histogram equalization, wavelet denoising has been applied as a post-processing technique. Denoising is one of the most important applications of wavelets [13]. Wavelet transform can be performed efficiently using discrete wavelet transform (DWT). Two-dimensional DWT leads to a decomposition of approximation coefficients at level j in four components: the approximation at level $j + 1$, and the details in three orientations (horizontal, vertical, and diagonal). Figure 4 describes the basic decomposition steps.

5 Results and Analysis

As mentioned in the block diagram, totally 6 methods are applied, and informational entropy of the image was extracted. Pomegranate images are segregated as diseased and healthy for the purpose of analysis. Tables 1 and 2 show the results obtained for each method for both diseased and healthy images of pomegranate. Figures 5 and 6 show the respective plots of comparisons. For the presentation of results, only 10 image samples are considered comprising both healthy and diseased ones.

From the table of results, following observations can be made:

Observation 1: There was a reduction in the informational entropy of the images after applying histogram equalization to the original images. Figure 7 illustrates the same.
Observation 2: The informational entropy was almost same or slightly reduced even after applying wavelet processing to the original images.
Observation 3: There is a reduction in the informational entropy, compared to the observation 2, after histogram equalization and wavelet processing on the original image.
Observation 4: There is no further improvement on the result even after applying spatial filtering to the original image.
Observation 5: There is a better contrast and improved informational entropy after applying wavelet denoising.
Observation 6: There is a better improvement with respect to noise reduction when spatial filtering is compared with wavelet denoising.

Table 1 Comparing informational entropy for samples with different methods (diseased images)

	Sample 1	Sample 2	Sample 3	Sample 4	Sample 5	Sample 6	Sample 7	Sample 8	Sample 9	Sample 10
Method-1	7.5906	7.16	7.5814	7.3047	7.1207	7.2194	7.204	7.2018	7.7363	7.3533
Method-2	6.5214	6.8963	6.5525	6.559	6.774	6.4935	6.6255	6.9322	6.7382	6.8388
Method-3	7.5904	7.1594	7.5782	7.3037	7.1178	7.2193	7.2019	7.2011	7.7358	7.3495
Method-4	6.4981	6.8869	6.5482	6.5404	6.7658	6.4921	6.6125	6.9244	6.7317	6.8259
Method-5	6.5071	6.8773	6.5593	6.5355	6.7374	6.4804	6.5766	6.9293	6.7283	6.7917
Method-6	7.5521	7.6691	7.5869	7.6527	7.6824	7.5049	7.6013	7.7632	7.7084	7.7384

Table 2 Comparing informational entropy for samples with different methods (healthy images)

	Sample 1	Sample 2	Sample 3	Sample 4	Sample 5	Sample 6	Sample 7	Sample 8	Sample 9	Sample 10
Method-1	7.355	7.262	7.3727	7.4534	7.6878	7.7134	7.3135	7.5824	7.2811	7.3089
Method-2	6.543	6.4615	6.5466	6.5078	6.5198	6.4456	6.4755	6.5294	6.5541	6.417
Method-3	7.3547	7.2615	7.3707	7.4442	7.6744	7.7011	7.3133	7.5805	7.2796	7.2979
Method-4	6.5227	6.4386	6.5396	6.4997	6.5027	6.4321	6.4566	6.5088	6.5519	6.4016
Method-5	6.7917	6.4315	6.5664	6.4989	6.489	6.4242	6.4609	6.4977	6.5489	6.434
Method-6	7.555	7.6103	7.6603	7.6138	7.6101	7.6019	7.5296	7.5905	7.6104	7.5344

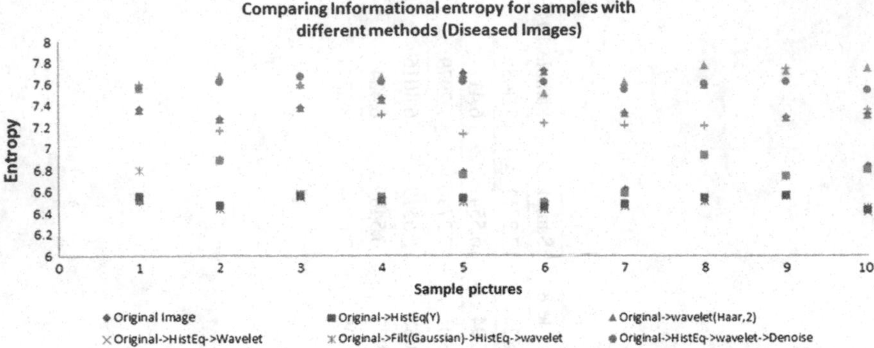

Fig. 5 Comparing informational entropy for samples with different methods (diseased images)

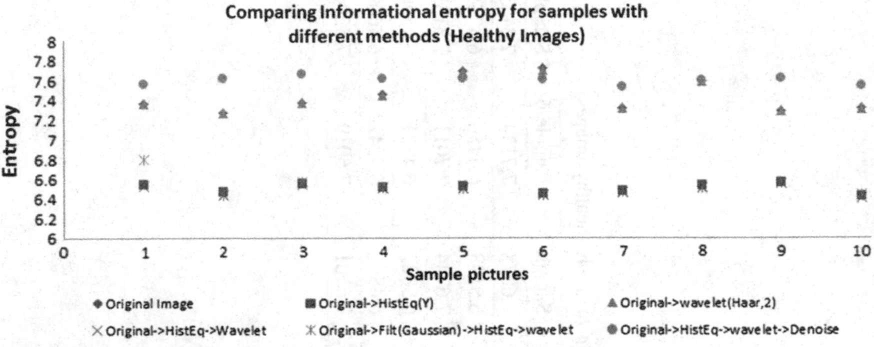

Fig. 6 Comparing informational entropy for samples with different methods (healthy images)

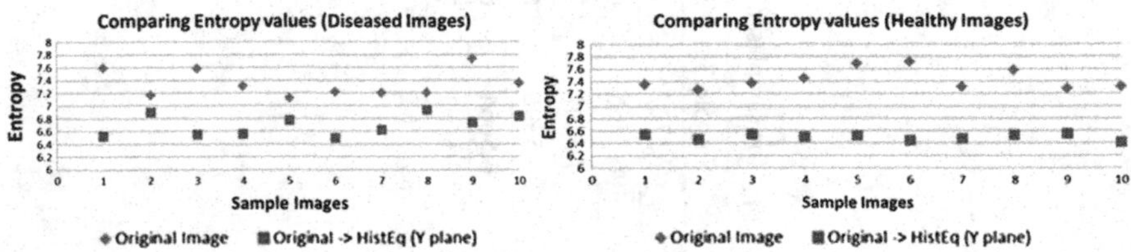

Fig. 7 Comparing informational entropy for samples before and after histogram equalization (diseased images and healthy images)

Inference: Histogram equalization improves the contrast of the image but reduces the informational entropy, which can be compensated with wavelet denoising processing which also yields a better denoised image.

6 Conclusion

In the present paper, an optimal method has been presented to effectively preprocess the images of pomegranate. There are 6 methods applied for each of the sample image. The problem with captured images was that there is an influence of the light source and noise. Hence, to reduce such effects, usually, histogram equalization is the method which enhances the contrast of the image. And spatial filtering can be applied to reduce the noise. But there is a loss of information in the image after histogram equalization. Hence, to compensate for the loss of information, a wavelet processing has been applied followed by denoising application which can effectively denoise the input image. The accuracy of the method is 90% which is a promising result to apply wavelet denoising in image processing applications for agricultural products.

References

1. Rafael C. Gonzalez, Richard E. Woods, and Steven L. Eddins. 2009. Digital Image Processing, 3rd ed. Pearson Education.
2. Krutsch, Robert and David Tenorio. 2011. Histogram Equalization, Application Note, Freescale Semiconductor, Document Number: AN4318.
3. Sonker, D., & Parsai, M. P. (2013). Comparison of Histogram Equalization Techniques for Image Enhancement of Grayscale images of Dawn and Dusk. *International Journal of Modern Engineering Research (IJMER)* 3 (4).
4. Rajput, Y., V.S. Rajput, A. Thakur, and G. Vyas. 2012. Advanced Image Enhancement Based on Wavelet and Histogram Equalization for Medical Images. *IOSR Journal of Electronics and Communication Engineering* 2 (6): 12–16.
5. Chandra, Ram, Sachin Suroshe, Jyotsana Sharma, R.A. Marathe, and D.T. Meshram. Pomegranate Growing Manual. ICAR-National Research Center on Pomegranate, Solapur-413255 (Maharashtra), India, Nov 2011.
6. Benagi, V.I. 2009. *Pomegranate—Identification and management of diseases, insect pests and disorders*. University of Agricultural Sciences (UAS) Dharwad, India.
7. POMEGRANATE: Cultivation, Marketing and Utilization. Technical Bulletin No. NRCP/2014/1, ICAR-National Research Center on Pomegranate, Solapur-413255 (Maharashtra), India.
8. Fu, J.C., H. Lien, and S.T.C. Wong. 2000. Wavelet-based Histogram Equalization Enhancement of Gastric Sonogram Images. *Computerized Medical Imaging and Graphics* 24 (2): 59–68.
9. Fuzeng, Y., W. Zheng, and Y. Qing. 2004. Methods for Contrast Enhancement of Agricultural Images Using Wavelet Transform [J]. *Transactions of The Chinese Society of Agricultural Engineering* 3: 030.
10. Yin, S.C., and S.L. Yu. 2013. Infrared Image Enhancement Algorithm Based on Wavelet Transform and Histogram Equalization [J]. *Laser and Infrared* 2: 025.
11. Yi-qiang, L.I.A.N.G. 2010. Application of a New Method of Wavelet Image Denoising in Agriculture Picking. *Journal of Anhui Agricultural Sciences* 4: 153.

12. Zhang, R., Y Huang, and Z Zhao. 2012, October. Texture analysis of ultrasonic liver images based on wavelet denoising and histogram equalization. In *5th International Conference on Biomedical Engineering and Informatics* (*BMEI*), IEEE, 2012, 375–378)..
13. Wavelet Toolbox™ 7 User's Guide, ©Copyright 1997–2015 by The MathWorks, Inc.

Analysis and Exploitation of Twitter Data Using Machine Learning Techniques

Ganeshayya Shidaganti, Rameshwari Gopal Hulkund and S. Prakash

Abstract In the present era, Internet is a well-developed technology that supports most of the social media analysis for various businesses such as marketing of a product, analysis of opinions of different customers, and advertising most of the brands. This gathered huge popularity among different users with a fresh way of interaction and sharing the thoughts about the things and materials. Hence, social media comprises of huge data that categorizes the attributes of Big Data, namely volume, velocity, and variety. This leads to the research work of this huge data related to different organizations and enterprise firms. To analyze the demands, customer's efficient data mining techniques are required. Nowadays, twitter is the one among the social networks which is dealing with millions of people posting millions of tweets. This paper exemplifies the data mining with machine learning techniques such as TF-TDF and clustering algorithms such as hierarchical clustering, *k*-means clustering, *k*-medoid clustering, and consensus clustering along with their efficiencies.

Keywords Twitter data · Machine learning technique · Consensus clustering
Big data · Social media · TF-IDF · *K*-medoid clustering

G. Shidaganti (✉) · R.G. Hulkund
Department of Computer Science & Engineering, M.S. Ramaiah Institute of Technology, Bangalore 560054, India
e-mail: ganeshayyashidaganti@msrit.edu

R.G. Hulkund
e-mail: h.rameshwari.109@gmail.com

S. Prakash
Department of Computer Science & Engineering, Dr. Ambedkar Institute of Technology, Bangalore 560056, India
e-mail: prakash.hospet@gmail.com

M.S. Reddy et al. (eds.), *International Proceedings on Advances in Soft Computing, Intelligent Systems and Applications*, Advances in Intelligent Systems and Computing 628, https://doi.org/10.1007/978-981-10-5272-9_13

1 Introduction

Social medias [1] such as twitter, Facebook, MySpace, LinkedIn, and many more are being popular in this era of Internet of everything. These microblogging sites are very advantageous to business firms, service providers, and customers. The service providers give the opportunity to enterprises to advertise their applications and products to customers through these sites. The interested customers can easily get the information about these things. This saves their valuable time. Even multiple jobs and requirements can also be posted in these sites. These social media has attracted data scientists to study the relational, social, and behavioral aspects between social sites and their implications on society. Social network analysis [2] provides opportunity to understand the interaction between individuals and group of people and communities of different networks.

Twitter is the microblogging site which is acquiring more and more popularity and growing faster. The users post their messages as tweets, and hence, per day millions of tweets are being posted. Users use this site to update what is there in their mind as status and discuss about the products and services with their relatives and friends who are staying far. This is an example for real-world scenarios like reviewing about the electronic goods, clothing, automobiles, movies to be watched, hotels, and restaurants. This site has become very useful to marketers as they will easily get to know about the customers' satisfaction related to their products and services.

Twitter sentiment analysis [3] is done to analyze the sentiments of users. Sentiments mean thoughts in positive, negative, or neutral forms. The emotion-rich data are gathered from twitter. This work includes the analysis of effectiveness of machine learning techniques on twitter corpus. This dataset is continuous. Dataset on movie reviews is discrete one. It is simple to implement machine leaning techniques on discrete datasets compared to continuous datasets. Due to limited number of characters posting, people end up with short form of words and use emoticons which give different perspective for word context.

In this paper, we have considered different clustering algorithms and other machine learning techniques. The organization of paper is as follows: Sect. 2 gives information about data preprocessing, and Sect. 3 represents experimental results.

2 Data Preprocessing

Since analysis of microblogging sites has got more importance during crises, the analysis of data is very important. Hence, data preprocessing [4] is very necessary in data mining. Therefore, the phrase "garbage in and garbage out" is specially meant for machine learning and data mining processes. During the collection of huge data, data get jumbled in different impossible forms which give informal meaning. These kinds of things are needed to be clarified to produce meaningful and tactic results from the corpus. It also improves the quality of the data.

The prediction of knowledge during initial phases of data training becomes difficult when redundant and irrelevant data or noise is present as a part of collected data. In this paper, the noisy data are referred to URLs and stop words of English literature. This leads to maximal wastage of time during data preparation and filtering. Data preparation phase is the second phase of Big Data life cycle which includes cleaning, normalization, transformation, feature extraction, and selection of data processing techniques. The final set of the process is the processed data ready for further actions without any inconsistency. The preprocessing follows:

A. Special characters removal: The emoticons in the text file appear like the set of special characters. In certain applications or tasks, emoticons are not needed, and hence, these characters are removed from the datasets.
B. Identifying uppercases: Slang words such as BTW which is meant as "by the way," tomorrow as 2MRW, LOL, ROFL have to be either replaced or removed forever from the datasets.
C. Alphabet lower casing: In the twitter [5] dataset, most of the words are written in capital letters to highlight those words. For example, instead of writing "hello," it could be represented as "HELLO." Therefore, it is very important part of the data preparation phase of life cycle. Before removing the cases, capital letters are identified. In microblogging, even irregular casing exists as "TwInLkIIngofSTARS."
D. Compression of word: Sometimes, few words are simply exaggerated. For an instance, happy is exaggerated as "hhhaaappyyyyyy." This word contains irrelevant letters which are absolutely not needed. To increase the accuracy, the identification of the sense of a sentence is essential.
E. Identifying pointers: Pointers refer to usernames and Hash tags. In twitter, character "@" is being used to point out a particular person in their posts. To differentiate words, "#" is used instead of white spaces like "#Happy#Journey#Ishan."
F. Synset [6] detection: Synset finding is done on the words such as "create," "creation," "created," "creating" which are relevant to the word "create." Therefore, these words when appear are considered as the word "create," which is the base word. This reduces the feature vector size while preserving the worthy key terms.
G. Link removal: The URLs that are downloaded as a part of dataset are not useful for sentiment analysis and applying machine learning [7] techniques. These do not contribute anything to data mining. Hence, links are considered as garbage.
H. Stop word removal: In any natural language processing tool, the most important task is identifying the stop words and removing them.
I. Spell checking [8]: Usage of acronyms has become the trend, and in most of the microblogging sites, the number of characters is limited to 140 words. Hence, the shortened words have to be modified to the original words with the help of English dictionary.

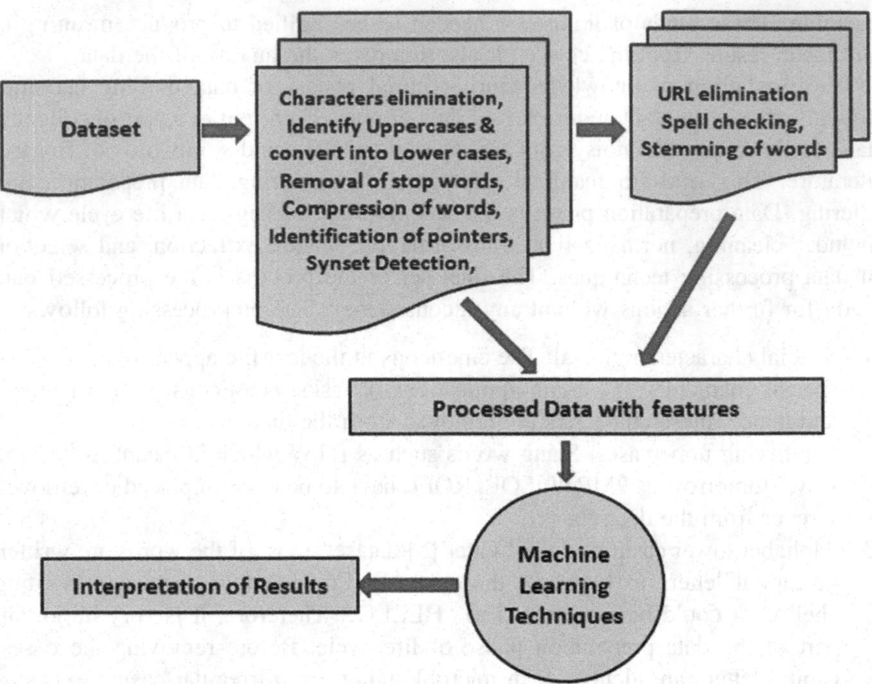

Fig. 1 Data preparation phase

J. Stemming of words: The term stemming is used in identifying the morphology of structure of given language's morphemes. It is also used to do information retrieval for the infected words to their word stems. In *R* programming, tm-package is used for stemming of words. For example, "engineer" is the root or stem word for "engineered," "engineering," and "engineers."

The data preprocessing steps are clearly shown in Fig. 1.

3 Machine Learning Techniques

As twitter data are unstructured data, to make it structured and apply some rules for further processing, machine learning comes into picture. Data refer to recorded data, whereas information refers to patterns underlying the data. To obtain the structural description of the data, the following techniques are used.

i. **TF-IDF**: Its elongated form is term frequency-inverse document frequency. This is used to check the number of times a word is repeated in a set of data. Based on the frequency of the word occurred in the different groups, categorization is done for an article. TF refers to how many times the word has occurred in an article. The term frequency for a word in an article means the ratio of the word count to the total number of words in the article. IDF describes the existence of a word in different documents as a common word between the documents. It is helpful in analyzing the different documents or article based on a single or multiple common words. For this paper, it is very helpful to analyze the tweets which share the same information based on the IDF terms.

ii. **Clustering**: Clustering is the technique used for statistical data analysis. It is needed in grouping the elements which are more similar to each other compared to other groups. In this paper, two clustering techniques are used.

A. **Hierarchical clustering**: To build the hierarchy of a statistical data, hierarchical clustering is used. The strategies for this are as follows:

- Agglomerative: It follows "bottom up" approach. Each element starts from its own cluster and pairs with other clusters which share near characters.
- Divisive: It follows "top down" approach. In this type, each element starts in one cluster are splits up as it moves further down the line.

In most of the information retrieval projects, agglomerative algorithms [4] are used rather than divisive algorithm. To do split and merge, greedy algorithm is applied. This paper work has used agglomerative algorithm. Metrics refers to the measurement of distance between the points. Some of the metrics are listed in Table 1.

In this paper, we are dealing with the Manhattan distance between the points.

A. ***k*-medoid clustering**: It is partition-based clustering algorithm. This clustering algorithm aims to distribute n observations into k clusters, in which each element belongs to the cluster of nearest mean. Euclidean distance is used as the metric. It uses PAM (Partitioning around Medoid) algorithm. PAM is faster than the exhaustive search because it uses greedy search. This algorithm follows the following procedure:

Table 1 Metrics for hierarchical clustering

Names	Formula
Euclidean distance	$\|\|a-b\|\| := \sqrt{\sum_i [\![(a\_i - b\_i)]\!]^{\wedge 2}}$
Manhattan distance	$\|\|a-b\|\| := \sum_i \|a_i - b_{i\|}$
Maximum distance	$\|\|a-b\|\| =:= \max\|a_i - b_i\|$

1. Initialize or randomly select the value of k.
2. Find association of each data point to its nearest neighboring clusters.
3. While the cost of configurations of data decreases,
4. For each medoid (m) data point and non-medoid data point (o),
   (i) Swap m and o and recalculate the cost.
   (ii) In the previous step, if the cost increases, then undo the swap that has happened in the previous step.

B. **Consensus clustering**: This clustering type is recommended for huge datasets. The main advantage of this approach is that these provide a final partition of data that is comparable to the best existing approaches, yet scale to extremely large datasets. Consensus clustering combines the advantages of many clustering algorithms.

4 Implantation with Results

Now, the preprocessed data can be used to interpret the results. Word cloud is used to show the importance of the words. The word cloud is formed from the term-document matrix. The representation of word cloud is shown in Fig. 2.

Fig. 2 Word cloud

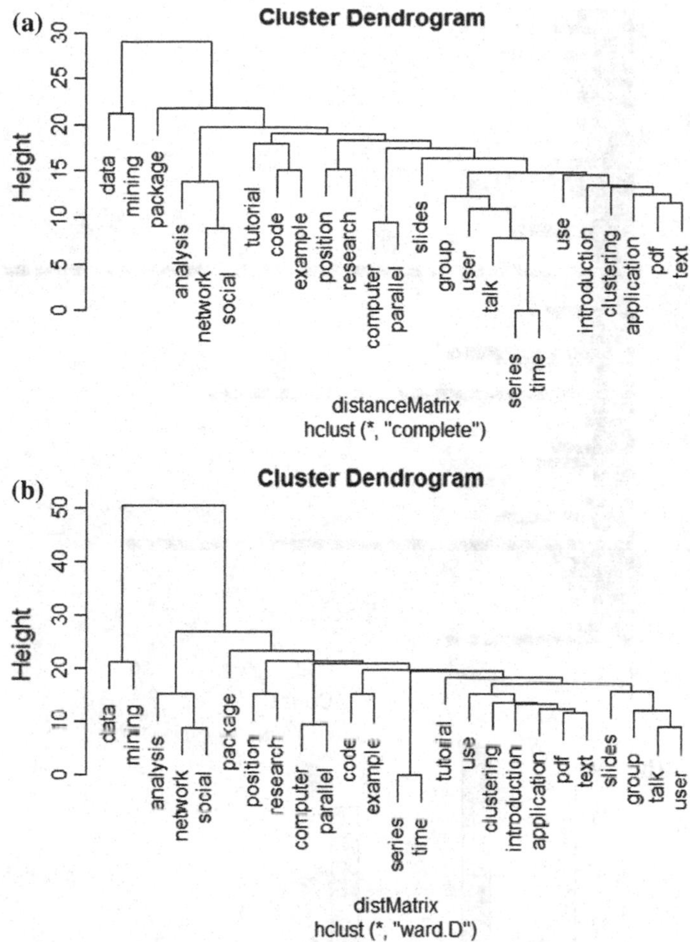

Fig. 3 Ordering of objects by hierarchical clustering, **a** with method as complete, **b** with method as "ward.D"

The analysis of hierarchical clustering shows that smaller clusters are generated. It also arranges the objects in certain orders. This is illustrated in Fig. 3.

TF-IDF [5] result gives the plot of number of counts versus terms in the dataset, as shown in Fig. 4, and Fig. 4b shows the bar plot of top few words which have occurred frequently.

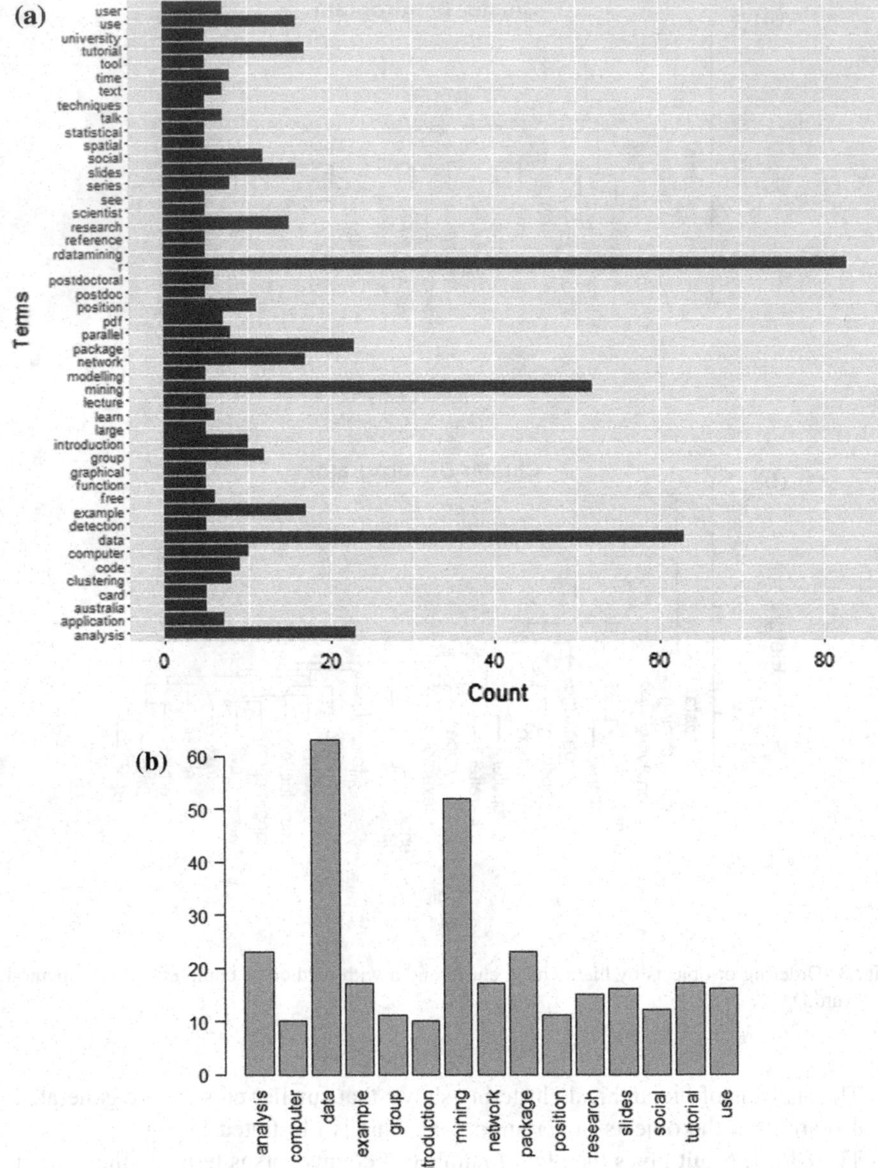

Fig. 4 **a** TF-IDF, and **b** bar plot of the terms which have occurred maximum no. of times

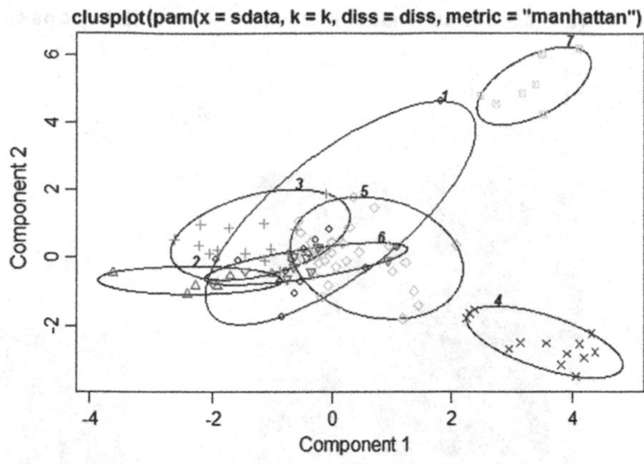

Fig. 5 *k*-medoid clustering

k-means and fuzzy [9] *c*-means(*c*-means centroid) minimize the squared error criteria and are computationally efficient. These algorithms do not require the user to specify the parameters.

There is no much difference between *k*-means and *k*-medoid clustering algorithms. *K*-medoid choose data points as centers (medoids). This algorithm is more robust to noise and outliers compared to *k*-means algorithm. It minimizes the sum of dissimilarities instead of sum of squared Euclidean distances. The common realization of *k*-medoid clustering is Partitioning around Medoid (PAM). The result for *k*-medoid algorithm is shown in Fig. 5.

The three disadvantages of these above-mentioned algorithms are as follows:

- Entities must be represented as points in *n*-dimensional Euclidean space.
- Objects have to be assigned to their respective clusters.
- Clusters must have same coordinates and must be of same shape.

To overcome these disadvantages, consensus clustering is implemented in this paper. The below graphs show the results of the same in Fig. 6.

The cumulative distribution function for the whole dataset is shown in Fig. 7.

The overall consensus clustering algorithm's cluster consensus is given in Fig. 8.

Figure 9 shows scatter plots of confidence and lift with respect to support.

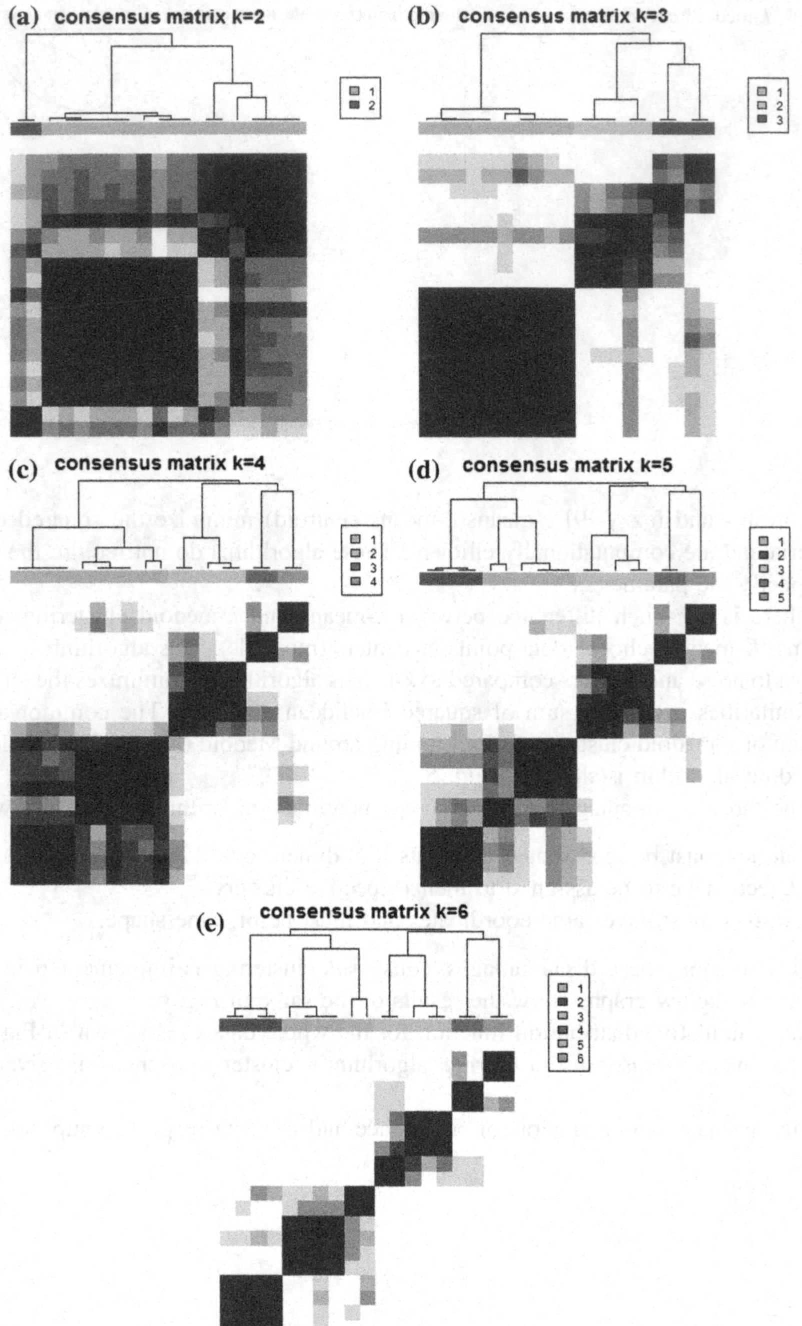

Fig. 6 Consensus matrix, **a** $k = 2$, **b** $k = 3$, **c** $k = 4$, **d** $k = 5$, and **e** $k = 6$

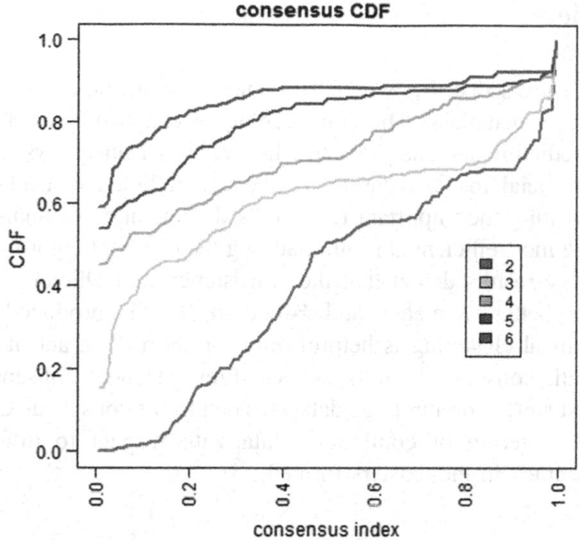

Fig. 7 CDF of consensus matrixes

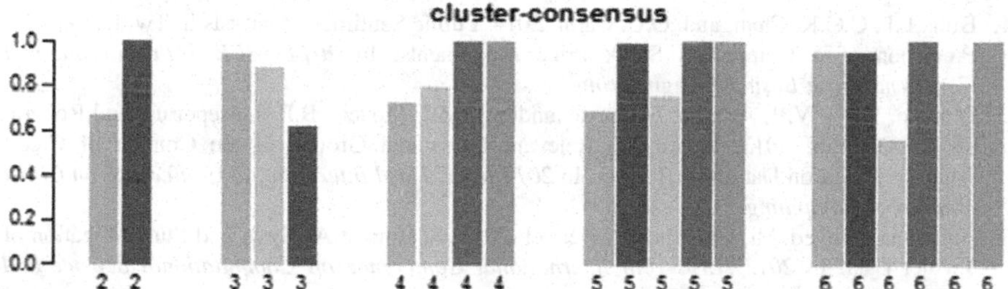

Fig. 8 Cluster consensus

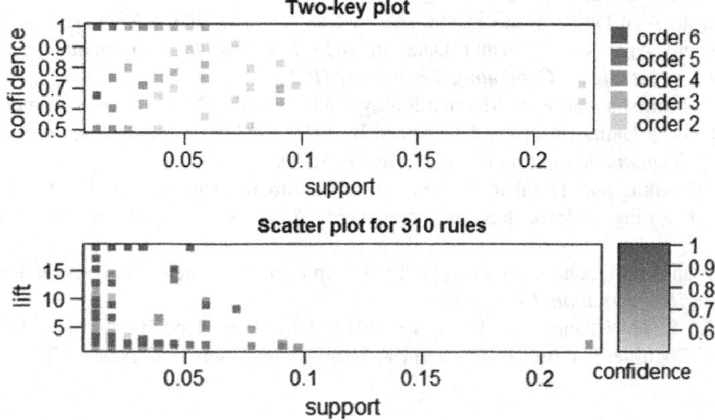

Fig. 9 Scatter plots of confidence and lift w.r.t support

5 Conclusion

This study has shown that the twitter data analysis with the use of different clustering techniques is beneficial. The same techniques can also be used in companies' stock market prediction and analysis and wherever Big Data analysis is required. In the analysis of social media datasets, we have concluded that TF-IDF finds its necessity in counting the important terms of a document. This analysis on twitter dataset gives the most efficient algorithm among the different algorithms mentioned in this work. The results depict that the consistency and efficiency of consensus clustering were better. *K*-means and *k*-medoid (PAM) produced almost same results. Hierarchical clustering is helpful only for short data, and it fails for large datasets. Overall, consensus results are satisfying. Hence, consensus clustering technique is best suited for any large dataset. The plot of consensus CDF graph, the probability of clustering of continuous data with respect to different clusters, esteems the accuracy in the clusters formed.

References

1. Bing, L.I., C.C.K. Chan, and, O.U. Carol. 2014. Public Sentiment Analysis in Twitter Data for Prediction of a Company's Stock Price Movements. In *2014 IEEE 11th International Conference on e-Business Engineering*.
2. Danyllo, W.A., V.B. Alisson, N.D. Alexandre, L.M.J. Moacir, B.P. Jansepetrus, and Roberto Felício Oliveira. 2013. Identifying Relevant Users and Groups in the Context of Credit Analysis Based on Data from Twitter. In *2013 IEEE Third International Conference on Cloud and Green Computing*.
3. Bahrainian, Seyed-Ali, and Andreas Dengel. 2013. Sentiment Analysis and Summarization of Twitter Data. In *2013 IEEE 16th International Conference on Computational Science and Engineering*.
4. Gokulakrishnan, Balakrishnan, Pavalanathan Priyanthan, Thiruchittampalam Ragavan, Nadarajah Prasath, and AShehan Perera. 2012. Opinion Mining and Sentiment Analysis on a Twitter Data Stream. In *The International Conference on Advances in ICT for Emerging Regions—ICTer 2012*, 182–188.
5. Bhuta, Sagar, Avit Doshi, Uehit Doshi, and Meera Narvekar. 2014. A Review of Techniques for Sentiment Analysis of Twitter Data. In *2014 International Conference on Issues and Challenges in Intelligent Computing Techniques (ICICT)*.
6. Kanakaraj, Monisha, and Ram Mohana Reddy Guddeti. 2015. NLP Based Sentiment Analysis on Twitter Data Using Ensemble Classifiers. In *2015 3rd International Conference on Signal Processing, Communication and Networking (ICSCN)*.
7. Gautam, Geetika, and Divakar Yadav. 2015. Sentiment Analysis of Twitter Data Using Machine Learning Approaches and Semantic Analysis. In *2015 IEEE International Conference*.
8. Venugopalan, Manju, and Deepa Gupta. 2015. Exploring Sentiment Analysis on Twitter Data. In *2015 IEEE International Conference*.
9. Liu, C.-L., T.-H. Chang, and H.-H. Li. 2013. Clustering Documents with Labeled and Unlabeled Documents Using Fuzzy Semi-K-Means. *Fuzzy Sets and Systems* 221: 48–64.

Real-Time Prototype of Driver Assistance System for Indian Road Signs

Karattupalayam Chidambaram Saranya and Vutsal Singhal

Abstract The paper presents a system for the detection and recognition of road signs in real time (tested for Indian road signs). The detection and recognition algorithm used is invariant to scale, angle, blur extent, and variation in lighting condition. Shape classification of road signs using Hu moments is done in order to categorize signs as either warning, mandatory, prohibitory or informational. Classified road signs are then matched to ideal road signs using feature extraction, and the matching is done with the help of Oriented FAST and Rotated BRIEF (ORB) descriptors. After recognition, the driver is given a feedback.

Keywords Driver assistance system · Traffic sign detection · ORB Feature extraction · Feature matching · Road signs

1 Introduction

Road sign detection and recognition is an integral part of unmanned vehicular navigation. It can also be of great assistance to differently abled people. Even in manned vehicles, in situations where a driver unknowingly missing a road sign might lead to accidents, an efficient driver assistance system would be of great importance. Hence, the development of robust real-time road sign detection and recognition system is crucial.

Road signs are usually of regular shape, size, and color for clear identification. Road signs in India can be classified as either warning, mandatory, prohibitory or informational. In informational signs, alphanumeric signs are separately considered. The road signs can be categorized into three different categories with respect to shape: circle (mandatory), triangle (warning), and octagonal (stop). All the circular

K.C. Saranya (✉) · V. Singhal
VIT University, Vellore, India
e-mail: saranya.kc@vit.ac.in

V. Singhal
e-mail: vutsalsinghal@gmail.com

M.S. Reddy et al. (eds.), *International Proceedings on Advances in Soft Computing, Intelligent Systems and Applications*, Advances in Intelligent Systems and Computing 628, https://doi.org/10.1007/978-981-10-5272-9_14

and triangular signs have red border with any background color (usually white). There is only one octagonal sign which is "stop."

2 Background

In past few years, many techniques have been developed to detect and recognize road signs. Most researchers have segmented the signs on basis of color, shape, or edge information to detect the road signs.

Siti et al. in [1] have used Hough Transform for road sign detection. In recognition phase, the ratio of areas of shape and symbol is used which fails when road signs having same symbols but different directions give ambiguous result (like left turn and right turn), and false results in case multiple signs are put together; algorithm is variant to rotation of road sign.

Chen et al. in [2] have used color-based segmentation method to establish ROI and then Harr wavelet transform for sign candidates. In recognition phase, SURF is used to find local invariant features. The recognition is performed by finding out the template image that gives the maximum number of matches.

Malik et al. in [3] have used color segmentation of road signs with red boundary after converting from RGB to HSV color space. Fuzzy Shape Descriptor, which uses area and center of mass, has been used to categorize images as circular or triangular shapes. In recognition phase, template matching has been used which is not scale and rotation invariant.

3 Implementation

Input from the camera is taken as video and processed frame by frame. As the frames per second is high (around 30), processing of consecutive frames can be neglected. Sequential flow (Fig. 1) implemented on development environment comprises Linux operating system with 4 GB of RAM and AMD A8-6410 APU 2.4 GHz quad-core processor.

3.1 Noise Removal

Noise can be amplified by the digital corrections of the camera. [4, 5] provide method of noise removal using non-local (NL) means. In this method, a pixel is taken, small window around it is considered, and the average value of similar windows in the entire image replaces the chosen pixel. NL means algorithm is defined by:

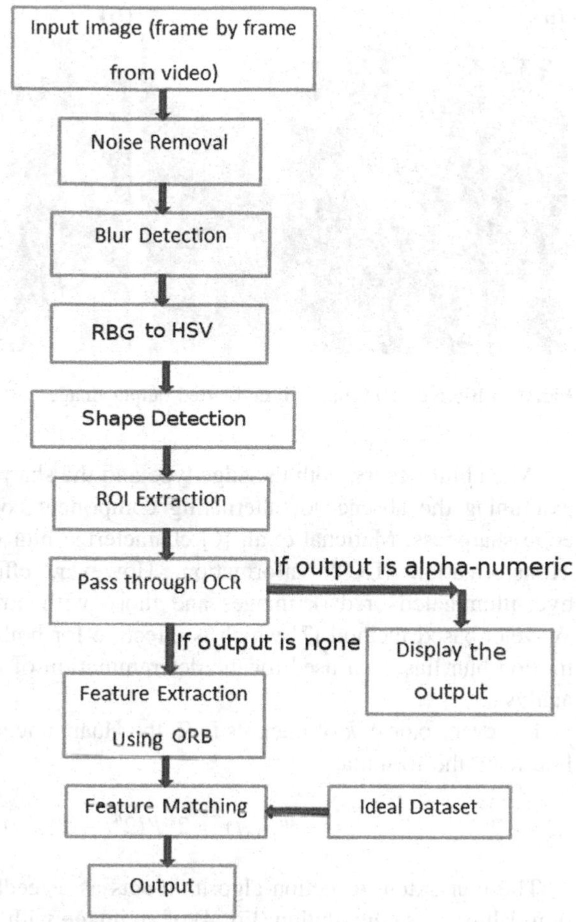

Fig. 1 Sequential flow

$$NL[u](x) = \frac{1}{C(x)} \int_{\Omega} e^{\frac{-(G_a^* |u(x+.)-u(y+.)|^2)(0)}{h^2}} u(y)\mathrm{d}y, \quad \text{where } x \,\varepsilon\, \Omega, \tag{1}$$

$$C(x) = \int_{\Omega} e^{\frac{-(G_a^* |u(x+.)-u(y+.)|^2)(0)}{h^2}} \mathrm{d}z \tag{2}$$

where C is the normalizing constant, G_a is the Gaussian kernel, and h acts as a filtering parameter.

3.2 Blur Detection and De-blurring

The extent of blur or focus measure of an acquired image is done before further processing. A straightforward method to detect blurriness is implemented here by taking into account the number of edges in the image.

Fig. 2 **a** Blurred input image, **b** de-blurred output image

When blur occurs, both the edge type and the sharpness of edge will change. By examining the absence of alternating component coefficients which indicate the edge sharpness, Marichal et al. [6] characterize blur extent using Discrete Cosine Transformation (DCT) information. However, efficiency is lost in case of over-illuminated or dark images and those with uniform background. So, Haar Wavelet based method [7] which is effective for both out-of-focus blur and linear motion blur has been used for the deteremination of blur extent in edge sharpness analysis.

For every pair *n*, *k* of integers in *Z*, the Haar function $\Psi_{n,k}$ is defined on the real line *R* by the formula:

$$\Psi_{n,k}(t) = 2^{\frac{n}{2}}\Psi(2^n t - k), \quad t \,\varepsilon\, R \tag{3}$$

The blur-extent detection algorithm acts as a feedback to the de-blurring function. Blurring is convolution (Eq. 4) of an image with a linear or circular kernel of a definite magnitude and direction (vector) which implies de-convolution of blurry image with the same kernel will give us the de-blurred or original image. In the most practical scenarios, kernel (vector) is unknown and blind de-convolution is implemented (Fig. 2).

$$h(z) = \int_R f(x)g(z - x)\mathrm{d}x \tag{4}$$

3.3 Shape Detection

Shape detection is performed using Hu moments [8], [9] in order to classify whether the road sign is circular, triangular, or octagonal. Hu moments are rotational and translational invariant. They consist of group of nonlinear centralized moment

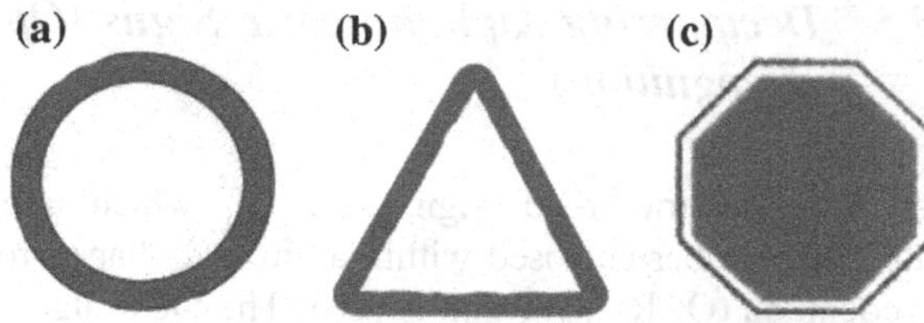

Fig. 3 Ideal shapes—**a** circle, **b** triangle, **c** octagon

expressions. The result is a set of absolute orthogonal moment invariants, which can be used for scale, position, and rotation-invariant pattern identification. Once the shape is detected (Fig. 3), it can then be matched with particular shaped ideal signs in the dataset.

3.4 Region of Interest

In this block, the RGB image is first converted to HSV color space; then, using color segmentation, a mask for red-colored border is created Now, contours are extracted for the given mask and sorted in descending order with respect to area enclosed. Certain number of such contours is plotted and extracted from the original image to obtain ROI with only the road sign present (Fig. 4).

Fig. 4 Mask and ROI for various inputs. **a** Input image 1, **b** mask of input image 1, **c** ROI of input image 1, **d** input image 2, **e** mask of input image 2, **f** ROI of input image 2, **g** input image 3, **h** mask of input image 3, **i** ROI of input image 3

3.5 Deciphering Alphanumeric Signs (Optical Character Recognition)

In alphanumeric road sign, (Fig. 5) which are usually characterized by alphabet/number enclosed within a circular-shaped road sign, an optical character recognition (OCR) algorithm is used. The OCR algorithm is based on Hu moments [11].

3.6 Feature Extraction and Matching ORB

The first feature extraction (or key points) is done, and then, feature matching of those key points with the key points of ideal road signs in the database is performed. Oriented FAST and Rotated BRIEF (ORB) [12] is used for feature extraction. ORB outperforms previously proposed schemes, such as SIFT [13], SURF [14] with respect to repeatability, distinctiveness, and robustness, and can be computed much faster.

3.6.1 ORB Descriptors

ORB uses an orientation compensation mechanism, which makes it rotation invariant, and it acquires the optimal sampling pairs to make it robust.

- Orientation Compensation

Use of measure of corner orientation using intensity centroid is done.

$$m_{pq} = \sum_{x,y} x^p y^q I(x, y) \tag{5}$$

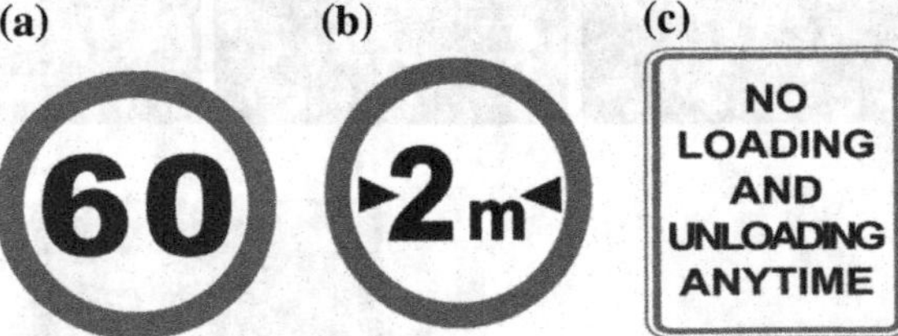

Fig. 5 Some alphanumeric signs [10]—**a** maximum speed 60 kmph, **b** prohibit entry for vehicles having overall width exceeding the specified limit, **c** no loading

With these moments, center of mass of the patch is calculated:

$$C = \frac{m_{10}}{m_{00}}, \frac{m_{01}}{m_{00}} \tag{6}$$

Now, a vector is constructed from the center to the centroid. The orientation of patch is given by:

$$\theta = a\tan 2(m_{01}, m_{10}). \tag{7}$$

- Learning Sampling Pairs

The sampling pair should possess high dissimilarity and high variance. High dissimilarity should be there so that each pair will have new information to add to the descriptor, thereby maximizing the amount of information the descriptor carries. High variance so as to make the features more distinguishable, as it responds differently to inputs.

3.6.2 Feature Matching

The extracted descriptors are now to be matched. A Fast Library for Approximate Nearest Neighbor (FLANN) [15] matcher object is used instead of usual brute force matcher. It takes two parameters. The first parameter specifies the distance to be used. The second parameter, if set true, returns only those matches with value (*m*, *n*) such that *m*th descriptor in set *X* has *n*th descriptor in set *Y* as the best match and vice versa, which implies the two features in both sets should match each other. It provides consistent result and is a good replacement of ratio test proposed in [13]. If enough good matches are found, then percentage matching is computed (Fig. 6) and the result is displayed.

Fig. 6 Matched images. **a** No entry sign, **b** turnabout sign, **c** no right turn sign, **d** no U turn sign

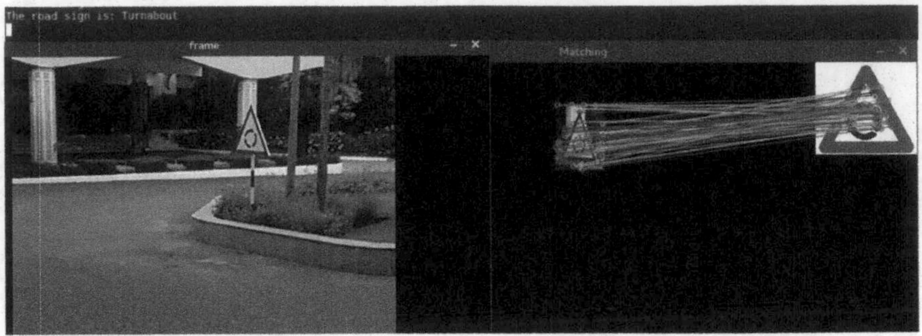

Fig. 7 Output of the proposed prototype for turn about sign

4 Result and Discussion

To evaluate the performance of the proposed system ,various videos of real time roads of 720x480 resolution were considered and processed using OpenCV library. The proposed algorithm is efficient and robust (Fig. 7). There were but some cases of failure in adverse conditions such as extreme fog or deteriorated sign. The algorithm is based on line-of-sight method. So direct visibility of sign is essential to yield desired output.

5 Conclusion

The prototype has proven itself to be capable of and robust for real-time purposes even in complex scenarios. In de-blurring phase, there is a scope for faster processing if consecutive frames are not considered as variation among consecutive frame can be assumed to be very less. In feature extraction and recognition phase, ORB [16] gives very robust output coupled with FLANN-based matcher.

References

1. Sallah, Siti Sarah Md, Fawnizu Azmadi Hussin, and Mohd Zuki Yusoff. 2010. Shape-Based Road Sign Detection and Recognition for Embedded Application Using MATLAB. In *2010 International Conference Intelligent and Advanced Systems (ICIAS)*.
2. Chen, Long, Qingquan Li, Ming Li, and Qingzhou Mao. 2011. Traffic Sign Detection and Recognition for Intelligent Vehicle. In *2011 IEEE Intelligent Vehicles Symposium (IV)*.
3. Malik, Rabia, Javaid Khurshid, and Sana Nazir Ahamd. 2007. Road Sign Detection and Recognition Using Color Segmentation, Shape Analysis and Template Matching. In *2007 International Conference on Machine Learning and Cybernetics*, vol. 6.

4. Buades, Antoni, Bartomeu Coll, and Jean-Michel Morel. 2011. Non-Local Means Denoising. *Image Processing On Line.*
5. Buades, A., B. Coll, and J.M. Morel. 2005. A Non-Local Algorithm for Image Denoising. *IEEE Computer Vision and Pattern Recognition.*
6. Marichal, X., W.Y. Ma, and H.J. Zhang. 1999. Blur Determination in the Compressed Domain Using DCT Information, *Proceedings of the IEEE ICIP'99.*
7. Tong, Hanghang, Mingjing Li, Hongjiang Zhang, and Changshui Zhang. 2004. Blur detection for digital images using wavelet transform, *IEEE International Conference on Multimedia and Expo ICME'04* Vol. 1. 2004.
8. Hu, Ming-Kuei. 1962. Visual pattern recognition by moment invariants, *IRE Transactions on Information Theory* 8.2 (1962): 179-187.
9. Gonzalez R.C., and R.E. Woods. 1977. Digital Image Processing. Reading: Line Detection Using the Hough Transform.
10. Ideal Road Sign Image. Source https://en.wikipedia.org/wiki/Road_signs_in_India
11. Chima, Y.C., A.A. Kassima, and Y. Ibrahimb. 1999. Character Recognition Using Statistical Moments. *Image and Vision Computing* 17 (3): S4.
12. Rublee, E. 2011. ORB: An Efficient Alternative to SIFT or SURF. *Computer Vision (ICCV).*
13. Lowe, D.G. 1999. Object Recognition from Local Scale-Invariant Features. *IEEE Computer Vision.*
14. Bay, Herbert, Tinne Tuytelaars, and Luc Van Gool. 2006. Surf: Speeded up robust features. *Computer vision–ECCV* (2006): 404-417.
15. Muja, Marius, and David G. Lowe. 2014. Scalable Nearest Neighbor Algorithms for High Dimensional Data. *Pattern Analysis and Machine Intelligence (PAMI)* 36.
16. Rublee, Ethan, Vincent Rabaud, Kurt Konolige, and Gary Bradski. 2011. ORB: An Efficient alternative to SIFT or SURF. In *2011 IEEE International Conference on Computer Vision (ICCV).* IEEE.

Application of Multivariate Statistical Techniques to Predict Process Quality

H.N. Navya and B.P. Vijaya Kumar

Abstract Almost all of the industrial processes are automated to collect large volumes of data that are produced during the manufacturing phase, and it is stored in the database as a backup. But, now the current trend is to understand the archived data and to obtain important information that is hidden among vast amounts of data. Such useful information obtained will further provide a detailed and better process knowledge that helps in improving the quality of the end product as per the standard specifications, and thereby, an increase in the profit can be expected. This paper focuses on two multivariate statistical techniques, principal component analysis (PCA) and partial least square (PLS), to predict the quality of an industrial process. A predictive model is built by extracting the important predictors that has a greater influence on the response quality. A case study is considered to evaluate the performance of the proposed techniques with respect to the specified quality and found that the parameters that influence the quality of product can be analyzed and predicted to better accuracy.

Keywords Multivariate statistical analysis · Principal component analysis
Partial least squares · Industrial process

1 Introduction

An industrial process consists of series of operations that help in the manufacturing of effective and quality end product on a large scale, and operations can be chemical, mechanical, or electrical. The complexity involved in an industrial process is usually high as it contains several tedious operations during the process of manufacturing. In today's competitive market place, predicting the quality of the

H.N. Navya (✉) · B.P. Vijaya Kumar
Department of ISE, M.S Ramaiah Institute of Technology, Bangalore, India
e-mail: navya992@gmail.com

B.P. Vijaya Kumar
e-mail: vijaykbp@yahoo.co.in

M.S. Reddy et al. (eds.), *International Proceedings on Advances in Soft Computing, Intelligent Systems and Applications*, Advances in Intelligent Systems and Computing 628, https://doi.org/10.1007/978-981-10-5272-9_15

product is an added advantage for the product to remain viable. Any tool that measures the performance of a process is a welcome addition and an integral to any manufacturing company.

Univariate statistical analysis explores one variable at a time; it basically describes the data and does not provide the relationship between variables. To predict the process quality, the different relationships between the variables should be explored. Hence, multivariate statistical analysis is the prime focus in this paper. Multivariate statistical techniques analyze more than one variable at a time and find the pattern simultaneously [1]. Applying multivariate techniques to an industrial process has several advantages such as improved process understanding, early fault detection, process optimization, online monitoring of quality [2]. This paper focuses on two multivariate statistical techniques PCA and PLS to analyze and explore the industrial process data set and build a model to identify the relationship between the predictors and response variables.

The rest of the paper is organized as follows. Section 2 discusses related work and their findings. Section 3 briefly explains the existing multivariate techniques used in our work. Our proposed model to predict process quality is described in Sect. 4. Section 5 consists of a case study on industrial process to evaluate the performance of the proposed technique, along with results. Finally, the concluding points are given in Sect. 6.

2 Related Work

Some of the related work pertaining to multivariate analysis, principal component analysis and partial least squares, are discussed in this section. In [3], multivariate statistical process control is applied to a high-speed continuous film manufacturing line. Here, the quality is improved by providing early warning of deviations in film quality. By monitoring the principal components, certain previously unidentified events were isolated and cause was identified. The capabilities of PCA that is fault detection and online monitoring were applied at Dofasco to detect the onset and to prevent the occurrence of mold breakouts in [4]. Multivariate analysis was used to analyze and interpret data from a large chemical process. PCA and PLS were used to identify correct correlations between the predictors and the output, to reduce the dimensionality of process data to build a model to predict output from known measurements in [5]. In online batch monitoring, PLS is applied to extract the variables that are more relevant to the quality of final product [6]. Multiway PLS was applied to an industrial fermentation process to identify the abnormal batches and hence improve the quality [7].

3 Multivariate Analysis Techniques

Multivariate statistical analysis is useful in the case where data is of high dimension. Since human vision is limited to 3 dimensions, all application above 2 or 3 dimensions is an ideal case for data to be analyzed through multivariate statistical analysis (MVA). This analysis provides joint analysis and easy visualization of the relationship between involved variables. As a result, knowledge that was unrevealed and hidden among vast amounts of data can be obtained.

A. **Principal Component Analysis (PCA)**

PCA is a powerful multivariate technique in reducing the dimension of data set and revealing the hidden relationship among the different variables without losing much of information [8]. The steps involved are:

1. Calculate the covariance matrix

Covariance is the measure of how one variable varies with respect to other variable. If the variables are more than two, then the covariance matrix needs to be calculated. The covariance matrix can be obtained by calculating the covariance values for different dimensions. The formula for obtaining the covariance matrix is

$$C^{n \times n} = \left(c_{i,j}, c_{i,j} = \operatorname{cov}\left(\mathrm{Dim}_i, \mathrm{Dim}_j\right)\right)$$

where

$C^{n \times n}$ is a matrix with n rows and n columns
i, j is the row and column indices, and covariance is calculated using below formula

$$\operatorname{cov}(X, Y) = \sum_{i=1}^{n} \left[(X_i - \overline{X})(Y_i - \overline{Y})\right] / (n - 1)$$

where

X_i is the value of X at ith position
Y_i is the value of Y at ith position
$\overline{X}$, $\overline{Y}$ indicates the mean values of X and Y, respectively.

2. Find eigenvectors for covariance matrix

Eigenvectors can be calculated only for square matrix, and not all square matrix has eigenvectors. In case $n \times n$ matrix contains eigenvectors, then total number of eigenvectors is equal to n. Another important property is that the eigenvectors are always perpendicular to each other. Eigenvalues are associated with eigenvector and describe the strength of the transformation.

3. Sort eigenvectors in decreasing order of eigenvalues

The highest eigenvalue corresponds to be the principal component which explains the maximum variance among the data points with minimum error. The first few principal components explain most of the variance, and eigenvalues with lesser values can be neglected with very little loss of information.

4. Derive the new data set

The original data will be retained back but will be in terms of the selected eigenvectors without loss of data.

B. **Partial Least Square (PLS)**

Partial least square is a method used to build the model and to predict the response variables from the large set of input predictors or factors. If the factors are few in number and not redundant, the multiple linear regressions can be a good approach. But, PLS is applied when the factors are many compared to the observations, and there is high collinearity between them. The general idea of PLS is to identify a set of orthogonal factors called latent variables which account for the maximum variance and have the best predictive power.

Let X be $n \times$ m matrix of predictors, where n is set of observations and m be the predictor variables. Let Y be $n \times p$ matrix of responses, where n is observations and p is response variables. The general model of PLS is

$$X = TP^T + E$$

$$Y = UQ^T + F$$

where

T and U are the projections (scores) of X and Y, respectively
P and Q are the loadings of X and Y, respectively
E and F are the errors associated with X and Y

The criterion is that the scores in X and scores in Y should have maximum variance. PLS determines the best subspace in X that explains Y.

4 Proposed Work

In order to build an accurate model, the samples should be collected from the production line over a considerate period of time. Once the data is collected, it should be preprocessed to increase the quality of the data. Then, multivariate statistical techniques are applied to build a prediction model. The next step is to validate the model offline by passing the training data set. If the model is not accurate enough, the model should be rebuilt. The final step is to feed the real-time process data to predict the process and the product quality. Figure 1 depicts the general steps in building a predictive model.

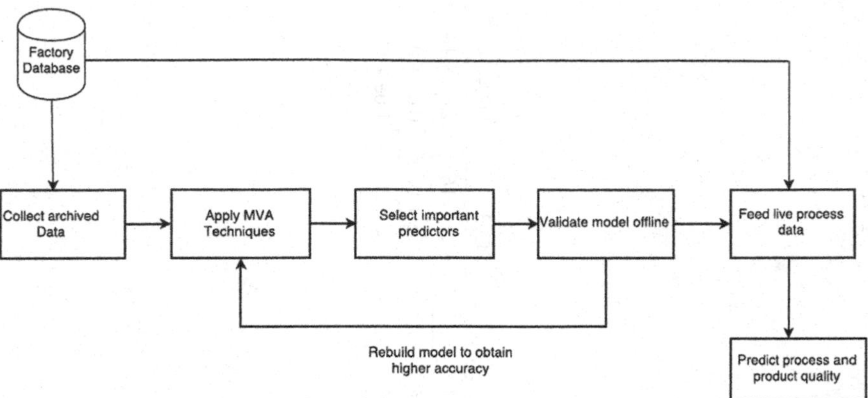

Fig. 1 General steps in building a predictive model

5 Case Study

The industrial process used for demonstration was an electrochemical machining process (ECM). ECM is used to produce workpiece by dissolving the metal in an electrolytic solution. This process is used in areas such as aerospace engineering, medical equipment, power supply industries, microsystem, and automotive industry [9]. Almost all metals, even nickel or titanium, can be electrochemically machined.

The data set consists of 105 predictor variables and 5 response variables. The response variables determine the quality of the end product. As the number of independent variables is more, multivariate statistical techniques are applied to explore the relationship between variables. Before the application of PCA, the data was prepared for analysis by identifying the duplicates, removal of outliers, and data conversion, and the number of predictors was reduced to 94.

5.1 PCA Application

A model for prediction is built by first applying PCA to the X matrix. PCA reduces the dimension and explores the hidden information. The first few principal components explain most of the variance. Table 1 indicates the summary of principal component analysis which explains the standard deviation between the principal components, proportion of variance, and cumulative proportion. It can be observed from Table 1 that 94 principal components were obtained, and as the number of components increases, the proportion of variance explained reaches 0.

Figure 2 indicates the number of components along X axis and cumulative variance along Y axis. The blue dot indicates the principal components, and value is the variance it explains. During experimentation, a variance of 80% was fixed, and

Table 1 Standard deviation, variance, and cumulative proportion associated with principal components

	PC1	PC2	PC3	PC4	PC5	PC6	PC7	PC8	…	PC93	PC94
Standard deviation	5.623	3.660	3.460	2.571	2.353	2.315	1.705	1.550	…	3.1e − 16	2.6e − 16
Proportion of variance	0.336	0.142	0.127	0.070	0.058	0.057	0.030	0.025	…	0.0e + 00	0.0e + 00
Cumulative proportion	0.336	0.479	0.606	0.676	0.735	0.792	0.823	0.849	…	1.0e + 00	1.0e + 00

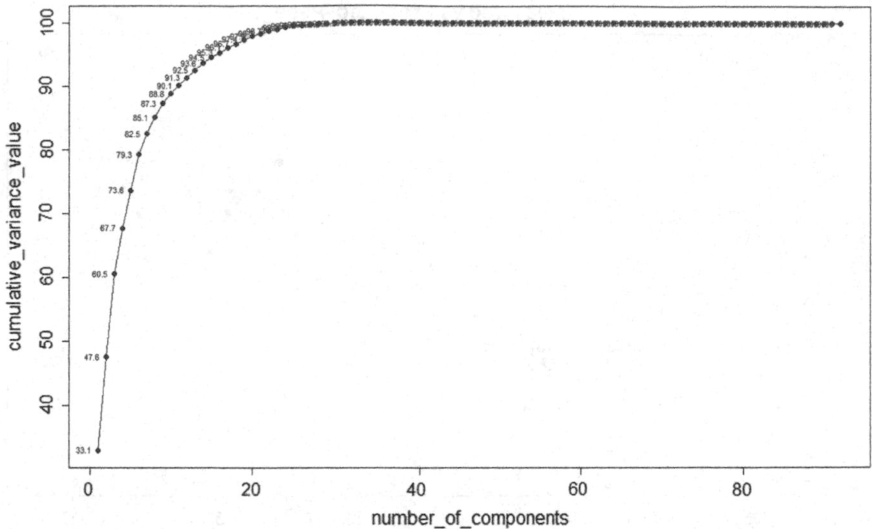

Fig. 2 Number of components required to explain variance

this was explained by the first 7 principal components. Also, it can be observed from the graph that the variance explained remains 100% after few components and hence not much data will be lost by selecting first few components.

5.2 PLS Application

Next, the partial least square technique was applied, and the number of components used to fit the model was the 7 principal components. Using the coefficient plot for PLS, the number of important predictors was reduced to 44 variables which best explains the output quality variables. A new set of data with same parameters was passed as test set to validate the model. The response values were predicted by the model and were similar to the training set.

The measured versus predicted plots were obtained for all the response quality variables. Figure 3 indicates that the predicted values are almost in line with the observed values for response variable 3.

5.3 Goodness Fit of Model

Goodness fit explains the variation between observed values and the values obtained from the model. A model fits in well only when there is a low discrepancy between measured and predicted. R^2 is a goodness fit measure for regression model

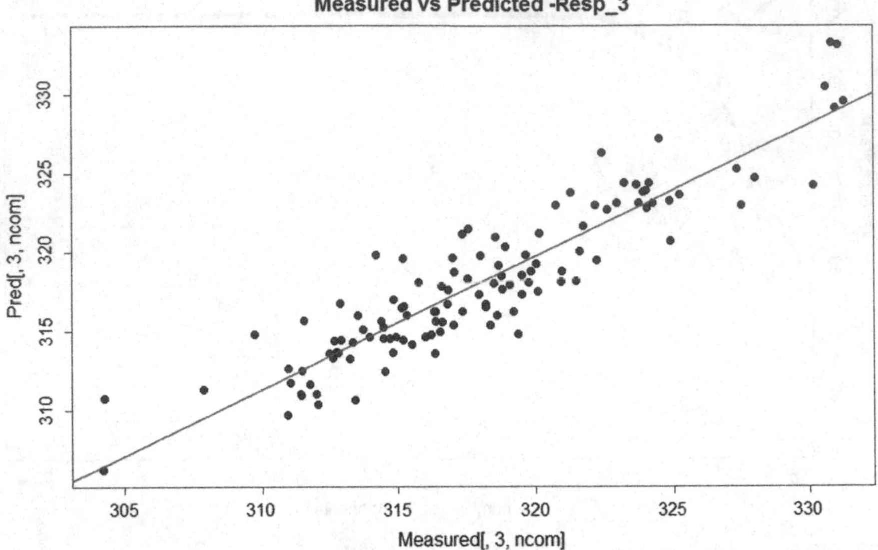

Fig. 3 Graph depicting actual values versus predicted values

which indicates how well the data fits the model. R^2 value of 1 indicates that the model perfectly fits, and value of 0 indicates the regression line does not fit the data. Figure 4 indicates the R^2 values for the response variables and the results are satisfactory for the ECM process.

Another test conducted to measure goodness fit is the root-mean-squared error of prediction which indicates the standard error obtained while predicting Y. Figure 5 indicates the RMSEP plot for all the 5 response variables, and the error decreases considerably for the first few components.

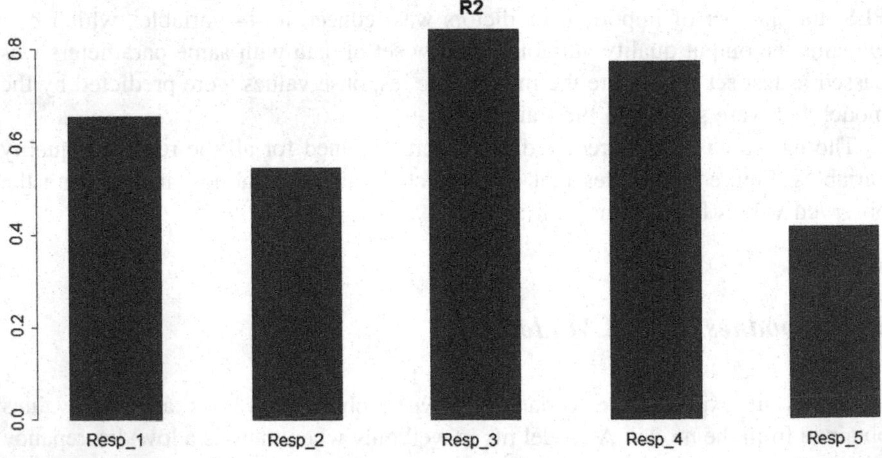

Fig. 4 R-squared plot for the model built

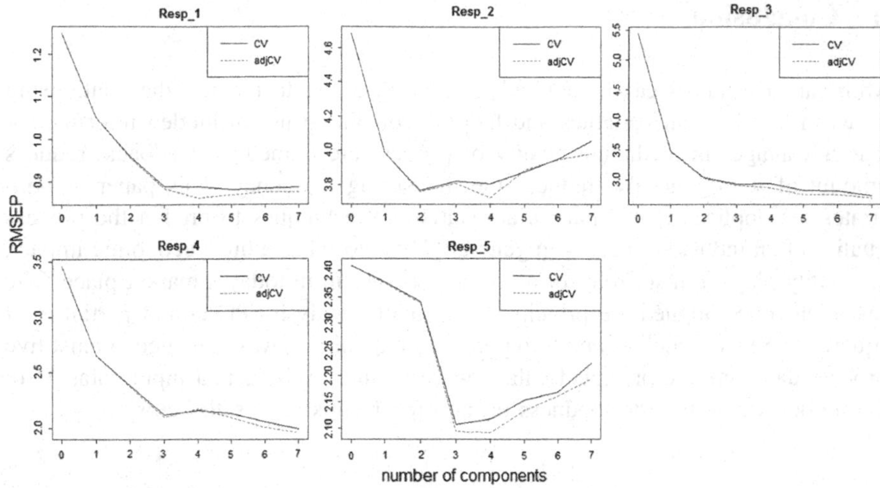

Fig. 5 Root-mean-squared error of prediction graph for the quality variables

Every second when a product is manufactured, its quality can be assessed based on certain standard values that have been set by the industry for the quality variables. The predicted plot depicts the normal, low, high, too low, and too high. If the predicted Y values lie within the normal line, it implies that the quality of the outcoming product has met the set standards. If the values are outside the mean, it indicates the product quality has been deviated, and immediate measures need to be taken. Figure 6 indicates the predicted plot with standard values, and blue dot indicates the product manufactured.

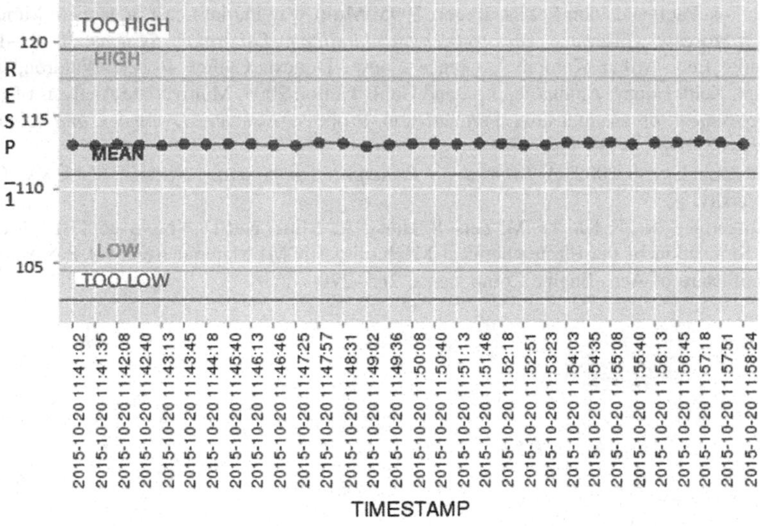

Fig. 6 Predicted plot depicting the quality of product against time of manufacturing

6 Conclusion

Multivariate analysis can be applied to large data sets to identify the relationship between the different variables and helps in obtaining useful hidden information. The advantages in predicting quality of process are immediate feedback, reduces amount of scrap, and also reduces the measuring frequency. This paper demonstrates the application of multivariate statistical techniques to predict the process quality of an industrial process in general. This work has a huge economic impact, as quality plays a major role for a product to sustain in today's market place. The two techniques applied are principal component analysis (PCA) and partial least square (PLS). A model was built to predict the quality, it was validated against live process data, and the predicted values were similar to the actual input data. Also, the model was tested for goodness fit, and the results were satisfactory.

References

1. Johnson, Richard Arnold, Dean W. Wichern, and others. 1992. *Applied Multivariate Statistical Analysis*, vol. 4. NJ: Prentice Hall Englewood Cliffs.
2. Kourti, Theodora, and John F. MacGregor. 1995. Process Analysis, Monitoring and Diagnosis, Using Multivariate Projection Methods. *Elsevier, Chemometrics and Intelligent Laboratory Systems*, 28: 3–21.
3. Weighell, M., E.B. Martin, and A.J. Morris. 1997. *Fault Diagnosis in Industrial Process Manufacturing Using MSPC*, 1–4. IET.
4. Dudzic, Michael, Vit Vaculik, and Ivan Miletic. 1999. *Applications of Multivariate Statistics at Dofasco*, 27–29. IEEE.
5. Kosanovich, K.A., and M.J. Piovoso. 1991. *Process Data Analysis Using Multivariate Statistical Methods*, 721–724. IEEE.
6. Nomikos, Paul, and John F. MacGregor. 1995. Multi-way Partial Least Squares in Monitoring Batch Processes. *Elsevier, Chemometrics and Intelligent Laboratory Systems*. 30: 97–108.
7. Chiang, Leo, Arthur Kordon, Lawrence Chew, Duncan Coffey, Robert Waldron, Torben Bruck, Keith Haney, Annika Jenings, and Hank Talbot. 2004. Multivariate Analysis of Quality Improvement of an Industrial Fermentation Process. *Elsevier, Dynamics and Control of Process Systems*, 443.
8. Smith, Lindsay I. 2002. *A Tutorial on Principal Components Analysis*, 65. USA: Cornell University.
9. Baumgartner, M., F. Klocke, M. Zeis, S. Harst, A. Klink, and D. Veselovac. 2013. Modeling and Simulation of the Electrochemical Machining (ECM) Material Removal Process for the Manufacture of Aero Engine Components, 265–270.

An Image Enhancement Technique for Poor Illumination Face Images

A. Thamizharasi and J.S. Jayasudha

Abstract Face recognition is used to identify one or more persons from still images or a video image sequence of a scene by comparing input images with faces stored in a database. The face images used for matching the image in the database has to be of good quality with normal lighting condition and contrast. However, face images of poor illumination or low contrast could not be recognized properly. The objective of the work is to enhance the facial features eyes, nose, and mouth for poor contrast facial images for face recognition. The image enhancement is done by first detecting the face part, then applying contrast-limited adaptive histogram equalization technique and thresholding to enhance the facial features. The brightness of the facial features is enhanced by using logarithm transformation. The proposed image enhancement method is implemented on AR database, and the face images appear visually good when compared to original image. The effectiveness of the enhancement method is compared by analyzing the histogram.

Keywords Image enhancement · Facial features · Illumination
Face images · Histogram

1 Introduction

Image processing techniques are used in various fields such as automated inspection of industrial parts and security systems, automated biometrics, i.e., iris recognition, fingerprint features and authentication, face recognition. There is a growing interest in biometric authentication, for use in application areas such as National ID cards, airport security, surveillance, site access. A wide variety of biometrics, such as

A. Thamizharasi (✉)
Manonmaniam Sundaranar University, Abishekapatti, Tirunelveli, Tamil Nadu, India
e-mail: radhatamil1@rediffmail.com

J.S. Jayasudha
SCT College of Engineering, Pappanamcode, Trivandrum, Kerala, India
e-mail: jayasudhajs@gmail.com

M.S. Reddy et al. (eds.), *International Proceedings on Advances in Soft Computing, Intelligent Systems and Applications*, Advances in Intelligent Systems and Computing 628, https://doi.org/10.1007/978-981-10-5272-9_16

automated recognition of fingerprints, hand shape, hand written signature, and voice, has been used. However, all of these systems need some cooperation from the operator. Face recognition and iris recognition are some of the noninvasive methods of identification of people.

Faces are the most common way people recognize each other. According to many researchers, it is not very convenient for computers to recognize individuals using faces. This is because human beings and computers possess different talents. The computers look at the face as a map of pixels of different gray (or color) levels.

In machine-based face recognition, a gallery of faces is first enrolled in the system and coded for subsequent searching. A probe face is then obtained and compared with each coded face in the gallery; recognition is noted when a suitable match occurs. The challenge of such a system is to perform recognition of the face despite transformations: changes in angle of presentation and lighting, common problems of machine vision, and changes also of expression and age, which are more special to faces. Computerized human face recognition has many practical applications. This includes face recognition in cluttered background like airport, bank ATM card identification, access control and security monitoring. The popular methods used for face recognition are Eigen faces [1], fisher faces [2], ICA [3], 2D PCA, and 2D LDA [4].

2 Related Works

The image enhancement techniques in spatial domain are logarithm transform, gamma transformation, contrast stretching, and histogram equalization (HE) [5]. The image enhancement technique in frequency domain is homomorphic filter [5]. It applies Fourier transform to the logarithm-transformed image.

The log transformations can be defined by the formula $s = c \log r + 1$, where s and r are the pixel values of the output and the input image and c is a constant. During log transformation, the dark pixels in an image are expanded as compared to the higher pixel values. The power law transformations are nth power and nth root transformation. These transformations can be given by the expression: $s = cr^\wedge\gamma$. This symbol γ is called gamma, due to which this transformation is also known as gamma transformation. Variation in the value of γ varies the enhancement of the images. Different display devices or monitors have their own gamma correction, that is why they display their image at different intensity.

The histogram equalization is an approach to enhance a given image. The approach is to design a transformation 'T' such that the gray value in the output is uniformly distributed in [0, 1]. It is also called histogram flattening. Histogram equalization can be described as method in which histogram is modified by spreading the gray level areas. It enhances the contrast of images by transforming the values in an intensity image or the values in the color map of an indexed image, so that the histogram of the output image approximately matches a specified histogram. This method usually increases the global contrast of images. Histogram equalization often produces unrealistic effects in photographs; however, it is very useful for scientific images like thermal, satellite, or X-ray image. Also histogram equalization can produce undesirable effects (like visible image gradient) when applied to images with low color depth.

The image enhancement technique in this work is based on the concepts of contrast-limited adaptive histogram equalization (CLAHE) and thresholding. These concepts are described below.

2.1 Contrast-Limited Adaptive Histogram Equalization (CLAHE)

In contrast-limited histogram equalization (CLHE), the histogram is cut at some threshold and then equalization is applied. It improves the local contrast of image. Contrast-limited adaptive histogram equalization (CLAHE) is an adaptive contrast histogram equalization method [6], where the contrast of an image is enhanced by applying CLHE on small data regions called tiles rather than the entire image. Each tile's contrast is enhanced, so that the histogram of the output region approximately matches the specified histogram [7]. The resulting neighboring tiles are then stitched back seamlessly using bilinear interpolation. The contrast in the homogeneous region can be limited so that noise amplification can be avoided.

2.2 Thresholding

Thresholding is the one of the simplest, computationally faster method which is used in image processing for image segmentation [5]. Given image '*f*', the object can be extracted from its background with threshold '*T*' and create image '*g*' using Eq. 1.

$$g(x,y) = \begin{cases} 1 & \text{if} f(x,y) > T \\ 0 & \text{if} f(x,y) \le T \end{cases} \tag{1}$$

The above method is called as global threshold. However, if image has noise or illumination, global threshold would not give good results and variable thresholding has to be applied [5]. An optimal global thresholding is proposed by Otsu [5].

3 Proposed Method

The image enhancement is done by following steps: The input color image is first converted into gray scale image. The gray scale image is then converted into black and white image (binary image) using Otsu's global threshold method. The isolated black pixels are the noise pixels, and they are removed from binary image. The first white pixel from the top row is the object pixel that shows the starting of the face image and the last white pixel is the end of the face image. The distance between first white pixel from top and last white pixel is face length. The distance from the first white pixel from left-hand side of the image and last white pixel is face width. The face length is the major axis of ellipse and face width is the minor axis of the ellipse. The center pixels 'x_0' and 'y_0' are calculated from face length and face width. A translation invariant elliptical mask is created over the binary mask 'BM' using the center pixels 'x_0' and 'y_0', major axis, and minor axis. The equation of ellipse is given in Eq. 2. The 'x' intercept 'a' is half of face length and 'y' intercept 'b' is half of face width.

$$\frac{(x - x_0)^2}{a^2} + \frac{(y - y_0)^2}{b^2} = 1 \tag{2}$$

An elliptical color and gray face image will be created by adding elliptical binary mask and original image 'I'.

The proposed method is implemented in AR database. The experimental work is done using AR database. The details of AR database are given below. The AR database [8] contains images of 100 persons taken in two different sessions. 50 are men and 50 are women [8]. In each session, photographs were taken in 13 different conditions. They are neutral expressions (anger, scream, and smile), illumination (right, left, and both sides) and occlusions (eye occlusion, eye occlusion with left illumination, eye occlusion with right illumination, mouth occlusion, mouth occlusion with left illumination, and mouth occlusion with right illumination). All the images are cropped to an image size of 160 × 120 pixels.

The image enhancement method is implemented using MATLAB R2013a [9]. Figure 1 shows the original image, and Fig. 2 shows the elliptical face image. The elliptical color face image is converted into gray scale image. Figure 3 shows the elliptical gray scale image. These steps give the detected face image as an ellipse. The next step is to highlight the facial features eyes, nose, and mouth.

The contrast-limited adaptive histogram equalization (CLAHE) is applied on the elliptical gray scale image to adjust the contrast variations in the image. The image obtained is 'f'. This image is then converted into binary image using threshold technique. The mean of the image 'f' is 'GT'. Image 'g' is obtained using Eq. 3 with image 'f' and threshold 'GT'.

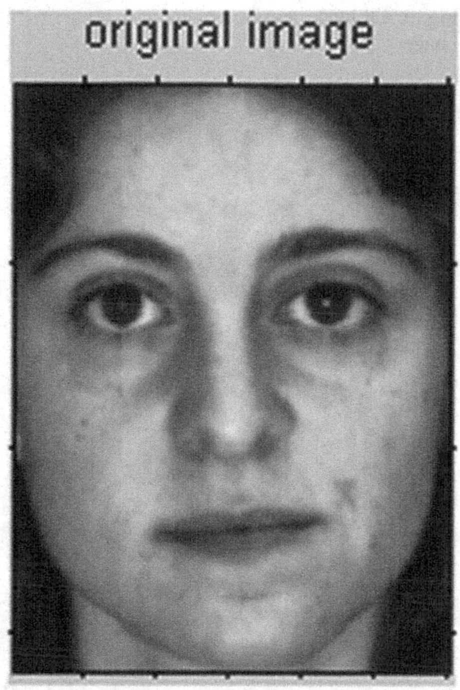

Fig. 1 Original color image

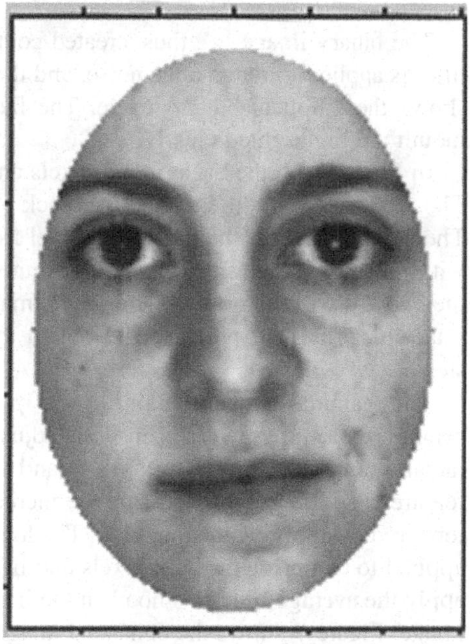

Fig. 2 Elliptical color face image

Fig. 3 Elliptical gray scale image

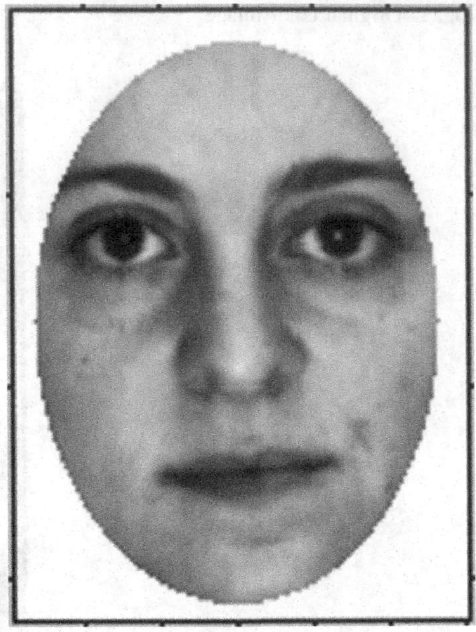

$$g(x,y) = \begin{cases} 1 & \text{if} f(x,y) > GT \\ 0 & \text{if} f(x,y) \leq GT \end{cases} \quad (3)$$

The binary image '*g*', thus, created contains salt and pepper noise. The median filter is applied to remove the noise, and the output image '*G*' is obtained. Figure 4 shows the elliptical binary image. The facial features eyebrows, eyes, nose, and mouth are highlighted clearly.

In Fig. 4, both the background pixels and facial features pixels are black pixels. The next step is to differentiate the black pixels in the face area and non-face area. The black pixels till the first white pixel from top, left, right, and bottom are filled with gray pixel intensity value 128. Figure 5 shows the output image obtained by the above step. The original gray scale image is shown in Fig. 6. The original gray scale image is now processed to obtain the enhanced image using the following steps.

The gamma transformation $g = f.^\wedge\gamma$ with γ equal to 1.1 is applied. The brightness of the gray scale image is adjusted by darkening the pixel intensities at facial features eyebrows, eyes, nose and mouth. The brightness in the non-facial features like foreground and chin are increased. The salt and pepper noise if any is removed by applying median filter. The logarithm transformation $c * \log(f + 0.8)$ is applied to compress the dark pixels and lighten the light pixels. The final step is to apply the average filter to smoothen the image. The image obtained is the enhanced image. Figure 7 shows the enhanced image. The original gray scale image has low contrast and the output image's facial features are enhanced.

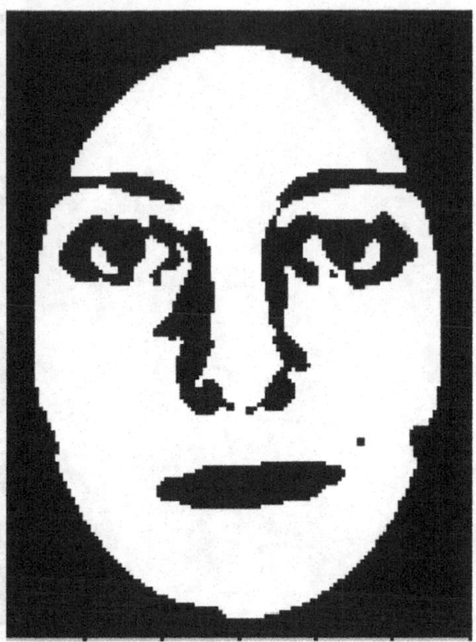

Fig. 4 Elliptical binary image

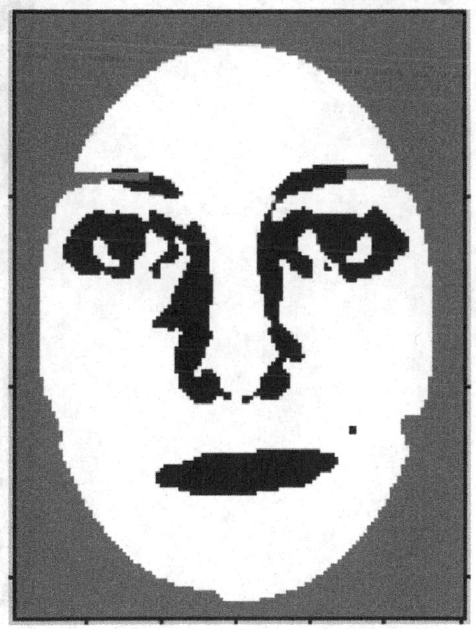

Fig. 5 Face segmentation

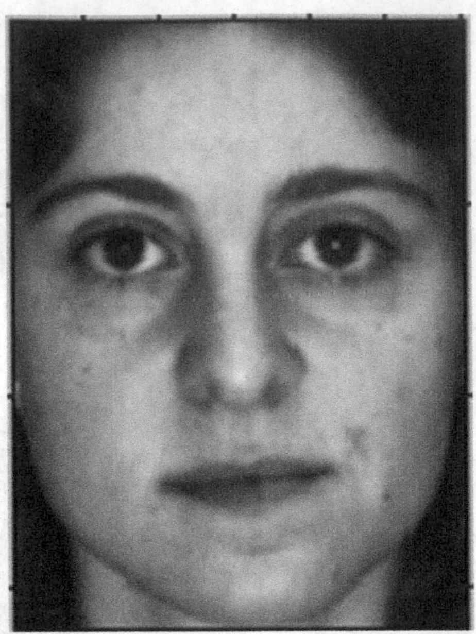

Fig. 6 Original gray scale image

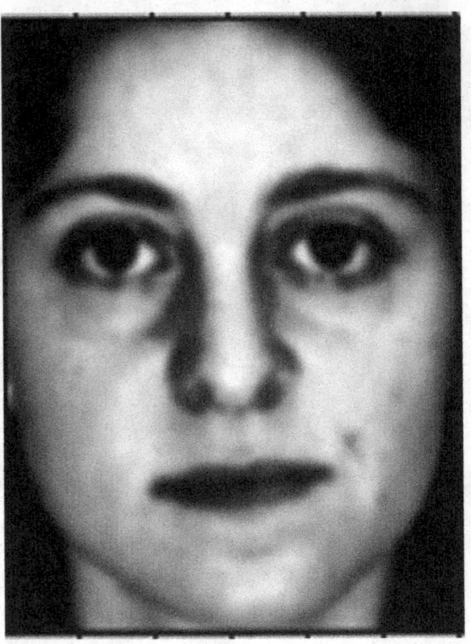

Fig. 7 Enhanced image

4 Results and Discussion

The result of the proposed method in Fig. 7 is compared with the results of histogram equalization image, CLAHE image, log transformation image, and gamma transformation image. The histogram-equalized image, CLAHE image, log transformation image, and gamma transformation image for the same image are shown in Figs. 8, 9, 10, and 11, respectively. The histogram equalization method is useful in images with backgrounds and foregrounds that are both bright or both dark. The enhanced image shown in Fig. 7 is visually good with good contrast when compared to images in Figs. 8, 9, 10, and 11.

However, the histogram analysis of the enhanced image and the existing methods are done here to compare the effectiveness of enhancement. Figure 12 shows the histogram of original gray scale image, Fig. 13 shows the histogram of the enhanced image, Fig. 14 shows the histogram of histogram-equalized image, and Fig. 15 shows the histogram of CLAHE image.

From the histogram analysis, it is found that in Fig. 12, the gray levels are at various levels. In Fig. 14, the histogram is equally spread and brightness is adjusted globally. The non-facial features are also enhanced here. From Fig. 15, it is found that in CLAHE method contrast is high. The histogram of enhanced image in Fig. 13 shows that the image contrast is good. The facial features and non-facial features are visually separable and the histogram also proves the same.

Fig. 8 Histogram equalized image

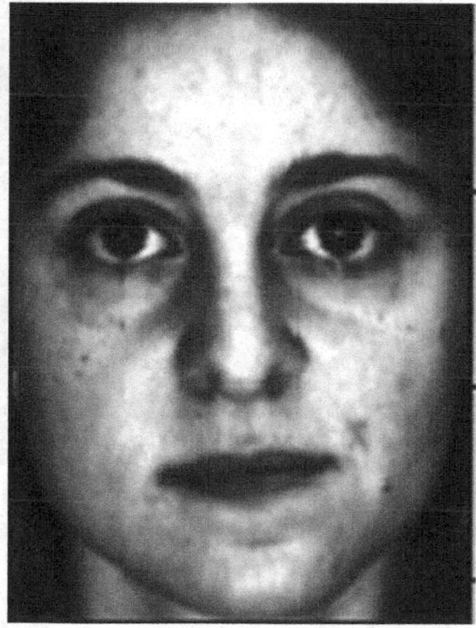

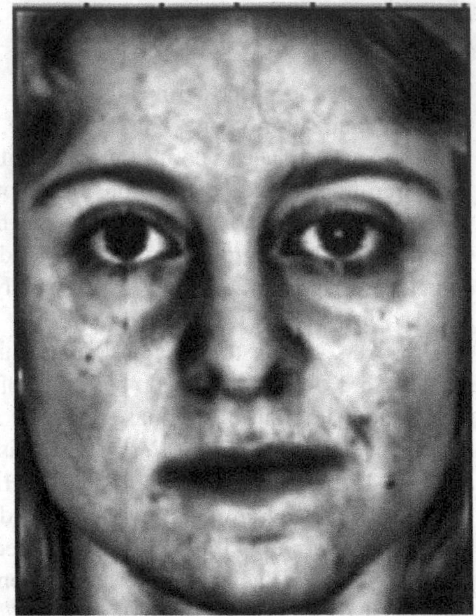

Fig. 9 CLAHE image

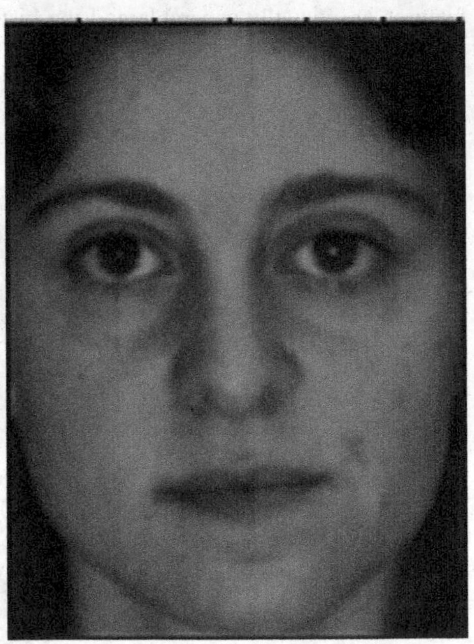

Fig. 10 Log transform image

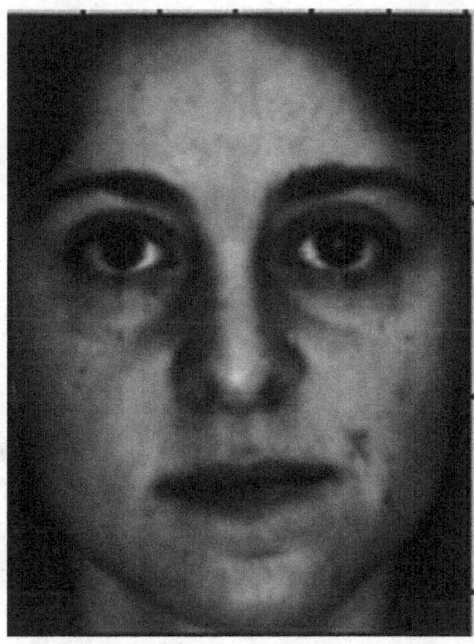

Fig. 11 Gamma transformation image

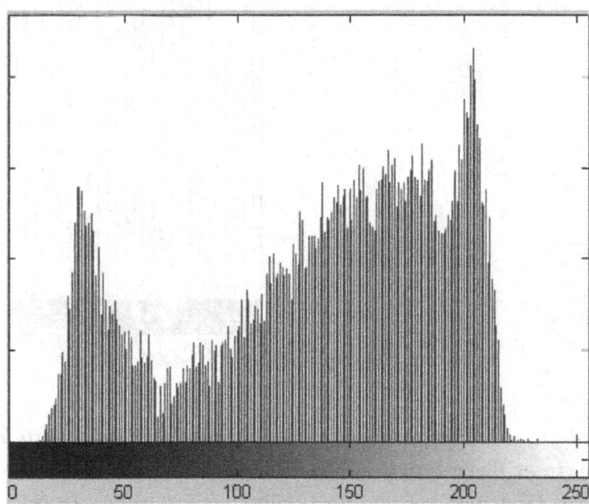

Fig. 12 Histogram of gray scale image

5 Conclusion

There are lots of image enhancement techniques available in the literature. The human face images under poor illumination could not be recognized effectively. An image enhancement particularly suitable for face images with poor illumination or contrast is proposed here. The facial features eyebrows, eyes, nose, and mouth are

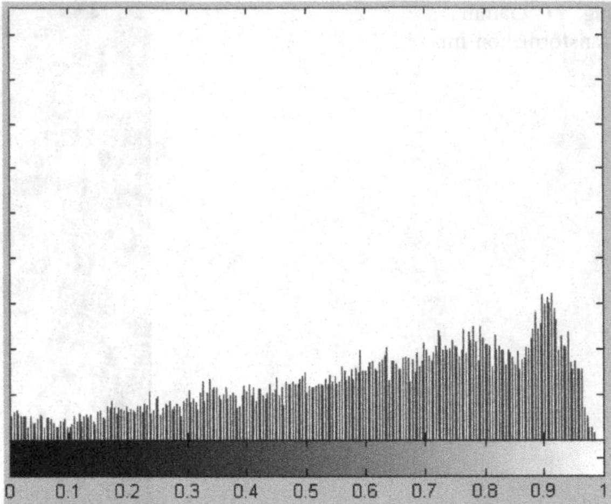

Fig. 13 Histogram of enhanced image

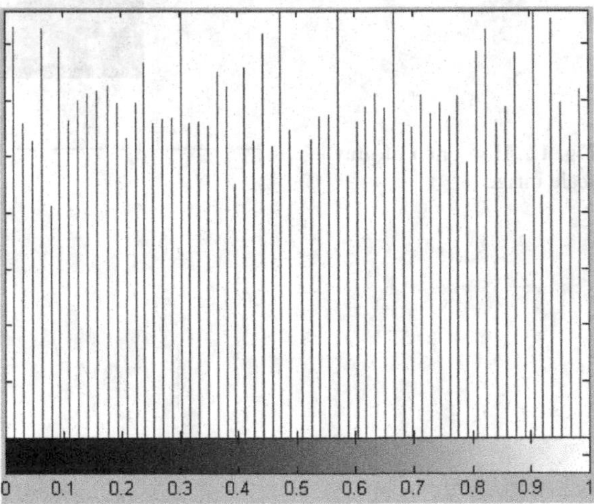

Fig. 14 Histogram of histogram-equalized image

highlighted from non-facial features, and this method could also be employed to extract these components individually. The extracted features could be compared for face part recognition. The brightness of facial features is darkened and the non-facial features are lightened. This process enhances the face image. The proposed method is compared with histogram equalization, CLAHE, logarithm, and gamma transformation methods using histogram. The enhanced method is visually good than other methods, and the histogram shows the image is of good contrast than the original gray scale image and other enhancement methods.

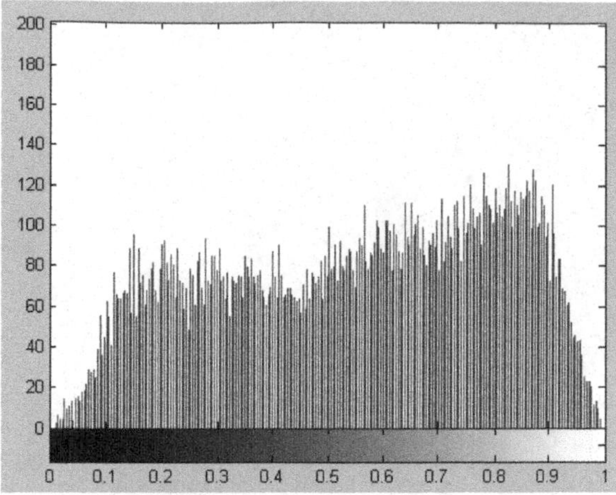

Fig. 15 Histogram of CLAHE image

References

1. Turk, M., and A. Pent land. 1991. Eigen Face for Recognition. *Journal of Cognitive Neuroscience* 3 (1): 71–86.
2. Belhumeur, P.N., J.P. Hespanha, and D.J. Krieg man. 1997. Eigen Faces Vs. Fisher Faces: Recognition Using Class Specific Linear Projection. *IEEE Trans. Pattern Analysis and Machine Intelligence* 19 (7): 711–720.
3. Draper, Bruce A., Kyungim Baek, Marian Stewart Bartlett, and J. Ross Beveridge. 2003. Recognizing Faces with PCA and ICA. *Computer Vision and Image Understanding* 115–137.
4. Wang, X., C. Huang, X. Fang, and J. Liu. 2009. 2DPCA vs. 2DLDA: Face Recognition using Two-Dimensional Method. *Proceedings of International Conference on Artificial Intelligence and Computational Intelligence* 2: 357–360.
5. Gonzalez, Rafael C., and Richard E. Woods. 1993. *Digital Image Processing*. Addison-Wesley.
6. Stark, J.A. 2000. Adaptive Image Contrast Enhancement Using Generalizations of Histogram Equalization. *IEEE Transactions on Image Processing* 9 (5): 889–894.
7. Zuiderveld, K. 1994. Contrast Limited Adaptive Histogram Equalization. In *Graphics Gems IV*, ed. P. Heckbert. Academic Press.
8. Martinez, A., and R. Benavente. 1998. *AR Face Database*. Computer Vision Centre, Technical Report 24.
9. Gonzalez, Rafael C., Richard E. Woods, and Steven L. Eddins. 2009. *Digital Image Processing Using MATLAB*, 2nd ed. ISBN 13:978-0982085400.

FPGA Implementation of Anti-theft Intelligent Traffic Management System

R. Suganya and K.R. Kashwan

Abstract The present-day traffic system in our nation provides an immobile traffic control plan, which is based on prior traffic counts. Besides increased crowd of vehicles on road, the theft of vehicles is also increased over the years. Due to this, emergency service vehicles may not achieve the target of being in time. This leads to the loss of human lives. To avoid the overcrowding issues, many systems have progressed using embedded systems. These, however, have a few problems for efficient controlling. To provide the solution for the conventional system, FPGA-based intelligent traffic system and vehicle theft detection system is proposed in this paper. It uses both software and hardware in combination with each other. FPGA are reconfigurable, and the program can be modified anytime according to the user requirements. The system functions faster than existing microcontroller systems already implemented. Extended memory option is the main option to store large database management. CAN protocol is used to connect a number of traffic junctions. XILINX software is used to carry out tests for simulations, and the results show that the system is efficient and practically usable.

Keywords FPGA · GSM · GPS · RFID · ZIGBEE

1 Introduction

Traffic management on the road has become a severe problem in today's time due ever-increasing traffic. There is a tremendous increase in the traffic density on the roads. When the number of vehicle increases, the traffic capacity is also increased. The more traffic results in many problems. Some of the problems are traffic jams

R. Suganya (✉) · K.R. Kashwan
Department of Electronics and Communication Engineering, Sona College of Technology (Autonomous), Salem 636005, Tamil Nadu, India
e-mail: rsuganyaslm@gmail.com

K.R. Kashwan
e-mail: kashwan.kr@gmail.com

M.S. Reddy et al. (eds.), *International Proceedings on Advances in Soft Computing, Intelligent Systems and Applications*, Advances in Intelligent Systems and Computing 628, https://doi.org/10.1007/978-981-10-5272-9_17

and accidents. Traffic signals control the traffic at each junction. Traffic norms and rules are followed to avoid accidents and to save the human life.

Traffic lights play a key role in traffic management. Traffic lights are the signaling devices that are placed on the intersecting points and are used to manage the flow of traffic on the road. Until the green light is enabled, the vehicle on red signal has to wait, if there is a little crowd or no traffic. This shows the dropping of a valuable time. Traffic control is a scheduled management of traffic on the roads in general and at intersecting points in specific. It is a matter of greater concern in larger cities. Several attempts are made to make traffic lights to operate in a chronological sequence. These traffic lights operate according to the accepted and legally valid rules of the traffic. Most of them use the sensor to calculate the density of traffic. This method has the limitations at the time of the peak hours of traffic on the road. However, the valuable time of the emergency vehicle is wasted at a traffic signal.

Thus, an advanced design for automatic control of traffic signals with routing taking place on vehicle density basis is preferred to achieve the objectives of efficient management of traffic. The traffic is regulated without any difficulty in a certain period of a time. Every traffic junction consists of controller to control the traffic process. Traffic junction is referred as a node. The nodes are controlled by the main server. Each node has GSM modem connected to the controller. The signal changes to green if there is traffic on any side of the road. It is not based on the round robin technique but it is controlled by nodes depending upon traffic density. The ambulance is allowed to pass through the road without waiting for the signal and it is said to be in ON state. For each node, the server supports a database. Every node has a different ID for addressing it. Whenever the ambulance passes through the traffic junctions or cross points, the similar action is executed and the ambulance is moved on a right path without any delay.

2 Literature Survey

The system allows the emergency vehicle by indicating a green signal to permit traffic and simultaneously indicating a red signal to the others sides. The drawback of this system may be interference due to traffic congestions. It provides the synchronization of green path of traffic indication. This system can find out the stolen vehicle through the green path traffic light [1]. This system does not need extra power to activate GPS. The length of vehicles in a green wave expands in size. Some of the vehicles cannot reach the green lights in time and must stop which is known as over-saturation [2]. The system uses the image processing techniques and beam interruption methods to avoid problems in RFID-controlled traffic. The drawback of this work is that it does not discuss what methods are used for communication between the emergency vehicle and the traffic signal controller. The RFID techniques work with more number of vehicles and lanes [3]. The main aim is to reduce the arrival time of ambulance. When the ambulance reaches the traffic signal or junction, the signal automatically changes to green even if it is currently red. The RFID is used to differ the emergency and non-emergency cases. The GPS and transceivers are used to communicate among the ambulance and traffic signal. It does not need of human power. The problem of this system is it needs all the information, such as beginning point and end point of the traveling [4]. This system provides the medical facilities as soon as possible without further delay. It is fully automated with the help of GPS. This system saves the human life [5]. From the literature, it is clear that conventional technologies are inadequate to solve the problems of congestion control, emergency vehicle clearance, and stolen vehicle detection. To implement this in real time, it is achieved by implementing on high-end FPGA. Other general references related to the literature are referred in [5–10]. The literature survey is mainly focused on the relevant topics of recent research interests in the area.

3 Design and Implementation Methodology

This section explains about working concept of the system, explanation of block diagram, and the stages of operation. The proposed system is split into three stages which are as follows. These three sections are interconnected for smooth working and achieving objectives.

- Automatic light signal changer
- Emergency congestion clearance
- Stolen vehicle detection system

3.1 Automatic Light Signal Changer

This is a control section which is developed with the help of an RFID tag and ZIGBEE. Whenever TAG comes into the range, the RFID reader reads the RF signal. The received RF signal is counted, and the network tracks the number of vehicles that has passed through for a specified time period. It then finalizes the congestion density. Consequently, it sets the consent duration time for particular path. Memory is established to store the counting value. From the above count, duration of the time signal is extended automatically further.

3.2 Emergency Congestion Clearance

Congestion clearance is managed by ZIGBEE transmitter and receiver. The transmitter side is concentrated for emergency vehicles and receiver side is used for

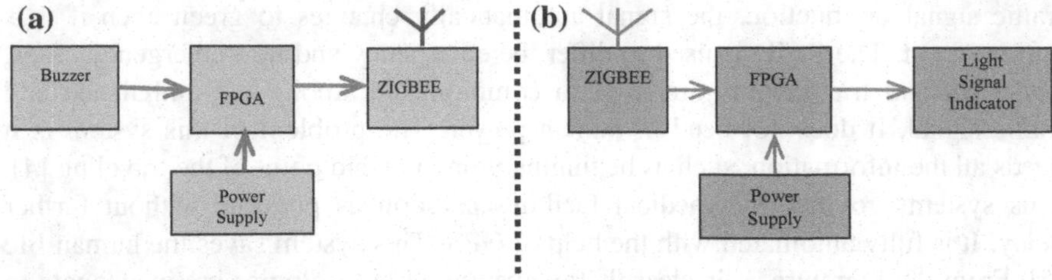

Fig. 1 **a** Transmitter block diagram, **b** receiver block diagram

establishing and managing connection points. In each instant of transmitter receiver meeting point, the buzzer is switched ON which helps the people in providing the path for emergency vehicles. Simultaneously, traffic light signal is changed to green until the ZIGBEE transmitter receiver signal is switched off at reaching the limit setup. Then traffic light signal is turned to normal green or red signals to restore normal procedure of traffic management. The transmitter side and receiver sides block diagrams are illustrated in Fig. 1a, b, respectively.

3.2.1 Stolen Vehicle Detection System

In this section, different RFID tags are explained that how these can be read by the RFID reader. The stolen RFIDs number is stored in the system. If a match is found in the toll booth, then the information is sent immediately to the control station or any other statutory authority through SMS using services of GPS networks. In addition, GPS tracker is placed in the vehicle to notice the exact point, if they missed the vehicle in traffic. This helps in tracking and relocating the lost vehicles. The architecture of block diagram for stolen vehicles is illustrated in Fig. 2 with transmitter and receiver sections.

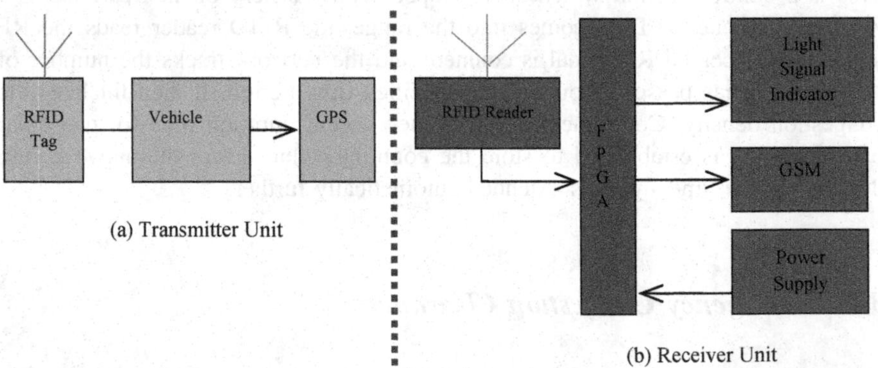

Fig. 2 Block diagram of detecting stolen vehicles. **a** Transmitter unit, **b** receiver unit

4 Results and Discussion

Three sections of coding are performed. The first section undergoes the baud rate setting case where it is set to a required range by the help of count value. The count value is defined by the ratio of input frequency to that of the required frequency. With the help of the count value for *on* and *off*, the system is conditioned accordingly. The transmitter section is used to obtain digital values from FPGA and then it presents further as an input to the GSM device where the signal is processed and send to phone in the form of a message. The next block is the receiver section, in which the RFID serial bits are converted to parallel bits for the processing of FPGA. These are processed to obtain the vehicle that is being stolen. The Xilinx simulation results are shown in Fig. 3, 4, 5, and 6.

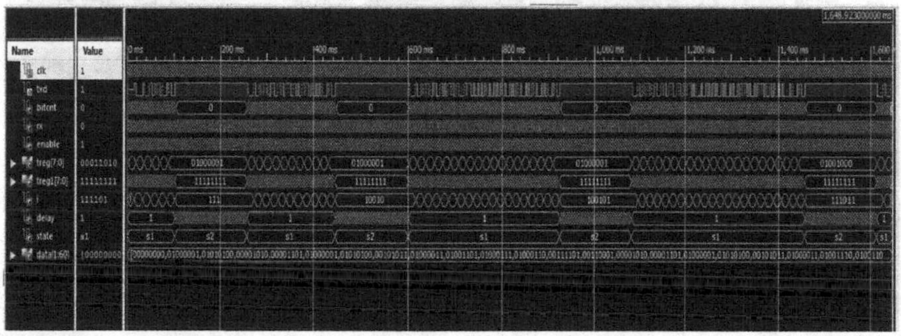

Fig. 3 Results of transmitter section block

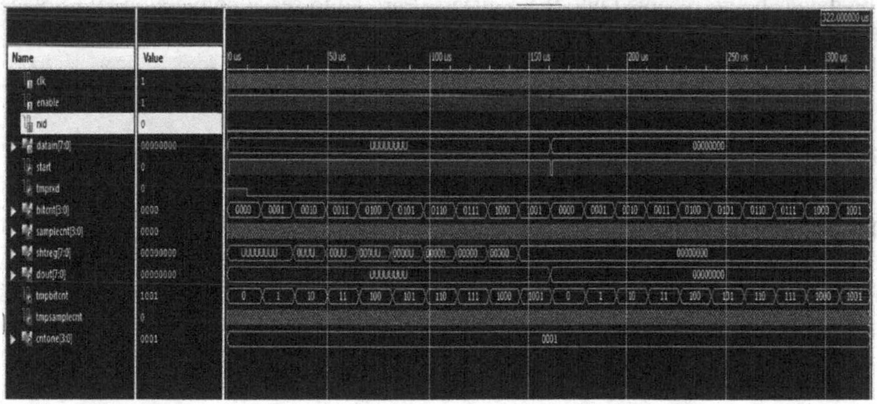

Fig. 4 Results of receiver section block

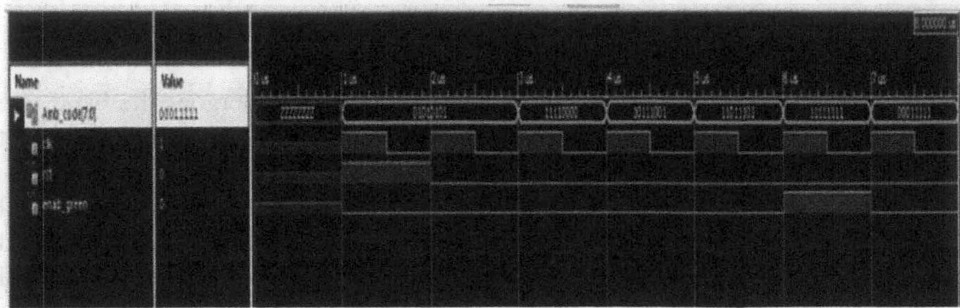

Fig. 5 Results of RFID receiver block

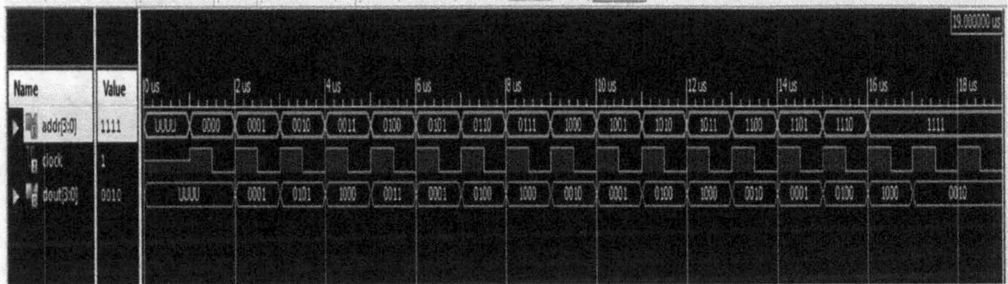

Fig. 6 Database storage block

4.1 Transmitter Section

In UART transmitter, the data that are stored in parallel registers are shifted as serial data for every clock cycle. For 8-bit data transfer, for each and every 8 clock cycles, the data is being received into the registers. Parallel to serial conversion takes place for compatibility usage.

4.2 Receiver Section

The UART receiver is a serial-to-parallel converter and hence the function of conversion from serial to parallel happens here. The receiver receives the serial data for every clock cycle and stores in the register. Once a specific group of data is accepted, it is transmitted in parallel mode to the array.

4.3 RFID Detection

At each road side signal, the RFID of the ambulance is saved in the memory. Whenever the receiver receives the corresponding ambulance RFID signal transmitted by the RF transmitter, the corresponding road signal is switched from red to green. This allows the ambulance to pass easily and without any delay at the road signals.

4.4 The Results of ROM Block

The ROM, a memory block with four-bit data storage facility, is used to store and supply data as and when required. For each clock cycle, the address of corresponding data is read. It is used only for storing data for future and further processing whenever is required. It neither modifies the data nor does any computation on its own. It supports database and supplies data for decision-making and computation.

5 Conclusion

An effective traffic congestion control and vehicle robbery detection system is implemented with the aid of wireless communication and FPGA boards. The algorithm is simulated on a system to verify the results. Further, it can easily be adopted for finding the quickest path with priority order. The emergency vehicles can reach their destination point within least possible time. It also reduces the complexity of computing environments. In common mode, the system is operated in such a way that the process is executed continuously for every two minutes time duration. Thus, the traffic jamming is eliminated even in the presence of multiple junctions on the path of course. The stolen vehicle is recovered from the thief by using a combination of RFID, GSM/GPS network, and FPGA. Previously reported research works are implemented using the embedded system which is not simple, and it takes long time. The entire database of such a system is complex. To locate a particular vehicle in large traffic is challenging. To manage such challenges, GPS tracker is employed to track the vehicle continuously. Every time RFID/TAG verifies the vehicle identification with the information in database. Once it analyzes the database, the report can be sent through SMS to the control room or to an authorized person to take suitable action. It helps us to seize the stolen vehicle if it comes in the reading range of any of the toll gates. The newly developed system operates much fast. It doesn't need human interventions for tracking and controlling. The system is experimentally verified for its functioning.

6 Future Improvement

Further improvement can be achieved for long-range RFID reader communication devices for field applications. The same concept can be easily adopted for health monitoring and management system in hospitals for timely and efficient treatment of patients.

References

1. Sundar, Rajeshwari, Santhoshs Hebber, and Varaprasad Golla. 2015. Implementing Intelligent Traffic Control System for Congestion Control, Ambulance Clearance, and Stolen Vehicle Detection. *IEEE Sensors Journal* 15 (2).
2. Mittal, A.K., and D. Bhandari. 2013. A Novel Approach to Implement Green Wave System and Detection of Stolen Vehicles. In *Proceedings of the IEEE 3rd International Advance Computing*, 1055–1059.
3. Sharma, S., A. Pithora, G. Gupta, M. Goel, and M. Sinha. 2013. Traffic Light Priority Control for Emergency Vehicle Using RFID. *International Journal of Innovation Engineering Technology* 2 (2): 363–366.
4. Hegde, R., R.R. Sali, and M.S. Indira. 2013. RFID and GPS Based Automatic Lane Clearance System for Ambulance. *International Journal of Advanced Electrical and Electronics Engineering* 2 (3): 102–107.
5. Goud, Varsha, and V. Padmaja. 2012. Vehicle Accident Automatic Detection and Remote Alarm Device. *IIRES* 1 (2): 49–54.
6. Sawwashere, Supriya, and Ashutosh Lanjewar. 2015 Automated Accident Intimation and Rescue System for Vehicle. *IJAIEM* 4 (5).
7. Varaprasad, G. 2013. High Stable Power Aware Multicast Algorithm for Mobile Ad hoc Networks. *IEEE Sensors Journal* 13 (5): 1442–1446.
8. Tawara, Katsunori, and Naoto Mukai. 2010. Traffic Signal Control by Using Traffic Congestion Prediction based on Pheromone Model. In *Proceedings of 22nd International Conference on Tools with Artificial Intelligence*.
9. Athavan, K., G. Balasubramanian, S. Jagadeeshwaran, and N. Dinesh. 2012. Automatic Ambulance Rescue System. In *Proceedings of 2012 Second International Conference on Advanced Computing and Communication Technologies*.
10. Sinhmar, Promina. Intelligent Traffic Light and Density Control Using IR Sensors and Microcontroller. *International Journal of Advanced Technology & Engineering Research (IJATER)*.

Gradation of Diabetic Retinopathy Using KNN Classifier by Morphological Segmentation of Retinal Vessels

Shreyasi Bandyopadhyay, Sabarna Choudhury, S.K. Latib, Dipak Kumar Kole and Chandan Giri

Abstract The extraction of blood vessels by morphological segmentation and the detection of the severity of diabetic retinopathy are proposed in this paper. The proposed algorithm extracts the finest vessels in the retina within a remarkably reduced computational time. The extracted blood vessel features being fed to KNN classifier determine the stage of Diabetic Retinopathy. The performance analysis is carried out which comes out to be of 94% along with the sensitivity (81.45%), specificity (96.25%), and accuracy (95.3%) defines the efficiency of the proposed system.

Keywords Morphological segmentation · Diabetic retinopathy · KNN classifier

S. Bandyopadhyay (✉) · S. Choudhury
Electronics and Communication Engineering, St. Thomas' College of Engineering & Technology, Kolkata, India
e-mail: s.bando.93@gmail.com

S. Choudhury
e-mail: sabarna.choudhury@gmail.com

S.K. Latib
Computer Science Engineering, St. Thomas' College of Engineering & Technology, Kolkata, India
e-mail: sklatib@gmail.com

D.K. Kole
Computer Science Engineering, Jalpaiguri Engineering College, Jalpaiguri, India
e-mail: dipak.kole@gmail.com

C. Giri
Information Technology, Indian Institute of Engineering Science & Technology, Shibpur, India
e-mail: chandangiri@gmail.com

M.S. Reddy et al. (eds.), *International Proceedings on Advances in Soft Computing, Intelligent Systems and Applications*, Advances in Intelligent Systems and Computing 628, https://doi.org/10.1007/978-981-10-5272-9_18

1 Introduction

The appearance and structure of blood vessels in an eye are important indicators for diagnoses of diseases such as diabetes, hypertension, and arteriosclerosis. People with diabetes are prone to be affected by diabetic retinopathy, a retinal disease causing serious loss of vision or damage to the retina in diabetic patients. Diabetic retinopathy resulting from high blood sugar affects blood vessels in the light-sensitive tissue called the retina that lines the back of the eye changing the configuration of retinal blood vessels. Diabetic retinopathy can be detected by analyzing the features of the retinal images. Vessels and arteries have many observable features, including diameter, color, tortuosity, and opacity. Microaneurysms, hemorrhages, cotton wool spots, and exudates are the various types of retinal abnormalities caused due to the damage of blood vessels in diabetic retinopathy. The retinal blood vessels are damaged due to the aging of the people and other factors. In its early stage, nonproliferative diabetic retinopathy, new abnormal blood vessels exponentially proliferate on the surface of the retina. These blood vessels which normally nourish the retina may swell and distort, thus losing their ability to transport blood. At the severe stage, proliferative diabetic retinopathy, growth factors secreted by the retina lead to formation of new blood vessels which are generally termed as abnormal, grow along the inside surface of the retina into the vitreous gel of the eye. These abnormal blood vessels are fragile, which makes them more likely to leak and bleed, ultimately leading to retinal detachment. Thus, the retinal blood vessel detection and segmentation [1] are one of the critical steps to detect and diagnose these abnormal vasculature structures. Ignoring these lesion symptoms can cause loss of vision because these symptoms are not exposed easily at an earlier stage and require diagnosis immediately. Early check-up and treatment of diabetic retinopathy in diabetes patients reduce the risk of damaging or losing the eyesight by 50%.

Palomera-Pérez et al. [2] have used feature extraction-based region growing algorithm to extract the blood vessels in the eye. Martinex-Perez et al. [3] proposed a semi-automatic method to identify the topological properties of retinal blood vessel. Chanwimaluang and Fan [4] and Gao et al. [5] identified efficient methods for automatic detection and extraction of blood vessels. Fraz et al. [6] used ensemble classifier to segment the vessels. The gradient vector field and Gabor transform were used as the feature vectors in ensemble classifier. Mendonça and Campilho [7] have proposed the algorithm of segmentation of blood vessels using vessel centerlines followed by the vessel filtering process. Numerous techniques [10, 11, 13] were used for the improvisation of the blood vessel contrasts. Vessel detection is also done by classification-based techniques [8], by Snake technique [9].

In this paper, we have presented a novel approach for the detection of retinal vessels and determination of the severity of the diabetic retinopathy by extracting features of the vessel (vessel pixel density). The methodology is adapted so as to decrease the computational time of vessel extraction, and the sensitivity, specificity, and accuracy of vessel detection are quite improved with a good correct rate percentage obtained by KNN classification. Section 2 deals with the methodology of the system. Section 3 describes our proposed work. The results and analysis and the performance evaluation have been discussed in Sects. 4 and 5, respectively.

2 Methodology

2.1 Preprocessing

A nonlinear median filtering is applied on an extracted green channel image, which has the highest contrast in an RGB image, to reduce salt and pepper noise and to preserve the soft edges as well. Contrast limited adaptive histogram equalization (CLAHE) is obtained twice to obtain a more enhanced image. CLAHE differs from ordinary adaptive histogram equalization in its technique to limit the contrasts which has to be applied for each neighborhood, and a transformation function is derived from the neighborhood to prevent the over amplification of noise contributed by only adaptive histogram equalization. The contrast amplification in the vicinity of a given pixel value is given by the slope of the function which is proportional to the slope of the neighborhood cumulative distribution function and therefore to the value of the histogram at that pixel value.

2.2 Morphological Segmentation

A morphologically flat structuring element was extracted from the twice histogram image. Top hat filtering is performed upon the gray scale image. Top hat filtering performs the morphological opening on the image and then subtracts the image from original image. Further, all the connected components below a certain threshold were removed to obtain the desired output. The mask of the original image was also with the mask image to remove the outer boundary of the retina and to give the resultant image a compact shape. Further morphological operations were performed to obtain the desired output of retinal vessels as explained in algorithm step 7 and 8.

2.3 KNN Classifier

The *K*-nearest neighbor (KNN) classifier [12] is a supervised method which requires feature vectors for each pixel and manually labeled images for training to classify image pixels as blood vessel or nonblood vessel pixels. The trained sets are feature vectors, each possessing a class label. The training phase of KNN classifier consists of only storing the feature vectors while labeling each of their classes.

While in the classification phase, the constant k is defined by the user, and the label which is most frequent among the k training samples nearest to that query point is assigned to the unlabeled vector. The best k can be selected by various heuristic techniques. In this algorithm, a vector is classified by majority votes polled by its neighbors, with the class being assigned to the most among its K-nearest neighbors measured by a distance function given by

$$\text{Euclidean Distance Function} = \sqrt{\sum_{i=1}^{k} (x_i - y_i)^2} \quad (1)$$

3 Proposed Work

The proposed work initiates with the acquisition of the retina pictures from the databases such as DRIVE, MESSIDOR. Before the preprocessing, the image (584 × 565) pixels' database may be resized if necessary to make the computation time faster. The preprocessing deals with green channel extraction, image enhancement, and de-noising. The morphological segmentation deals with different morphological operations to obtain the desired retinal vessels. Features of the vessels were obtained to obtain the stage of diabetic retinopathy. Here, we obtained the pixel density of the retinal vessels. These features are fed to the KNN classifier. The class

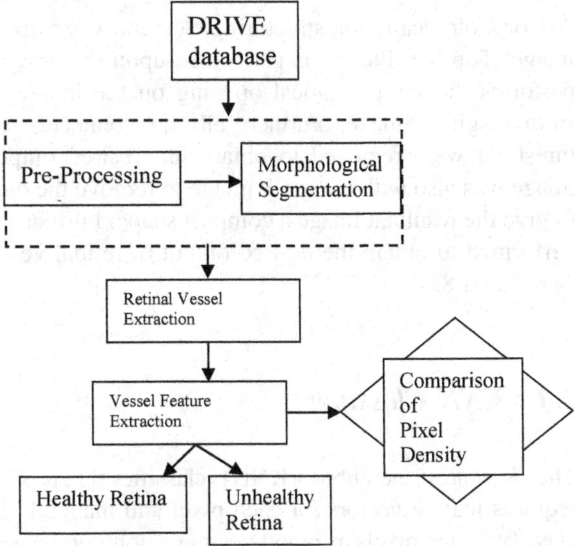

Fig. 1 Work flow process

obtained from the KNN classifier is used to determine the severity of diabetic retinopathy. Other factors such as exudates, hemorrhages are also responsible for the diabetic retinopathy. Here, by the observation of the growth of the vessels in the retina, it can be inferred whether the condition of the retina is good or unhealthy. Besides this, the stage of diabetic retinopathy will also be detected by Fig. 1 that illustrates the proposed work, which is followed by the algorithm.

3.1 Algorithm

1. Extracting the green channel image from original rgb image.
2. Resizing the image to limit the computational time.
3. Non-linear median filtering is performed.
4. Contrast Limited Adaptive Histogram Equalization is exercised twice to enhance the image.
5. Flat structuring element was extracted and top hat filtering was performed.
6. Thresholding and Morphological segmentation is done.
7. Multiplication of the binary converted image of this obtained image with the mask of original image is done.

   a. Find the rows and columns of the gray scale image of the original input image.
   b. for i is from 1 to no of rows
      for j is from 1 to no of columns
         when mask image pixel(i, j)=0
         morphologically segmented image(i, j) = 1;
      end operation
   end operation.

8. Vessel extraction is obtained by further morphological operations on masked image by (a) and (b) operations.

   (a) It sets a pixel to 1 if it is found that five or more pixels in its 3-by-3 neighborhood are 1s; otherwise, it sets the pixel to 0. Now, the isolated pixels to be removed to get a more accurate retinal vessel ending. Isolated pixels (individual 1s that are surrounded by 0s) are removed.
   (b) In the first subiteration, delete pixel p if and only if the conditions $H1$, $H2$, and $H3$ are all satisfied. In the second subiteration, delete pixel p if and only if the conditions $H1$, $H2$, and $H3'$ are all satisfied.

   Condition $H1$:

$$X_H(p) = 1$$

where

$$X_H(p) = \sum_{i=1}^{4} b_i$$

$$b_i = \begin{cases} 1 & \text{if } x_{2i-1} = 0 \\ 0 & \text{otherwise} \end{cases}$$

x_1, x_2, …, x_8 are the values of the eight neighbors of p, starting with the east neighbor and numbered in counter-clockwise order.

Condition $H2$:

$$2 \leq \min\{n_1(p), n_2(p)\} \leq 3 \tag{2}$$

where

$$n_2(p) = \sum_{k=1}^{4} x_{2k} \vee x_{2k+1} \tag{3}$$

$$n_1(p) = \sum_{k=1}^{4} x_{2k-1} \vee x_{2k} \tag{4}$$

Condition $H3$:

$$(x_2 \vee x_3 \vee \overline{x_8}) \wedge x_1 = 0 \tag{5}$$

Condition $H3'$:

$$(x_6 \vee x_7 \vee \overline{x_4}) \wedge x_5 = 0 \tag{6}$$

These two subiterations together make up one iteration of the algorithm by which fine retinal vessels detected are made smooth. Finally, desired retinal vessels are obtained.

9. Features (vessel pixel density) are obtained from the extracted image to classify the type of diabetic retinopathy.

Vessel Pixel Count = No. of Nonzero pixels in the final output image of retina vessel extraction.

Vessel Pixel Density =(Vessel Pixel Count)/Total no. of non-zero pixels in the original binary image.

4 Results and Analysis

The processed algorithm produces the desired output of retinal vessel. The stepwise processed pictures are shown in Fig. 2.

Further morphological operations (described in the algorithm step 8) were performed to smoothen the image and get a better output as shown in Fig. 3. The

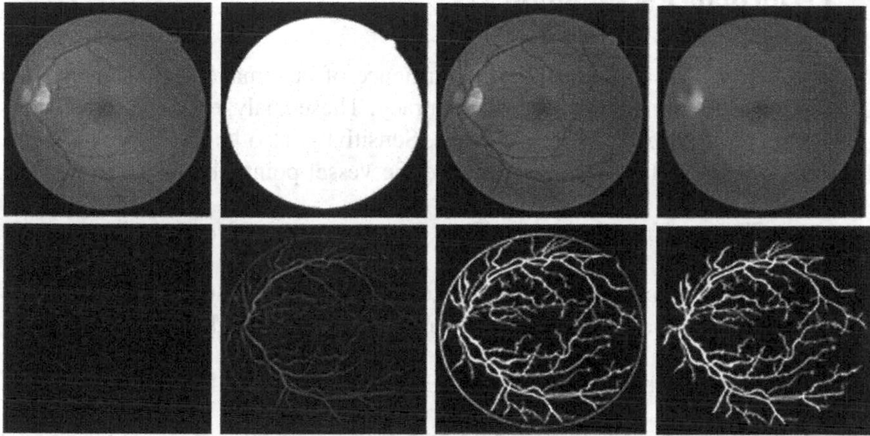

Fig. 2 Starting from top left, moving clockwise **a** original image, **b** mask of the image, **c** extracted green channel, **d** filtered image, **e** subtracted image, **f** adaptive histogram equalized image, **f** blood vessel extracted image, **g** image after being multiplied with masked image

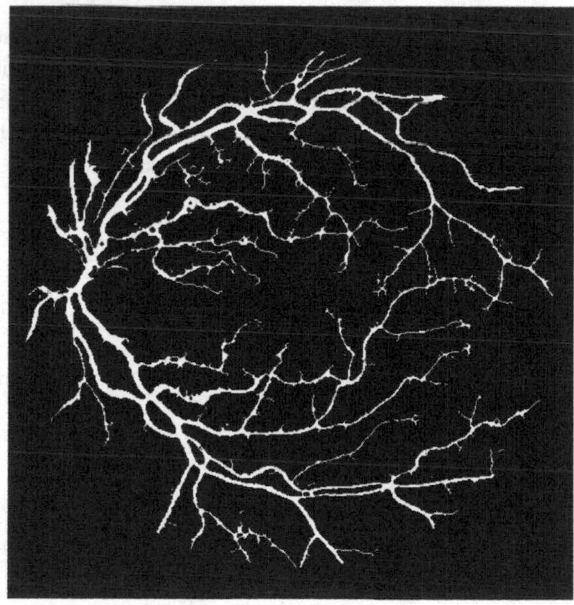

Fig. 3 Final retinal vessel detection

extraction of vessels was obtained with 81.45% sensitivity, 96.27% specificity, and 95.3% accuracy within a very reduced computational time. We have trained KNN classifier by using 25 normal and 50 abnormal images. These images are classified by KNN classifier as normal or abnormal. After feature extraction, KNN classifier is applied to determine the stage of diabetic retinopathy whose performance is of 94%.

5 Performance Evaluation

In this paper, we have analyzed the performance of our image segmentation with the help of sensitivity, specificity, and accuracy. These analyzers are the indicators of the number of properly classified pixels. Sensitivity, also known as true positive rate (TPR), gives us how much perfect are the vessel points that is given by:

$$\mathrm{TPR} = (\text{True Positives})/(\text{True Positives} + \text{False Negatives}) \tag{7}$$

Specificity refers to the true negative rate (TNR), which stands for the fraction of pixels erroneously classified as vessel points. Specificity stands out as:

$$\mathrm{TNR} = (\text{True Negatives})/(\text{True Negatives} + \text{False Positives}) \tag{8}$$

Accuracy is the measurement of the tiniest details in the obtained segmented image when compared to the original input image. It basically depends on how good is the signal-to-noise ratio.

$$\text{Accuracy} = (\mathrm{TN} + \mathrm{TP})/(\mathrm{TN} + \mathrm{TP} + \mathrm{FN} + \mathrm{FP}) \tag{9}$$

Table 1 gives us the results of the stage declaration of diabetic retinopathy. Three cases are shown here. Both normal and diabetic retinopathy affected retinal images were given as inputs. The proposed system attains efficacy in finding the stage of diabetic retinopathy by attaining the class by KNN classifier. The performance parameters are shown in Tables 2 and 3.

Table 1 Results

Input image type	Pixel density	Stage of diabetic retinopathy
Healthy	12.69	Normal
Unhealthy	16.08	Mild diabetic retinopathy
Unhealthy	33.78	Severe diabetic retinopathy

Table 2 Performance parameters

Database	Average performance parameters	
	Accuracy (average) (%)	Computational time (average) (s)
DRIVE	95.3	12.07

Table 3 Evaluation chart

Algorithm	Sensitivity (%)	Specificity (%)	Accuracy (%)
Palomera-Pérez et al. [2]	–	–	92.6
Espona et al. [9]	66.34	96.8	93.16
Marín et al. [1]	69.44	98.19	95.26
Budai et al. [10]	58.00	98.20	93.86
Espona [13]	74.36	96.15	93.52
Chaichana et al. [11]	78.16	95.75	–
Soares et al. [8]	–	–	94.66
Proposed	81.45	96.27	95.3

6 Conclusion

The extraction of vessels is important for further investigation of diabetic retinopathy. Moreover, the growth of vessels can determine the stage and severity of diabetic retinopathy.

Acknowledgements Thanks to DRIVE database for making the retina vessels publicly available.

References

1. Marín, D., A. Aquino, M.E. Gegúndez-Arias, and J.M. Bravo. 2011. A New Supervised Method for Blood Vessel Segmentation in Retinal Images by Using Gray-level and Moment Invariants-based Features. *IEEE Transactions on Medical Imaging* 30 (1): 146–158.
2. Palomera-Pérez, M.A., M.E. Martinez-Perez, H. Benítez-Pérez, and J.L. Ortega-Arjona. 2010. Parallel Multiscale Feature Extraction and Region Growing: Application in Retinal Blood Vessel Detection. *IEEE Transactions on Information Technology in Biomedicine* 14 (2): 500–506.
3. Martinez-Perez M.E., A.D. Hughes, A.V. Stanton, S.A. Thom, N. Chapman, and A.A. Bharath. Retinal Vascular Tree Morphology: A Semi-automatic Quantification. *IEEE Transactions on Biomedical Engineering.*
4. Chanwimaluang T, and G. Fan. An Efficient Algorithm for Extraction of Anatomical Structures in Retinal Images. In *Image Processing. International Conference.* Barcelona, Spain.
5. Gao, X.W., A. Bharath, A. Stanton, A. Hughes, N. Chapman, and S. Thom. 2000. Quantification and Characterisation of Arteries in Retinal Images. *Comput Methods Programs Biomed.*
6. Fraz, M.M., P. Remagnino, A. Hoppe, et al. 2012. An Ensemble Classification-Based Approach Applied to Retinal Blood Vessel Segmentation. *IEEE Transactions on Biomedical Engineering* 59 (9): 2538–2548.
7. Mendonça, A.M., and A. Campilho. 2006. Segmentation of Retinal Blood Vessels by Combining The Detection Of Centerlines and Morphological Reconstruction. *IEEE Transactions on Medical Imaging* 25 (9): 1200–1213.
8. Soares, J.V.B., J.J.G. Leandro, R.M. Cesar Jr., H.F. Jelinek, and M.J. Cree. 2006. Retinal Vessel Segmentation Using the 2-D Gabor Wavelet and Supervised Classification. *IEEE Transactions on Medical Imaging* 25 (9): 1214–1222.

9. Espona, L., M.J. Carreira, M. Ortega, and M.G. Penedo. 2007. *A Snake for Retinal Vessel Segmentation*. LNCS 4478: 178–185.
10. A. Budai, R. Bock, A. Maier, J. Hornegger, and G. Michelson. 2013. Robust Vessel Segmentation in Fundus Images. *International Journal of Biomedical Imaging* 2013, Article ID 154860, 11 p.
11. Thanapong, Chaichana, Wiriyasuttiwong, Watcharachai, Reepolmaha, Somporn, Pintavirooj Chuchart and Sangworasil Manas. 2007. Extraction Blood Vessels from Retinal Fundus Image Based on Fuzzy C-Median Clustering Algorithm. In *IEEE Fourth International Conference on Fuzzy Systems and Knowledge Discovery (FSKD 2007)*.
12. Niemeijer, M., J. Staal, B. van Ginneken, M. Long, and M.D. Abramoff. 2004. Comparative Study of Retinal Vessel Segmentation Methods on a New Publicly Available Database. In *Proceedings of the SPIE medical imaging*, vol. 5370, 648–656.
13. Espona, L., M.J. Carreira, M.G. Penedo, and M. Ortega. 2008. Retinal Vessel Tree Segmentation using a Deformable Contour Model. 2008 *IEEE*.

Design of F-Shape Fractal Microstrip Patch Antenna for C- and X-Band Applications

Dilveen Kaur and Jagtar Singh Sivia

Abstract In this paper, F-shape fractal microstrip patch antenna is designed using roger RT/duroid and FR4 glass epoxy substrate materials. Proposed antenna is fed by microstrip line feed. Simulations of proposed antenna are done up to third iteration using Ansoft HFSS software. F-shape fractal notation in antenna design makes the patch antenna flexible in terms of generating resonant frequency and bandwidth, as an iteration number of fractal increases. Performance parameters of the antenna such as bandwidth, radiation pattern, return loss, and VSWR are compared for both the substrate materials. It is found that rogers RT/duroid as a substrate provides less return loss and more gain. Vector Network Analyzer (VNA) is used for the measurement of return loss and VSWR of the antenna. Simulated and measured results show good agreement with each other. The measured results of proposed antenna with rogers RT/duroid substrate exhibit a gain of 7.2261, 5.0946, 2.9631, 8.3167, and 1.2998 dB at resonant frequencies 5.6, 7.72, 9.34, 11.26, and 13.58 GHz, respectively.

Keywords F-shape fractal · Patch antenna · Return loss · Gain
VSWR · Microstrip line feed

1 Introduction

A fractal is an antenna which uses the fractals. Fractals are self-similar shapes so as to increase the length and perimeter of the material. Fractal is basically an irregular shape. To reduce the size of patch, fractal geometries are used. Fractal geometries have the properties of space filling and self-similarity. For the design of multiband fractal microstrip patch antenna, the self-similarity property is used [1].

D. Kaur (✉) · J.S. Sivia
ECE, YCoE, Punjabi University GKC, Talwandi Sabo, Bathinda, Punjab, India
e-mail: dilveenkr@yahoo.com

J.S. Sivia
e-mail: jagtarsivian@gmail.com

M.S. Reddy et al. (eds.), *International Proceedings on Advances in Soft Computing, Intelligent Systems and Applications*, Advances in Intelligent Systems and Computing 628, https://doi.org/10.1007/978-981-10-5272-9_19

The self-similarity property means that the design is subdivided into parts and the part is a reduced size or the smallest portion of the whole shape [2]. There are many different shapes of fractal antennas like Sierpinski Gasket, Minkowski Hilbert Curve, and Koch Curve [1]. For the antenna to radiate effectively and to increase the bandwidth of antenna, fractal geometry is used. The fractal geometry can be designed using Iterative Function System (IFS). In IFS, there are a number of affine transformations. The starting part of the fractal geometry is called initiator, which can be divided into equal parts [3]. This is the first iteration [4]. The size of a radiating patch increases as the number of iterations increases but the resonant frequency diminishes [2]. Fractal antennas are inspired by nature [5]. Fractals are widely used to design complex natural objects like cloud boundaries, galaxies, mountain ranges, and many more [6]. Fractals have jagged shapes. So these discontinuities help to improve the bandwidth and effective radiation of antennas [7]. Broadband operation, gain, space filling, multiple scale properties, self-similar pattern, mechanical simplicity, and robustness are the special properties of fractal antennas [2 ,8]. Microstrip patch antennas have radiating elements on one side of a dielectric substrate, and these are fed using different feeding techniques: the microstrip line, coaxial probe, aperture coupling, and proximity coupling [1, 9]. The feed line and the antenna input impedance should be properly matched. For good impedance matching between line and patch, different feeding techniques are used [4, 9]. In this paper, proposed antenna is fed with microstrip line feed as this feed makes impedance matching simple to achieve and easy to fabricate.

2 Antenna Design

The design of F-shape fractal antenna starts with rectangular patch geometry whose dimensions are calculated using Eqs. (1)–(3).

The width of microstrip patch is calculated as [4, 9, 10]:

$$w = \frac{c}{2f_0\sqrt{\frac{\varepsilon_r+1}{2}}} \tag{1}$$

The effective dielectric constant is determined using the following equation [4, 9, 10, 11]:

$$\varepsilon_{\text{reff}} = \frac{\varepsilon_r+1}{2} + \frac{\varepsilon_r-1}{2}\left[1+12\frac{h}{w}\right]^{1/2} \tag{2}$$

The effective length (L_{eff}) of the patch is calculated using the following equation [4, 9]:

$$L_{eff} = \frac{c}{2f_0\sqrt{\varepsilon_{reff}}} \quad (3)$$

Following are the steps for the design of proposed antenna:

Step 1: For rogers RT/duroid substrate with dielectric constant 2.2, thickness 1.6 mm, and resonant frequency 7.5 GHz, the length and width of a patch are found to be 12.45 and 16 mm, respectively. This iteration is known as the zeroth iteration which is shown in Fig. 1a.

Step 2: Two F-shapes along the horizontal length of patch are obtained by applying the concept of fractal geometry. The dimensions of various cuts to form F-shapes are shown in Fig. 1a. This geometry is known as first iteration geometry.

Step 3: Two more small F-shapes in a patch are obtained using similar concept, and various cut dimensions are shown in Fig. 1a. The geometry thus obtained is known as second iteration geometry.

The fabricated geometries of proposed antennas with rogers RT/duroid and glass epoxy as substrate materials are shown in Fig. 1a, b, respectively.

Fig. 1 Fractal iterations of the proposed antenna. **a** Antennas fabricated with rogers RT/duroid substrate material. **b** Antennas fabricated with FR4 epoxy substrate material

3 Results and Discussions

Results of the proposed antennas with rogers RT/duroid and FR4 glass epoxy substrates are given below.

3.1 Rogers RT/Duroid as a Substrate

Figure 2 shows the simulated and measured return loss versus frequency plots of the proposed antenna using rogers RT/duroid substrate for all iterations. Comparison of simulated and measured results of proposed antenna in terms of performance parameters such as resonant frequency, return loss, gain, and VSWR are shown in Table 1. From this table, it is clear that for zeroth iteration, antenna has a return loss less than −10 dB at resonant frequencies 7.22 and 9.44 GHz. Thus, zeroth iteration works at these two frequencies, and gains of antenna at these frequencies are 7.1410 and 5.0810 dBi, respectively.

Likewise, first iteration works at three frequencies 5.70, 7.72, and 9.44 GHz having gains of 7.2639, 4.9289, and 2.5940 dBi, respectively. In the second iteration of F-shape patch, the antenna resonates at 5.60, 7.72, 9.34, 11.26, and 13.58 GHz with a return loss of −13.9944, −14.0891, −13.7137, −12.6360, and −15.0583 dB and having gains of 7.2261, 5.0946, 2.9631, 8.3167, and 1.2998 dBi, respectively. Figure 2 shows the simulated and measured return loss versus frequency plots for different iterations of proposed fractal antenna.

Figure 3 shows the variation of VSWR versus frequency for the proposed fractal F-shaped antenna for different iterations. This antenna has VSWRs 1.498, 1.492, 1.519, 1.609, and 1.429 at resonant frequencies 5.6, 7.72, 9.3, 11.2, and 13.5 GHz, respectively.

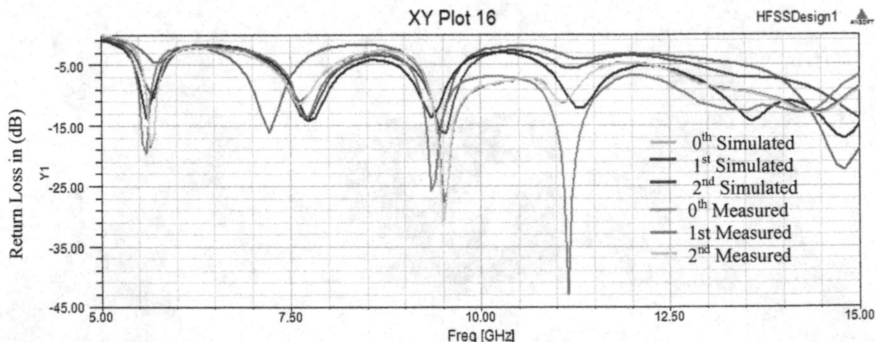

Fig. 2 Simulated and measured return loss versus frequency for different fractal iterations with rogers RT/duroid substrate

Table 1 Simulated and measured results of antenna designed with rogers RT/duroid substrate

Iteration number	Simulated results				Measured results		
	Resonant frequency (GHz)	Return loss (dB)	Gain (dBi)	VSWR	Resonant frequency (GHz)	Return loss (dB)	VSWR
0th iteration	7.2222	−16.2533	7.1401	1.3639	7.65	−11.4297	1.73
	9.4444	−15.4561	5.0810	1.4060	9.5	−30.9716	1.06
1st iteration	5.7071	−10.7073	7.2639	1.9084	5.65	−18.6651	1.26
	7.7273	−14.4979	4.9289	1.4643	7.6	−11.1444	1.77
	9.4444	−16.6403	2.5940	1.3453	9.5	−28.7114	1.08
2nd iteration	5.6061	−13.9944	7.2261	1.4989	5.55	−18.0052	1.29
	7.7273	−14.0891	5.0946	1.4922	7.7	−13.8834	1.51
	9.3434	−13.7137	2.9631	1.5196	9.35	−25.8371	1.11
	11.2626	−12.6360	8.3167	1.6091	11.15	−44.5149	1.01
	13.3859	−15.0583	1.2998	1.4291	13.5	−12.5791	1.61

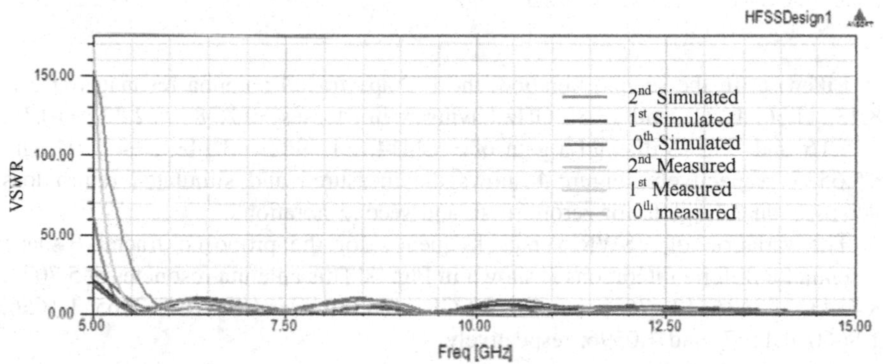

Fig. 3 Simulated and measured VSWR versus frequency for different iterations of proposed fractal antenna with rogers RT/duroid substrate

3.2 *FR4 Epoxy as a Substrate*

The comparison of simulated and measured results of the proposed antenna with FR4 epoxy substrate in terms of the performance parameters such as return loss, gain, and VSWR are shown in Table 2.

Zeroth iteration of this antenna resonates at 5.20 and 13.68 GHz with a return loss of −14.0525 and −28.5633 dB and having gains of −1.3273 and −2.5234 dBi, respectively. First iteration of antenna resonates at 5.70, 8.23, 11.36, 12.47, and 13.98 GHz having return losses −17.22, −15.10, −12.76, −23.71, and −19.15 dB and gain of −2.9338, 1.6307, −2.9680, −4.3053, and −5.6425 dBi, respectively.

Table 2 Simulated and measured results of antenna designed with FR4 epoxy substrate

Simulated results					Measured results		
Iteration number	Resonant frequency (GHz)	Return loss (dB)	Gain (dBi)	VSWR	Resonant frequency (GHz)	Return loss (dB)	VSWR
0th iteration	5.2020	−14.0525	−1.3273	1.4948	5.5	−8.5295	2.20
	13.6869	−28.5633	−2.5234	1.0775	14.3	−7.4438	2.20
1st iteration	5.7071	−17.2224	−2.9338	1.3193	5.85	−8.8502	2.13
	8.2323	−15.1043	1.6307	1.4263	8.4	−13.6191	1.53
	11.3636	−12.7608	−2.9680	1.5978	11.5	−10.7243	1.82
	12.4747	−23.7120	−4.3053	1.1395	12.65	−37.8358	1.03
	13.9899	−19.1541	−5.6425	1.2478	14.45	−9.1747	2.07
2nd iteration	5.7071	−17.7820	−6.4414	1.2965	5.9	−9.9256	1.94
	8.4343	−22.3280	−1.8962	1.1656	8.45	−23.9585	1.14
	11.3636	−10.3950	−3.1483	1.8660	11	−11.7587	1.70
	12.6768	−25.1688	−4.4004	1.1167	12.85	−19.5771	1.23
	13.8889	−30.7651	−5.6525	1.0596	14.15	−11.3359	1.74

Likewise, in the second iteration, the F-shape fractal antenna resonates at 5.7, 8.43, 11.36, 12.67, and 13.88 GHz having return losses −17.78, −22.32, −10.39, −25.16, and −30.76 dB and a gain of −6.4414, −1.8962, −3.1483, −4.4004, and −5.6525, respectively. Figure 4 shows the measured and simulated return loss versus frequency plots for zeroth, first, and second iterations.

The variation of VSWR versus frequency for the proposed fractal F-shape antenna for different iterations is shown in Fig. 5. This antenna resonates at 5.7071, 8.4343, 11.3636, 12.6768, and 13.8889 GHz and provides VSWRs 1.2965, 1.1656, 1.8660, 1.1167, and 1.0596, respectively.

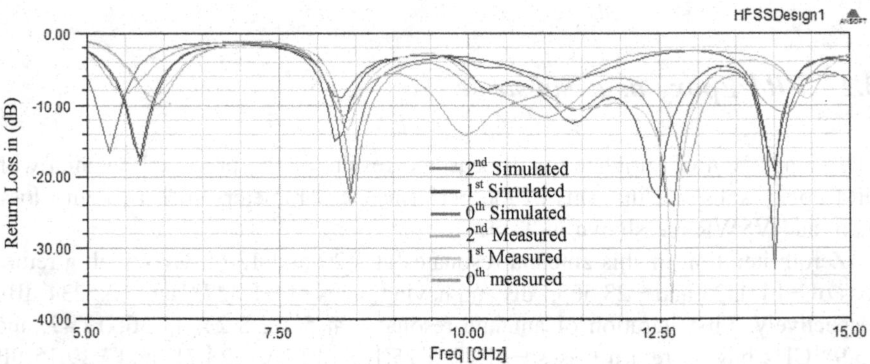

Fig. 4 Simulated and measured return loss versus frequency for different fractal iterations with FR4 epoxy substrate

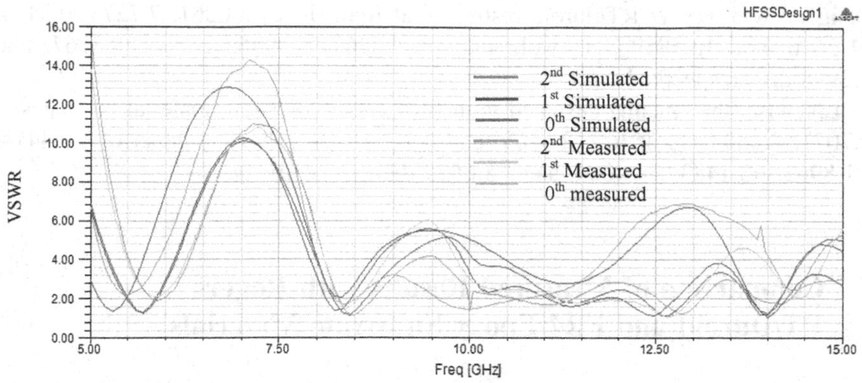

Fig. 5 Simulated and measured VSWR versus frequency for different fractal iterations with FR4 epoxy substrate

4 Comparison of Simulated Results with Rogers RT/Duroid and FR4 Epoxy Substrate Materials

The comparison of simulated results with rogers RT/duroid and FR4 epoxy substrate materials for different iterations and at different frequencies is shown in Table 3. The simulated return loss, gain, and VSWR of the designed antenna with different substrates are compared. From the table, it is clear that the antenna

Table 3 Comparison of simulated results with rogers RT/duroid and FR4 epoxy substrate materials

Iteration number	Rogers RT/duroid				FR4 epoxy			
	Resonant frequency (GHz)	Return loss (dB)	Gain (dBi)	VSWR	Resonant frequency (GHz)	Return loss (dB)	Gain	VSWR
0th iteration	7.2222	−16.2533	7.1401	1.3639	5.2020	−14.0525	−1.3273	1.4948
	9.4444	−15.4561	5.0810	1.4060	13.6869	−28.5633	−2.5234	1.0775
1st iteration	5.7071	−10.7073	7.2639	1.9084	5.7071	−17.2224	−2.9338	1.3193
	7.7273	−14.4979	4.9289	1.4643	8.2323	−15.1043	1.6307	1.4263
	9.4444	−16.6403	2.5940	1.3453	11.3636	−12.7608	−2.9680	1.5978
	–	–	–	–	12.4747	−23.7120	−4.3053	1.1395
	–	–	–	–	13.9899	−19.1541	−5.6425	1.2478
2nd iteration	5.6061	−13.9944	7.2261	1.4989	5.7071	−17.7820	−6.4414	1.2965
	7.7273	−14.0891	5.0946	1.4922	8.4343	−22.3280	−1.8962	1.1656
	9.3434	−13.7137	2.9631	1.5196	11.3636	−10.3950	−3.1483	1.8660
	11.2626	−12.6360	8.3167	1.6091	12.6768	−25.1688	−4.4004	1.1167
	13.3859	−15.0583	1.2998	1.4291	13.8889	−30.7651	−5.6525	1.0596

designed with rogers RT/duroid resonates at frequencies 5.6061, 7.7273, 9.3434, 11.2626, and 13.5859 GHz with gain of 7.2261, 5.0946, 2.9631, 8.3167, and 1.2998 dBi, respectively.

Likewise, the antenna designed with FR4 epoxy substrate material resonates at 5.7071, 8.4343, 11.3636, 12.6768, and 13.8889 GHz having gains of −6.4414, −1.8962, −3.1483, −4.4004, and −5.6525 dBi, respectively.

5 Comparison of Measured Results with Rogers RT/Duroid and FR4 Epoxy Substrate Materials

The comparison of measured results of the designed antenna with both the substrate materials are shown in Table 4. The measured return loss and VSWR are compared. The antenna proposed with rogers RT/duroid substrate material resonates at 5.55, 7.7, 9.35, 11.15, and 13.5 GHz with return losses −18.0052, −13.8834, −25.8371, −44.5149, and −12.5791 respectively.

Likewise, the antenna designed with FR4 epoxy substrate material resonates at 5.9, 8.45, 11, 12.85, and 14.15 GHz with return losses −9.9256, −23.9585, −11.7587, −19.5771, and −11.3359 dB, respectively.

Table 4 Measured results of the designed antenna with rogers RT/duroid and FR4 epoxy substrate materials

Rogers RT/Duroid				FR4 epoxy		
Iteration number	Resonant frequency (GHz)	Return loss (dB)	VSWR	Resonant frequency (GHz)	Return loss (dB)	VSWR
0th iteration	7.65	−11.4297	1.73	5.5	−8.5295	2.20
	9.5	−30.9716	1.06	14.3	−7.4438	2.20
1st iteration	5.65	−18.6651	1.26	5.85	−8.8502	2.13
	7.6	−11.1444	1.77	8.4	−13.6191	1.53
	9.5	−28.7114	1.08	11.5	−10.7243	1.82
	–	–	–	12.65	−37.8358	1.03
	–	–	–	14.45	−9.1747	2.07
2nd iteration	5.55	−18.0052	1.29	5.9	−9.9256	1.94
	7.7	−13.8834	1.51	8.45	−23.9585	1.14
	9.35	−25.8371	1.11	11.00	−11.7587	1.70
	11.15	−44.5149	1.01	12.85	−19.5771	1.23
	13.5	−12.5791	1.61	14.15	−11.3359	1.74

6 Conclusion

F-shape fractal microstrip patch antenna using two different substrate materials is designed in this paper. Simulated and measured results of the antenna are compared. It is concluded that the antenna designed with rogers RT/duroid as a substrate material provides better performance paramters as compared to FR4 epoxy substrate as it provides high gain, low return loss, and VSWR. The designed antennas can be used in C and X-band applications for satellite links, wireless communication and microwave links.

Acknowledgements The authors are grateful to Miss Garima Saini, assistant professor, Electronics and Communication Department, National Institute of Technical Teacher Training and Research (NITTTR), Chandigarh, for providing the antenna measurement facilities. The authors would also like to thank the Rogers RT/Duroid authority for donating the Rogers RT/duroid material.

References

1. Nagpal, A., and S.S. Dhillon. 2013. Multiband E-shaped Fractal Microstrip Patch Antenna with DGS for Wireless Applications. *IJETAE International Journal of Emerging Technology and Advance Engineering* 2 (12): 241–244.
2. Kumar, M. 2014. Design and Analysis of Minkowski Fractal Antenna Using Microstrip Feed. *International Journal of Application or Innovation in Engineering & Management (IJAIEM)*, 3 (1).
3. Werner, D.H., P.L. Werner, and K.H. Church. 2001. Genetically Engineered Multiband Fractal Antennas. Electronics Letters 37 (19): 1150–1151.
4. Yadav, S., and P. Jain. 2014. Analysis and Design Rectangular Patch with Half Circle Fractal Techniques. In *International Conference on Advances in Computing, Communications And Informatics (ICACCI)*, 2014, 978-1-4799.
5. Bayatmaku, N., P. Lotfi, and M. Azarmanesh. 2011. Design of Simple Multiband patch Antenna for Mobile Communication Applications Using New E-shape Fractal. *IEEE Antennas and Wireless Propagation Letters* 10.
6. Soliman, A., and D. Elsheakh. 2011. Multi-Band Printed Metamaterial Inverted-F Antenna (IFA) for USB Applications. *IEEE Antennas and Wireless Propagation Letters* 10: 25–28.
7. Ojaroudi, M., and N. Ojaroudi. 2014. Ultra-Wideband Small Rectangular Slot Antenna with Variable Band-Stop Function. *IEEE Transactions of Antennas and Propagation* 62 (1).
8. Deshmukh, A.A., and G. Kumar. 2005. Compact Broadband E-shaped Microstrip Antenna. Electronics Letters 41 (18): 989–990.
9. Balanis, C.A. 2008. Microstrip Antennas. In *Antenna Theory, Analysis and Design*, 3rd ed., 811–876. New York: Wiley.
10. Pharwaha, A.P.S., J. Singh, and T.S. Kamal. 2010. Estimation of Feed Position of Rectangular Microstrip Patch Antenna. *IE Journal-ET*, 91.
11. Sivia, J.S., A. Singh, and T.S. Kamal. 2013. Design of Sierpinski Carpet Fractal Antenna Using Artificial Neural Networks. *International Journal of Computer Applications* 68 (8): 5–10.

Face Image Detection Methods: A Survey

Varsha Kirdak and Sudhir Vegad

Abstract Face detection is an important part of face recognition systems. In this paper, we presented various methods of face detection, which are commonly used. These methods are local binary pattern (LBP), Adaboost, support vector machine (SVM), principal component analysis (PCA), hidden Markov model (HMM), neural network-based face detection, Haar classifier, and skin color models. Each method is summarized along with their advantages and disadvantages.

Keywords Face detection · Adaboost method · Local binary pattern (LBP)
Principal component analysis (PCB) · Hidden Markov model (HMM)
Support vector machine (SVM) · Skin segmentation

1 Introduction

The face detection simply means to determine that the given input is a face image or not, regardless of the size, position, background, etc. The current evolution of computer technologies has boosted in this era; for example, computer vision contributes in face recognition and video coding techniques. Face detection in computer vision involves segmentation, extraction, and verification of faces. Face detection is considered as primary steps toward face recognition. In recent years with the development of artificial intelligence, Internet of Things, e-commerce, and other computer applications, face detection and recognition gain much more importance [1].

Applications of face detection include as follows: **Surveillance**: Surveillance in the form of CCTVs can be proved effective in gathering evidences, e.g., criminal

V. Kirdak (✉)
GTU PG School, Ahmedabad, India
e-mail: varsha.kirdak100@gmail.com

S. Vegad
ADIT, New V V Nagar, Vitthal Udyognagar, India
e-mail: svegad@gmail.com

M.S. Reddy et al. (eds.), *International Proceedings on Advances in Soft Computing, Intelligent Systems and Applications*, Advances in Intelligent Systems and Computing 628, https://doi.org/10.1007/978-981-10-5272-9_20

evidences or during exams. Face detection is the best biometric for video data too. **General identity verification**: Smart cards are most often used in today's world for maintaining high security, e.g., employee id, driving license, e-registration, voter id card. **Criminal justice systems**: For gathering criminal evidences, forensic analysis and postevent investigation can be surely benefitted by face detection even if the culprit is wearing mask or cloth for distraction. **Image database investigations**: Searching in large number of image databases for any kind of identification purposes like missing children, licence holders, account holders. **Multimedia** environments with adaptive human–computer interfaces; i.e., it is a robust way as well as proved beneficial in various domain areas.

Computational powers and availability of recent sensing is increasing results in human–computer interactive applications such as face detection, which gradually includes authentication, verification, tracking of facial images. Face detection goal is to determine the input contains any face images and provides results regardless of expressions, occlusions, or lighting conditions [2]. Face detection is a first step in human interaction systems which include expression recognition, cognitive state, tracking, surveillance systems, automatic target recognition (ATR), or generic object detection/recognition. Face detection challenges include as follows: out-of-plane rotation: frontal, 45°, profile, upside down, presence of beard, mustache, glasses, etc. Facial expressions occlusions by long hair, hand, in-plane rotation image conditions include as follows: size, lighting condition, distortion, noise, and compression.

Face detection applications in areas such as content-based image retrieval (CBIR), video coding, video conferencing, crowd surveillance, and intelligent human–computer interfaces are gaining popularity. Basically, face detection falls under two categories that are local feature-based ones and global methods. Approaches based on features include: geometrical method, color-based or texture-based method, motion-based method [3].

Figure 1 shows face detection approaches in graphical representation, and detail description is as follows [4].

2 Face Detection Techniques

Generally, prior to basics include following methods:

Knowledge-Based Methods: It encodes what constitutes typical face, e.g., the relationship between facial features. It includes knowledge about some features that could fall into face localization like distance of eyes and mouth or mouth is below nose, etc., thus reducing computational time for detection. It includes top-down approaches and bottom-up approaches.

Feature Invariant Approaches: This approach aims to find structural features of a face that exist even when pose, viewpoint, or lighting conditions vary. Therefore useful in locating face faster, and correct identification of faces is possible even in adverse conditions and include skin color and texture.

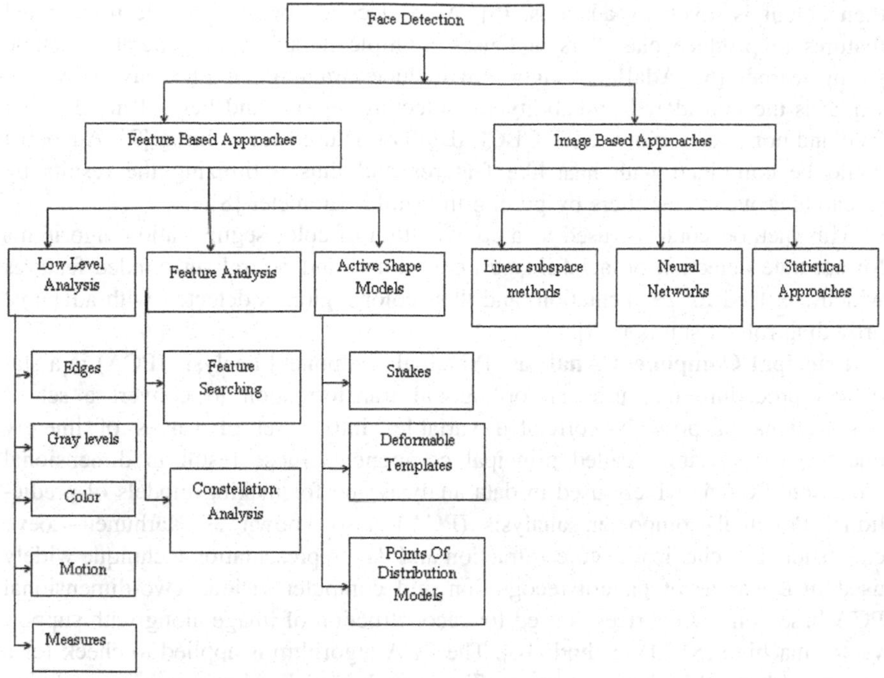

Fig. 1 Face detection approaches

Template Matching: Here, several standard patterns are stored in databases to describe the face as a whole or the facial features separately. So after comparing the input images and stored images, output is provided as match result considering the template of facial images. This technique is used for localization as well as detection of facial images and includes predefined templates.

Appearance-Based Methods: The models are learned from a set of training images that capture the representative variability of faces. Basically, templates are from trained databases and are used for detection purposes, and it includes distribution-based methods, support vector machine (SVM), hidden Markov model.

3 Face Detection Methods

Adaboost: AdaBoost is abbreviated for "Adaptive boosting." It is a machine learning meta-algorithm by Yoav Freund and Robert Schapire who won the global prize in 2003 for their work. It can be used in combination with learning algorithms to improve their performance and accuracy. The output of the other learning algorithms ("weak learners") is combined into a weighted sum that represents the final output of the boosted classifier. Blob filtering results identify ROI region, and then output is given by adaboost [5]. Also, accurate prediction rule uses visual features to produce classifiers and uses multiple iterations to generate a single strong learner [6]. AdaBoost algorithm includes training samples having weight which is the considered probability for selecting regions and helpful in detecting face and non-face regions. MIT-CBCL database is used for training [7]. Adaboost could be combined with haar-like features and thus optimizing the results by researching weak classifiers by putting threshold parameter [8].

This method could be used as a combination of color segmentation algorithms for accurate detection of facial images, edge lines, and central surrounded features which are used for segmentation, and then color region is detected with adaboost providing robust solution [9].

Principal Component Analysis: Principal component analysis (PCA) is a statistical procedure that uses an orthogonal transformation to convert a set of observations of possibly correlated variables into a set of values of linearly uncorrelated variables called principal components those result in dimensional reduction. PCA is widely used in data analysis and for making models of predictions. Principal component analysis (PCA), also known as Karhunen–Loeve expansion, is a classical feature extraction and data representation technique widely used in the areas of pattern recognition and computer vision. Two-dimensional PCA based on 2D matrices is used for reconstruction of image along with support vector machine (SVM) method [10]. The PCA algorithm is applied to check for a face at each pixel in the input image. This search is realized using crosscorrelation in the frequency domain between the entire input image and eigenvectors. A few low-frequency discrete cosine transform (DCT) coefficients are used and then perform PCA on the enhanced training images set, and ORL database is used [11]. Initially, the eigen space is created with eigenvalues and eigenvectors then the input images are classified based on Euclidian distance. Classifier is used for decision-making based on feature vectors; ORL face database is used [12].

Hidden Markov Model: A hidden Markov model (HMM) is a statistical Markov model in which the system being modeled is assumed to be a Markov process with unobserved (hidden) states. Direct states are not visible, but the output is clearly shown. These models are especially known for their application in temporal pattern recognition such as speech, handwriting, gesture recognition, for example, Gaussian distribution. The observation vectors used to classify the states of the HMM are obtained using the coefficients of the Karhunen–Loeve transform (KLT). This method is useful in detecting gray scale images as well. MIT database is used, and both face detection and face recognition are using HMM model [13]. It consists of Markov chain of hidden, invisible, and limited states and a probability density function. Markov chain includes probability distribution function of initial state and probability matrix of transferred state. Probability density function is related to those states [14, 15]. HMM could be used as an expression classifier also. First relative features are extracted based on relative displacement of face muscles [16].

Support Vector Machine (SVM): SVM is a learning model in machine learning that is used for analysis of data and can perform nonlinear classifications.

SVM classifier includes sub-gradient descent and coordinates descent. The SVM algorithm maps the test images into higher-dimension transform space where a hyper-plane decision function is made based on kernel function. Three kernel functions are polynomial, Gaussian radial basis function (RBF), and sigmoidal. ORL database is used [17]. Integration of feature analysis, class modeling, and SVM are used for detection of faces. Input images are represented using 1-D Haar wavelet representation, and its amplitude projections, and then using classification rules face and non-face images are classified. Distribution-based distance and SVM give a novel approach in detection [18].

Skin color segmentation: As name indicates, this method uses skin color detection and segmentation methods for fast retrieval of image detection. Lighting compression is carried out followed by noise removal on YCbCr model then euler computations then put bounding box ratio and eccentricity ratio, thus separating facial and non-facial regions [19]. Color image threshold is provided based on centroid segmentation. Three-dimensional color space is transformed to 2D, and then hexagonal distribution is done, and threshold value is fixed based on centroid segmentation for detection of faces [20]. A robust detector is used which combines skin color, heuristic rules, and geometric features. Pixel-based detection strategy is used and localization of regions of eyes and lips are performed that results in a robust way of detecting faces [21]. Skin segmentation includes locating objects of facial components like eyes, lips. Skin representation is carried out in the chrominance plane and based on rotation of Cg and Cr axis, and boundary box is used for face segmentation. YCgCr model is used, and threshold maximum and minimum is based on Gaussian distribution, thus reducing the processing time [22].

Local binary pattern: Local binary pattern (LBP) is a feature used for classification in computer vision, especially for texture classification. Image texture features extraction is used in fields of image retrieval, texture examination, face recognition, image segmentation, etc. Every pixel has texture value that is useful in finding the key points of the image and then forming a color-texture feature selection [23]. Gradient local binary pattern feature is more discriminative and effective than histogram of orientated gradient (HOG), histogram of template (HOT), and semantic local binary patterns (S-LBP). Combining gradient information and texture information together could result in gradient local binary pattern. INRIA dataset is used [24]. A feature-centric cascade method detector makes use of local binary pattern along with the combination of Haar features. After that, feature-centric detection method is used. Multi-view face detection is possible using this particular method [25].

Neural Network: Firstly, face detection is carried out using a polynomial neural network (PNN) and face recognition using pseudo-2D hidden Markov models; thus, detection of images is even possible in clustered images [26]. Eigen-face algorithm is combined with neural network. Training sets are prepared by eigen algorithm, and then matched blocks are used for detecting the facial image constitute in template-based face detection method [27]. Neural network when combined with adaboost algorithm provides faster processing speed and less computation time.

Table 1 Advantages and disadvantages of various face detection methods

Method	Advantages	Disadvantages
Adaboost method	No prior knowledge is needed	Complex weak classifier, sensitive to noisy data, slow training
PCA	Dimensionality reduction, eigenvalues are used as features	High-dimensional space not having good results, lighting condition varies performance
HMM	Easy to determine hidden states, better compression	Algorithm is expensive and time-consuming
SVM	Nonlinear classification	Training and testing takes time
Skin color segmentation	Detect face and non-face regions	Lighting condition is affecting the results
LBP	Texture classification	Use is restricted
Neural network	High detection rates, less computationally expensive	Less accurate results compared to other methods

Adaboost cascade classifier uses Haar-like features; thus, detection rates are increased drastically [28]. The advantages and disadvantages of above discussed methods are shown in Table 1.

4 Conclusion

The purpose of this paper is to identify face detection issues and challenges and compare various methods for face detection. There is a significant advancement in this area as it is beneficial in real-world application products. Various face detection techniques are summarized, and finally methods are discussed for face detection, their features, advantages, and disadvantages.

There is still a good scope for work to get efficient results by combining or improving the selection of features for detection of face images regardless of intensity of background color or any occlusion.

References

1. Norway, and Boon Kee. 2001. Low Face Detection: A Survey, April 17, 2001.
2. Gupta, Varsha, and Dipesh Sharma. 2014. A Study of Various Face Detection Methods. *IJARCCE* 3 (5).
3. Lu, Yongzhong, Jingli Zhou, and Shengsheng Yu. 2003. A Survey of Face Detection, Extraction and Recognition, vol. 22. University of Science and Technology Computing and Informatics.
4. Bakshi, Urvashi, and Rohit Singhal. 2014. A Survey on Face Detection Methods and Feature Extraction Techniques.

5. Niyomugabo, Cesar, Kwon Jae Hong, and Tae Yong Kim. 2014. *A Fast Face Detection Method Based On Skin Color and Adaboost*. IEEE ISCE.
6. Meir, R., and G. Rätsch. 2003. An Introduction to Boosting and Leveraging. Springer, Berlin.
7. Li, Wan, and Chen Pu-chun. 2010. Face Detection Method Based on Skin Color and AdaBoost Algorithm. In *International Conference on Computational and Information Sciences*. IEEE.
8. Zhang, Huaxun, Yannan Xie, and Cao Xu. 2011. A Classifier Training Method for Face Detection Based on AdaBoost. In: *International Conference on Transportation, Mechanical, and Electrical Engineering (TMEE)*. IEEE.
9. Ma, Songyan, and Tiancang Du. 2010. Improved adaboost face detection. In *International Conference on Measuring Technology and Mechatronics Automation*. IEEE.
10. Zhang, Xiaoyu, Jiexin Pu and Xinhan Hu. Face Detection Based on Two Dimensional Principal Component Analysis and Support Vector Machine. In *Proceedings of the 2006 IEEE International Conference on Mechatronics and Automation*.
11. He, Jia-Zhong, Qing-Huan Zhu, and Ming-Hui Du. 2006. Face Recognition using PCA on Enhanced Image for Single Training Images. In *Proceedings of the Fifth International Conference on Machine Learning and Cybernetics, Dalian*, 13–16 August 2006. IEEE.
12. Gomathi, E., and K. Baskaran. 2010. Recognition of Faces using Improved Principal Component Analysis. In *Second International Conference on Machine Learning and Computing*, 2010. IEEE.
13. Nefian, Ara V., and Monson H. Hayes. 1998. *Faces Detection and Recognition using Hidden Markov Models*. IEEE.
14. Nefian, A.V., and M.H. Hayes. 1998. Face Detection and Recognition Using Hidden Markov Models. In *Proceedings of the International Conference on Image Processing*.
15. Samaria, F., and S. Young. 1994. HMM Based Architecture for Face Identification. *Image and Computer Vision* 12.
16. Boruah, Dhrubajyoti, Kandarpa Kumar Sarma, and Anjan Kumar Talukdar. 2015. Different Face Region Detection Based Facial Expression Recognition. In *2nd International Conference on Signal Processing and Integrated Networks (SPIN)*.
17. Shavers, Clyde, Robert Li, and Gary Lebby. 2006. An SVM Based Approach to Face Detection. In *Proceedings of the 38th Southeastern Symposium on System Theory Tennessee Technological University Cookeville, USA*, March 2006.
18. Shih, Peichung, and Chengjun Liu. 2005. Face Detection Using Distribution-based Distance and Support Vector Machine. In *Proceedings of the Sixth International Conference on Computational Intelligence and Multimedia Applications (ICCIMA'05)*.
19. Yadav, Shalini, and Neeta Nain. 2015. Fast face detection based on skin segmentation and facial features. In *11th International Conference on Signal-Image Technology & Internet-Based Systems*.
20. Zhang, Qieshi, Sei-ichiro Kamata, and Jun Zhang. 2009. Face Detection and Tracking in Color images Using Color Centroids Segmentation. In *Proceedings of the 2008 IEEE International Conference on Robotics and Biometrics*, Bangkok, Thailand, February 2009.
21. Zhang, Qieshi, Sei-ichiro Kamata, and Jun Zhang. 2009. Face Detection and Tracking in Color images Using Color Centroids Segmentation. In *Proceedings of the 2008 IEEE International Conference on Robotics and Biomimetics, Bangkok, Thailand*, February 2009.
22. Yi, Daokui Qu, and Fang Xu. 2010. Face Detection Method Based on Skin Color Segmentation and Facial Component Localization. In *2nd International Asia Conference on Informatics in Control, Automation and Robotics*.
23. de Dios, Juan José, and Narciso Garcia. 2004. Fast Face Segmentation in Component Color Space. In *International Conference on Image Processing (ICIP)*.
24. Meir, R., and G. Rätsch. 2003. *An Introduction to Boosting and Leveraging*. Berlin: Springer.
25. Jiang, Ning, Jiu Xu, Wenxin Yu and Satoshi Goto. 2013. Gradient Local Binary Patterns for Human Detection. IEEE.
26. Yan, Shengye, Shiguang Shan, Xilin Chen, and Wen Ga. 2008. *Locally Assembled Binary (LAB) Feature with Feature-Centric Cascade or Fast and Accurate Face Detection*. IEEE.

27. Xu, R-Qiong, Bi-Cheng Li, and Bo Wang. 2003. Face Detection and Recognition using Neural Network and Hidden Markov Models. In *IEEE International Conference on Neural Networks & Signal Processing Nanjing, China,* December 2003.
28. Tsai, C.C., W.C. Cheng, J.S. Taur, and C.W. Tao. 2006. Face Detection Using Eigenface and Neural Network. In *IEEE International Conference on Systems, Man, and Cybernetics*.

Recommendation Engine to Improvise Curriculum of OBE by Using Comparative Analysis of Institutional Curriculum with Industrial Requirements

S. Likitha Korad and A. Parkavi

Abstract Nowadays, industries require people with better job skills compared to earlier. Institutions are moving towards OBE in order to improve the knowledge skills of the students and to meet the industrial expectations. In this paper, authors have devised a recommendation system to perform the comparative analysis of the institutional curriculum data with industrial requirement data. Based on the comparative analysis, recommendations to improve the curriculum can be made to the institutions, so that the industry requirements are met.

Keywords OBE · Recommendations · Token matching

1 Introduction

The major challenge in education is to elevate skill set required in an educational course. This skill set will change with trends and technology. The recent trends in industry have impacted the education system. This is because the industry always seeks out for relevant skill set in the students. The educational course is directly dependent on the curriculum provided by the institutions.

The complexity of curriculum decision-making includes planning, applying and evaluating the curricular courses in the education. The courses that match the trending skill sets need to be identified by the institutions and be included in the curriculum. By doing so, students can enhance their skill sets and the job placements in the institute will increase as well. After graduation, gaining a job is the most important thing. This, however, depends on the student's skills gained during their graduation. The job profile gained by them depends less on their graduation grade but more on the skills gained. There is a clear difference between the needs of

S. Likitha Korad (✉) · A. Parkavi
M.S. Ramaiah Institute of Technology, Bangalore, India
e-mail: liki.s.korad@gmail.com

A. Parkavi
e-mail: parkavi.a@msrit.edu

M.S. Reddy et al. (eds.), *International Proceedings on Advances in Soft Computing, Intelligent Systems and Applications*, Advances in Intelligent Systems and Computing 628, https://doi.org/10.1007/978-981-10-5272-9_21

the industry and the education garnered by the student. This paper primarily stresses on the skills required by the industry and considers the data collected from over 500 companies all over the world, in the order of their turnover.

The application is designed for the institutions, who can check the skills present in the curriculum of the college and also the industries that are majorly aligned with these companies. The curriculum can be refined according to the industry skills specified by the companies in their websites.

These companies prefer candidates who possess the skills that are required by the company. The students who have acquired these skills in their courses will be placed in the industry. For institutions, this provides scalable data which can be used to tally the percentage of industry requirement that will match. This matched percentage will gloss over the recommendations for the change in the curriculum.

2 Related Works on Educational Data Analysis

2.1 Social Network Model for Educational Data Analysis

Social networking along with the Web 2.0 technologies has brought significant changes to the manner in which people interact with each other, by sharing personal information like pictures, videos and opinions. However, this constitutes informal learning where people learn various aspects of life. Extending this power to the formal learning enhances the present educational methodologies. Social networking model in education can be used to describe the roles of the students, and the analysis of the students' work can be based on this model. Social network model can be used powerfully to perform the educational data analysis. The activity consisted of seven people, and the Flickr photographs of the participants were posted and the corresponding comments were also given as per the activity. At the end of the activity in the five phases, the most used phases were sharing/comparing and negotiation/co-construction. This concluded that the network-based technologies can be used for educational purposes and can be collaborated to exchange the information [1].

Due to the advent of Web 2.0 technology, the development of digital games is changing and heading to the functions of multi-player social interaction, such as MMORPG games. Few relevant researches have investigated the cognitive issues of interaction in this kind of games. Currently, many researches are investigating the learning performances of educational games, motivation of learners and the design issues of these interactive games [2]. The calculation of these included the lag sequential calculation and social network analysis to detect the players. The calculations are in four procedural types like frequency transition matrix, conditional probability matrix, expected-value matrix and adjusted residuals table, for the behavioural patterns of the players. This frequency transition matrix will infer the behavioural patterns of the players or an individual for certain period of time. Using these calculations of many matrices, we can also infer the player's interaction with the other players. This will give the social network analysis of the individual with the society. Using this analysis, the framework that is proposed will help the students to do the work using the social network and also the teachers can assign the problem tasks that are related to the educational scenarios [3].

Evolution of the Web to Web 2.0, where now people are producers and consumers of information, has allowed the Web to become a huge database which is in the state of constant expansion. In this social world, vast amount of information is shared using the social media networks such as Facebook, Twitter. This information may be of multiple kinds. Other contributions of social media network are the explicit relationship that exists between participants and implicit relationship through different characteristics of each media network. These relationships can be used to find relevant information to extend the analysis of networks beyond simple statistics [4, 5]. Many social media networks provide the APIs which help in sharing the information, query the information to extract the essential, out of interest. The combination of the technologies between the SNA and the Web technologies was used by the OER's discovery process. There are three types of graphs. A Hashtag graph gives the updated overview of the project as the end result. Two networks were taken in the network of users: The first condition is the directed network where the network involves the retweet and the second is similar, but the name of the user is the name of the destination node recovery of OER [6].

2.2 *Microblog for Educational Data Analysis*

Microblog is a miniature blog which is originated from Twitter. It is a new type of social network platform which uses the broadcast system of sharing real-time message by concerned mechanism. Microblog has 4 aspects, the study of both network structure of microblog and its features, the study of microblog users, hot spot mining of microblog and the trending analysis, and also its application study in the fields such as business and scientific research. Social network analysis is a research technique which involves the social relations, and current hot research technique, which is concerned about network structure of virtual learning community. Microblog is kind of a potential informal social organization. The knowledge is communicated and dynamically shared in the microblog circle enormously. This has formed microblog network concerned to knowledge which is centred on microblog user. Microblog knowledge network about educational technique is dependent on opinion leaders, and these leaders are spread amongst famous people and experts in the industry. Interaction of this microblog knowledge network [7] is less optimistic than what is expected through social network analysis. There is frequent communication between famous people, leaders and the expertise in the circle of microblog about 'education technique' and also between expertises in the industries through research findings. Communication status in the circle of microblog about 'educational technique' would improve to some extent and will help the people struck in that issue, and also if it is done with strengthening subjective wishes of communication and activity guidance, it will emerge globally [8].

2.3 Recommendation System Using Social Networks

The syllabus of the curriculum can be modified by giving it to the website that includes both teaching professors of an organization and industry people who will tell the present required tools or knowledge-based subjects. They have teachers group, and the semi-structured information is converted to structure by giving the outcomes of the course and plotting it on the graph. Using LinkedIn Alumni's skills set, profiles can be collected and put into the recommendation system [9]. Educational videos can be identified using the Facebook and meet expertise. An Android app is developed to get such types of educational videos. The Android app is extracting the information from the Facebook [10].

The syllabus of the curriculum can be modified by giving it to the website that includes both teaching professors of an organization and industry people who will tell the present required tools, or knowledge-based subjects. They have teachers group, and the semi-structured information is converted to structure by giving the outcomes of the course and plotting it on the graph. Using LinkedIn Alumni's skills set, profiles can be collected and put into the recommendation system. The Alumni's profiles in the LinkedIn, where the skills are specified in profiles. These skills are extracted and are compared with the subject skills that are present in the institutional curriculum. If any of the skills that are not present in the syllabus of the institution are recommended for the institute by the recommendation systems. [11].

3 Recommendation Engine Designed to Improve Curriculum Contents Using Comparative Analysis

In this paper, the authors have attempted giving recommendations for modifying the curriculum. To facilitate this, the authors have designed a system using Java that scans the contents of courses from educational institutions' Web pages and compares it with the skill requirement Web pages of the industry. Then, recommendations will be made by the system to convey the modifications required in the curriculum [12–14] (Fig. 1).

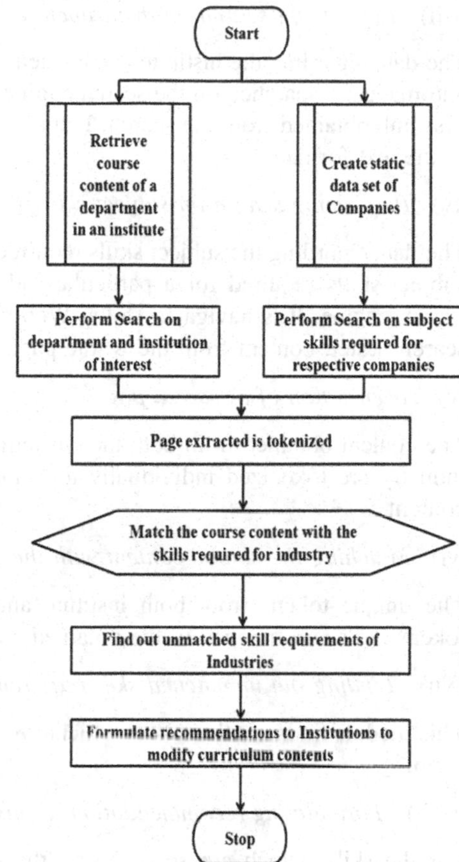

Fig. 1 Recommendation system to improve course content

3.1 Methodology

(i) *Retrieval of course content of a department in an institute*

At the start, it is required to extract the data from a particular institute along with the branch details. For that, the authors have decided to aggregate the data from the institute of interest to know about the skills required for the students of the current generation.

(ii) *Creation of static data set of companies*

In this paper, the authors have considered about 500 companies, using one website. That website has listed the companies from all around the world, in ascending order of its ranking.

(iii) *Performing search on department and institution of interest*

The data regarding the institute is collected. The syllabus for a particular institute is automatically searched on the search engine. It is navigated to the Web page of the first link obtained from the search. Later, content from the whole page is extracted to the text format.

(iv) *Performing search on subject skills required for respective companies*

The data regarding the subject skills required for the institute is to be collected. The subject skills required for a particular industry is automatically searched on the search engine. It is navigated to the Web page of the first link obtained from the search. Later, content from the whole page is extracted to the text format.

(v) *Tokenization of extracted page*

The content obtained from both the curriculum of the institutes and the skills from industry are tokenized individually to obtain the unique words amongst the text contents.

(vi) *Matching the course content with the skills required for industry*

The unique tokens from both institute and industry are merged. These merged tokens are again mapped to the obtained word frequency.

(vii) *Finding out unmatched skill requirements of industries*

This module finds out the tokens which are unmatched (i.e. unique) and then groups them.

(viii) *Formulating recommendations to institutions to modify curriculum contents*

For the skills which are specified by the company as requirements and are not matched with institutions' course contents, the authors have designed their system to provide recommendations. The unmatched skill set of the industry with institutions will be given as recommendations to institutions, so the course contents can be changed to improve the OBE of institutions.

4 Implementation Results

The authors have developed recommendation engine using Java programming. First, they have developed a code which accepts the input from the user in the form of the institution and department name for which the comparative analysis is done, for industry requirement.

The input from the user is taken as shown in Fig. 2, and the syllabus for the department is retrieved using search engine through the system developed by the authors.

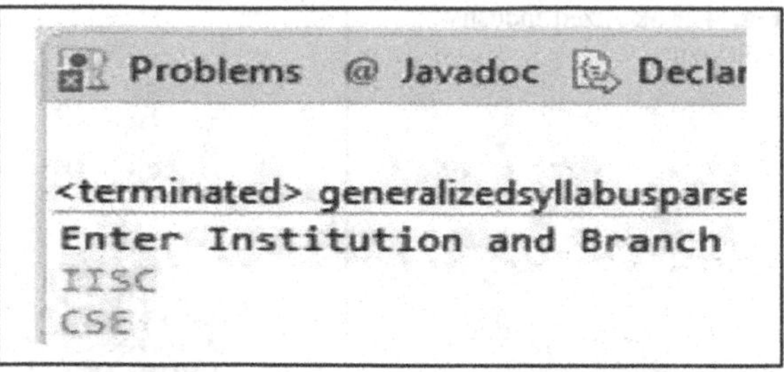

Fig. 2 Input from the user to search for institution curriculum contents

Fig. 3 Tokenized form of the course contents of specific department

From the institute's Web page which has the departmental courses, contents are tokenized and the repeated tokens are eliminated as shown in Fig. 3. Then, the names of the industries are to be entered by the user, with which the institution data has to be compared.

The system developed by the authors will retrieve the job requirements mentioned in the Web page of the companies. The system will tokenize the job requirement skills specified in the Web page as shown in Fig. 4.

Then, the system will compare the tokenized contents of institute data with industry data. Tokens of the institute that match with the industry are displayed as shown in Fig. 5.

The unmatched tokens of the industry with institutions are shown in Fig. 6. Based on this, the recommendations to improve course content can be made to the institutions.

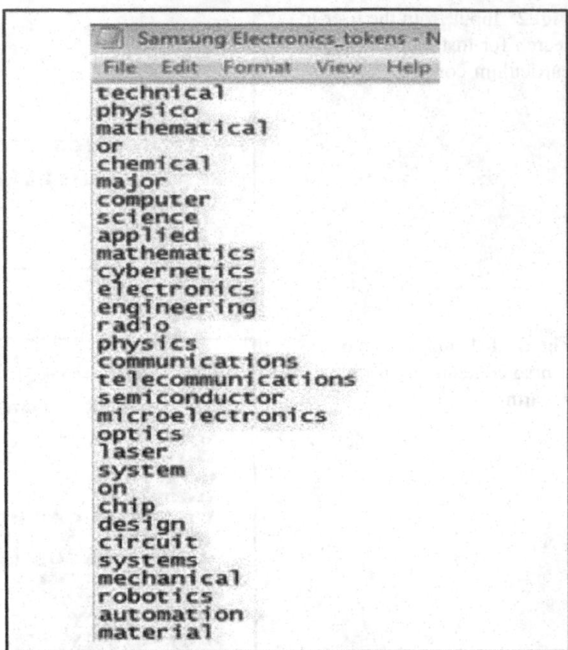

Fig. 4 Tokenized industry requirement

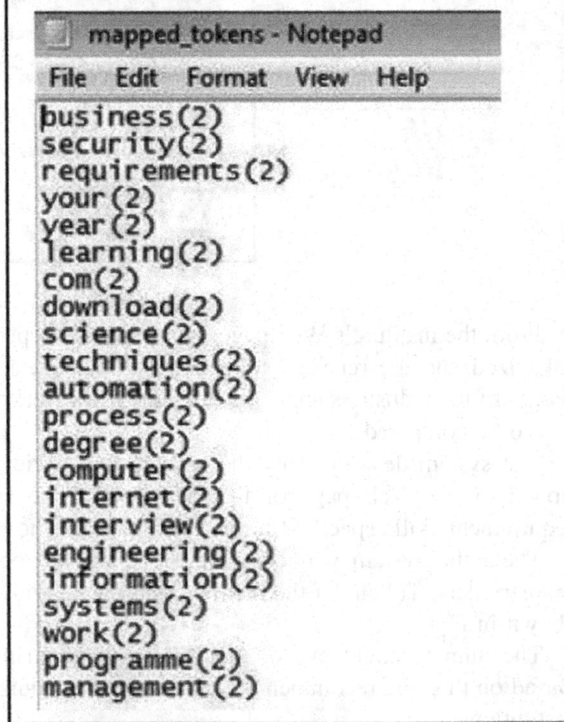

Fig. 5 Mapped tokens of institute with the industry

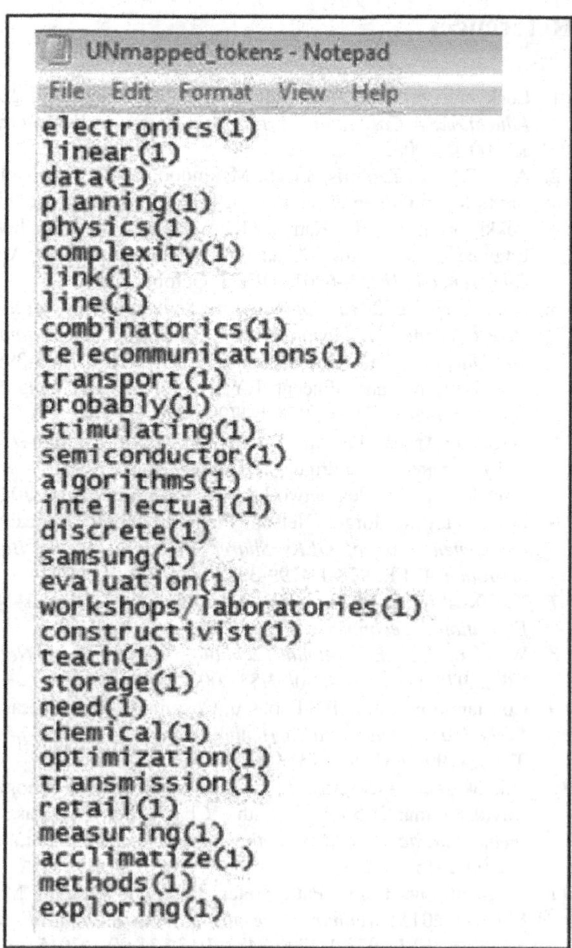

Fig. 6 Unmatched tokens of institute and industry

5 Conclusion

In this paper, the authors have designed a recommendation engine through which the outcome of education can be improved by comparing curriculum content with industry requirements. After comparing the curriculum content with industry requirements, the ones which are not matched are suggested to be included in the curriculum course content of the institution.

References

1. Lockyer, Lori, and John Patterson. 2008. *Integrating Social Networking Technologies in Education: A Case Study of a Formal Learning Environment*. IEEE. 978–0-7695-3167-0/08 $25.00 © 2008.
2. Ang, C.S., P. Zaphiris, and S. Mahmood. 2007. A Model of Cognitive Loads in Massively Multiplayer Online Role Playing Games. *Interacting with Computers* 19 (2): 167–179. Pokkiyarath, M., R. Raman, K. Achuthan, and B. Jayaraman. 2014. Preparing Global Engineers: USA-India Academia & Industry Led Approach. *Frontiers in Education Conference (FIE)*, 1–62014. IEEE, October 2014.
3. Hou, Huei-Tse. 2010. *Applying Lag Sequential Calculation and Social Network Analysis to Detect Learners' Behavioral Patterns and Generate Automatic Learning Feedback-Scenarios for Educational MMORPG Games*. IEEE. 978-0-7695-3993-5/10 $26.00 © 2010.
4. Sun, Beiming, and Vincent T.Y. Ng. 2014. *Analyzing Sentimental Influence of Posts on Social Networks*. IEEE. 978-1-4799-3776-9/2014.
5. Ferraz de Arruda, G., A.L. Barbieri, P. Martín Rodriguez, Y. Moreno, L. da Fontoura Costa, and F. Aparecido Rodrigues. 2014. The role of centrality for the identification of influential spreaders in complex networks. ArXiv e-prints, April 2014.
6. Lopez-Vargas, Jorge, Nelson Piedra, Janneth Chicaiza, and Edmundo Tovar. 2014. *Recommendation of OERs Shared in Social Media Based-on Social Networks Analysis approach*. IEEE. 978-1-4799-3922-0/14/$31.00 ©2014.
7. Yu, Xiao Hua. 2010. Using Microblogging's Social Networking in Education. *Modem Educational Technology*. 12: 97–101.
8. Wei, Xu. 2013. *Educational Technology Circle: Social Network Analysis on Sina Micro Blog*. IEEE. 978-1-4673-4463-0/13/$31.00 ©2013.
9. Constantinov, Calin, P.Ş Popescu, Cosmin Marian Poteras, and Mihai Lucian Mocanu. 2015. *Preliminary results of a curriculum adjuster based on professional network analysis*. IEEE. 978-1-4799-8481-7/15/$31.00 ©2015
10. Palaiokrassas, Georgios, Athanasios Voulodimos, Kleopatra Konstanteli, Nicholas Vretos, David Salama Osborne, Efstathia Chatzi, Petros Daras, and Theodora Varvarigou. *Social Media Interaction and Analytics for Enhanced Educational Experiences*. IEEE. 10897801/ $26.00 2015.
11. Constantinov, Calin, Paul Ştefan Popescu, Cosmin Marian Poteras, and Mihai Lucian Mocanu. 2015. *Preliminary Results of a Curriculum Adjuster Based on Professional Network Analysis*. IEEE. 978-1-4799-8481-7/15/$31.00 ©2015.
12. Johnson, L., S. Adams Becker, M. Cummins, V. Estrada, A. Freeman, and H. Ludgate. 2013. *NMC Horizon Report: 2013 Higher Education Edition*. Austin, Texas: The New Media Consortium. NMC Horizon Report: 2013 Higher Education Edition, 2013.
13. http://nlp.stanford.edu:8080/parser/index.jsp.
14. Pokkiyarath, M., R. Raman, K. Achuthan, and B. Jayaraman. 2014. Preparing global engineers: USA-India academia & industry led approach. In *Frontiers in Education Conference (FIE)*, 1–6. IEEE, October 2014.
15. http://www.ling.helsinki.fi/kit/2009k/clt234/docs/NLTK-0.9.8-ch06.html.

A Survey of Optimization of Rate-Distortion Techniques for H.264 Scalable Video Coding

G.S. Sandeep, B.S. Sunil Kumar, G.S. Mamatha and D.J. Deepak

Abstract Rate-Distortion optimization is a process of improving a video quality during video compression. This will mainly concentrate on amount called rate and is a measure of distortion against data required to encode the video. The main usage of video encoders is to improve quality in encoding image, video, audio, and others to decrease the file size and to improve the quality. The classical method of making encoding decisions for the video encoder is to choose the result that yields highest quality output. Scalable video coding (SVC) is a process in which encoding and decoding can be done on a bit stream. The resulting video has lower temporal or spatial resolution or a reduced fidelity, while retaining reconstruction quality that is close to the original. SVC reduces bit rate necessarily required to represent a given level of perceptual quality and also improves coding efficiency. The main advantage of this Rate-Distortion is applied to measure the quality obtained from the video signals. Some of the rate-distortion models are prediction, approximation, Lagrangian multiplier, empirical, and parametric. Among these Lagrangian multiplier provides good rate-distortion optimization. The MSE- or PSNR-based quality metrics are usually used to assess the visual quality in the cases that the spatial resolution and frame rate are fixed. Our main aim is to demonstrate gain in PSNR with optimized value of bit rates. Finally, we conclude by presenting the efficient techniques to achieve improved quality of video.

Keywords Bit rate · CGS · FGS · MGS · MSE · Peak Signal-to-Noise Ratio Rate Distortion · Scalability · Scalable Video Coding

---

G.S. Sandeep (✉) · B.S. Sunil Kumar · D.J. Deepak
G M Institute of Technology, Davangere, India
e-mail: sandeepgs@gmit.ac.in

B.S. Sunil Kumar
e-mail: sunilkumarbs@gmit.ac.in

D.J. Deepak
e-mail: deepakdj@gmit.ac.in

G.S. Mamatha
R.V. College of Engineering, Bengaluru, India
e-mail: mamatha.niranjan@gmail.com

M.S. Reddy et al. (eds.), *International Proceedings on Advances in Soft Computing, Intelligent Systems and Applications*, Advances in Intelligent Systems and Computing 628, https://doi.org/10.1007/978-981-10-5272-9_22

1 Introduction

Rate-Distortion is a quality metric for measuring both deviation of source and cost of each bit stream in a possible set of decision outcomes. These bits are measured mathematically by multiplying the cost of each bit with Lagrangian value and are a set of values representing relationship between bit cost and quality level. To minimize true Rate-Distortion cost in hybrid coding, a proper designing of framework is required that jointly designs motion compensation, quantization, and entropy in H.264. Some of the rate-distortion algorithms are available and are named as: firstly, graph-based algorithm for soft decision quantization requires motion compensation and step sizes. Secondly, iterative algorithm uses residual coding in H.264 with only motion compensation. Some of the innovative features of H.264/SVC combine temporal, spatial, and quality scalabilities into a single multilayer stream. The provision of scalability in terms of picture size and reconstruction quality in these standards comes with a considerable growth in decoder complexity and a significant reduction in coding efficiency. The spatial and temporal scalabilities have bit stream to represent content of the source with reduced picture size and frame rate, respectively. The usage of SVC in spatial scalability with arbitrary resolution ratios gives a value and is a ratio of picture size for complete bit stream to the included sub-stream.

In the year 2007, the scalable extension of the H.264 codec was invented named as SVC. In the year 2010, a joint collaboration by ITU-T VCEG and ISO MPEG derives a new model for SVC. To increase flexibility of video encoder at application layer for packet level, we need to use bit rate adaptation. Compression efficiency has been improved by applying inter-layer estimation for the video frames.

In recent years, the video traffic in Internet forms a significant increase of information. In 2011, this share of internet video was 51% of all consumer Internet traffic and is expected to rise to 55% in 2016. These high numbers do not include video exchanged through peer-to-peer sharing. Finally, the sharing of video on the overall Internet traffic is expected to reach 86% in 2016. The important aspect related to video quality is a measure of bit rate of the video carried over the network. When this measure is too low, means the visual quality of the decoded video may be degraded. On the other hand, when this is too high, it may lead to video freezes or longer waiting times or loss of data.

The aim of this paper is to survey the different papers to identify the different scalability measure and that uses prediction parameters and list out the peak signal-to-noise ratio values that are related with the rate-distortion parameters. In this work, we are focus on the measures of Rate-Distortion and statistical values are obtained from different authors and finally concluded that by using Lagrangian rate-distortion optimization, we get good Rate-Distortion results and is measured with PSNR values. From the analysis, we can identify that Rate-Distortion mainly depends on CGS and MGS scalability parameters, and it will produce an acceptable value of PSNR with lesser bit rates.

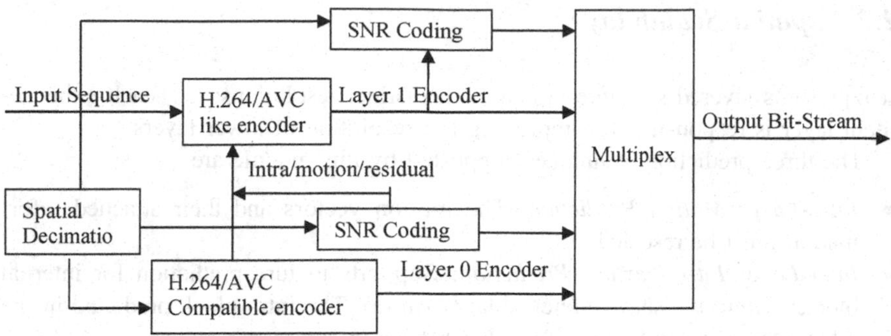

Fig. 1 Simplified SVC encoding structure with two spatial layers

2 Overview of SVC

Figure 1 shows the architecture of the SVC encoder. The architecture proposes a more flexible design to improve the functional efficiency of encoder. Basically, the prediction of blocks using motion-compensated and intra prediction within the same layer are introduced. However, dependency between layers is carried out by inter prediction section, which again utilize the motion vector, similar structures in a frame, so as to improve the ratio of compression in the encoding process.

The base layer is designed with coarse grain scalability to code it as first. The enhancement layer uses fine grain scalability to improve the resolution which may increase the complexity of hardware design.

There are three types of scalability in SVC: (a) Temporal scalability, (b) Spatial scalability, and (c) SNR scalability.

2.1 Temporal Scalability

By using frame rates and bit-stream subsets, we can be able to represent video. Encoded video streams have three distinct types of frames: I (intra), P (predictive) or B (bi-predictive). I frames only explore the spatial coding within the picture, whereas both P and B frames do have interrelation with different pictures. While in P frames inter-picture predictive coding is performed based on one preceding reference and B frames consist of a combination of inter-picture and bi-predictive coding.

2.2 Spatial Scalability

It represents layered structure videos with distinct resolutions; i.e., each enhancement layer is responsible for improving the resolution of lower layers.

The three prediction techniques supported by this module are

- *Inter-Layer Motion Prediction*: The motion vectors and their attached information must be rescaled.
- *Inter-Layer Intra Texture Prediction*: Supports texture prediction for internal blocks within the same reference layer (intra). The intra block predicted in the reference layer can be used for other blocks in superior layers.
- *Inter-Layer Residual Prediction*: In SVC, the inter-layer residual prediction method can be used after the motion compensation process to explore redundancies in the spatial residual domain.

2.3 SNR Scalability

We legitimize transporting the complementary data in different layers in order to produce videos with distinct quality levels. In SVC, SNR scalability is implemented in the frequency domain.

The H.264/SVC standard supports three distinct SNR scalability modes;

(a) *Coarse Grain Scalability*: Each layer has an independent prediction procedure, and consecutive layers have the same resolution.
(b) *Medium Grain Scalability*: It increases efficiency by using a more flexible prediction module, whereas both base and enhancement layers can be referenced.
(c) *Fine Grain Scalability*: It employs an advanced bit-plane technique where different layers are responsible for transporting distinct subset of bits corresponding to data information.

3 Rate Distortion Done by Different Authors

By calculating peak signal-to-noise ratio (PSNR) values of the frames of a video sequence as objective video quality measure. Here we consider a frame with pixels (8bit per pixel), and now Calculating PSNR from mean squared error (MSE).

$$\mathrm{MSE} = \frac{1}{Nx \cdot Ny} \sum_{x=0}^{Nx-1} \sum_{y=0}^{Ny-1} [F(x,y) - R(x,y)]^2 \quad (1)$$

$$\text{PSNR} = 10 \cdot \log_{10} \frac{255^2}{\text{MSE}} \tag{2}$$

In Sun et al. [1], authors have analyzed sub-bit-plane technique of scalable video coding for fine granular SNR and also checks that MSE-based distortion rate (D-R) function should be linear within a FGS layer and finally achieve good coding efficiency with reduced complexity by using model-based quality layer assignment algorithm.

In Mansour et al. [2], authors have analyzed rate-distortion prediction model for medium grain scalable, and it allows for video encoder to predict size and distortion of video frame. Here, cross-layer optimization capabilities are used to achieve best picture quality.

In Thomas et al. [3], authors describe secure scaling for bit streams at intermediate nodes, and it mainly supports medium grain scalability and coarse grain Scalability and finally incorporates FGS packets and produces more robust pictures.

In Sun et al. [4], authors analyzed the rate-distortion function of SVC FG coding and provide a solution using approximation model of the rate-distortion function. This solution consists of integer transform coefficients and extracts properties of Rate-Distortion using Gaussian model and apply approximation to obtain good quality picture.

In Li et al. [5], authors describe Lagrangian multiplier for rate-distortion optimization. It directly supports multilayer scenario, but is not efficient because correlation between the layers is not considered. To overcome this, finally a new selection algorithm for rate-distortion optimization was proposed.

In [6], authors have made an assumption for fixed display scenario at constant rate of the frame, and it consists of one CGS and another MGS for creating quality layers for bit stream. Author proposed hierarchical coding which yields dyadic decomposition of temporal layers.

In [7], authors proposed a scheme named rate-distortion optimization with ROI slices to improve coding efficiency. It discards background slices, and it uses base layer information to generate prediction signal of EL. And lastly, author derives Lagrange multiplier for performance improvement.

In [8], authors have proposed new rate control scheme for principle component analysis of scalable video coding. An improvement has made and create a new model named TMN8 for rate control on all the frames like P, B, I. This scheme is more accurate in case of bit rate and gains much peak signal-to-noise ratio.

In [9], authors also proposed updated rate control scheme for spatial scalability for scalable video coding. The adaptive approach is used to decide a quantization parameter and by applying transform coefficients on spatial scalability. A new model called Cauchy Density is used to derive natural logarithm domain Rate-Distortion model this matches spatial scalability characteristics this in turn will provide an excellent performance over the earlier rate control schemes.

In [10], authors have analyzed empirical model for Rate-Distortion, and it checks for channel rates and capacity. By utilizing optical rate prediction architecture with cross-layer design side information, a new empirical model has been created to get good video distortion method. Finally, author uses low-density parity-check approach to estimate the performance of video Rate-Distortion.

In [11], authors have analyzed that medium grain scalability is more superior than coarse grain scalability in case of rate-distortion complexity, and it also improves error resilience. By applying inter-layer prediction; it improves rate-distortion complexity of the scalable video coding. On usage of low-delay group of picture-structure, mainly in handheld devices, reduces decoding complexity also the structural delay.

In [12], authors have described adaptive pre-filtering system for SVC computational complexity-rate-distortion analysis; as a result the visual presentation quality at decoder side is improved with using limited resources. SVC preprocessing will provide efficient coding for desired region of interest.

In [13], authors have analyzed large-scale Rate-Distortion and rate variability distortion characteristics of CGS and MGS approach and found that CGS achieves low bit rate compared to single-layer encoding. From the analysis, the traffic variability of CGS is lower compared to single-layer streams. Similarly, MGS can achieve slightly higher Rate-Distortion efficiency than single-layer encoding. By using hierarchical B frame structure of MGS layer, a high Rate-Distortion performance is achieved.

In [14], authors have developed a model named as parametric Rate-Distortion for medium grain scalability and that uses spatial and temporal complexity of video sequences with variable bandwidth that will improves the quality of video sequences.

In [15], authors have analyzed optimal bit-stream extraction scheme for SVC, and it produces scalable layer representation for multicasting network with varying bandwidth. This scheme also extracts the optimal path for multicast nodes and also helps in reducing optimal path extraction for less capable devices through path truncation. Finally, it gives good PSNR and MSE.

In [16], authors have analyzed rate-distortion function of SVC FGS pictures by using Gaussian model, and it proves distortion-rate curve will be concave. By using sub-bit-plane technology, distortion rate is inferred to be linear under MSE. To find drift error, author uses effective distortion model in SVC. A new virtual group of picture concept and new priority setting algorithm is designed to achieve good optimal Rate-Distortion performance.

In [17], authors have proposed rate-distortion model in scalable wavelet video coding to find efficiency. Generally, Lagrangian multiplier provides optimized solution for mode decision and rate-constrained motion estimation. This operates on multiple bit rates and open loop structure. Finally, proposing motion prediction gain metric to measure efficiency and always improved.

In [18], authors have analyzed rate-distortion optimization scheme to utilize transform coefficients in spatial SVC for solving l1-regularized least squares problems and produce larger PSNR in reducing bit rate. Finally, this will improve the coding efficiency.

4 Results and Analysis

An exhaustive survey is made over 12 papers published in international conferences or in journals of Scopus index. The very purpose of this survey is to identify the best technique to produce highest bit rate of transmission; at the same time it should ensure an appreciable quality (PSNR) of video. Table 1 enlists the different techniques and their performance parameters. The observation indicates all the techniques proposed over the timeline are not efficient. This can be observed in Fig. 2. However, few papers showed an aggressive performance in bit rate results retaining a video quality near to the original. Such techniques are identified and are listed in Table 2. The graph of the results in these techniques clearly shows an improvisation

Table 1 Comparison of PSNR versus bit rate for different rate-distortion techniques

Ref. paper no.	Rate-distortion techniques	PSNR (dB)	Bit rate (kbps)
10	Empirical model	29.87	379
11	Inter-layer prediction	30.45	320
15	Bit-stream extraction scheme	31.05	135
14	Parametric Rate-Distortion	31.60	246
18	Inter-layer residual prediction	31.80	190
09	Rate control scheme	32.38	132
13	Large-scale Rate-Distortion and rate variability distortion	35.00	200
05	Lagrangian multiplier	35.10	300
01	Sub-bit-plane technique	35.35	192
17	Motion prediction gain metric	36.50	253
16	Bit-stream extraction scheme	37.53	384
12	Adaptive pre-filtering system	40.40	183

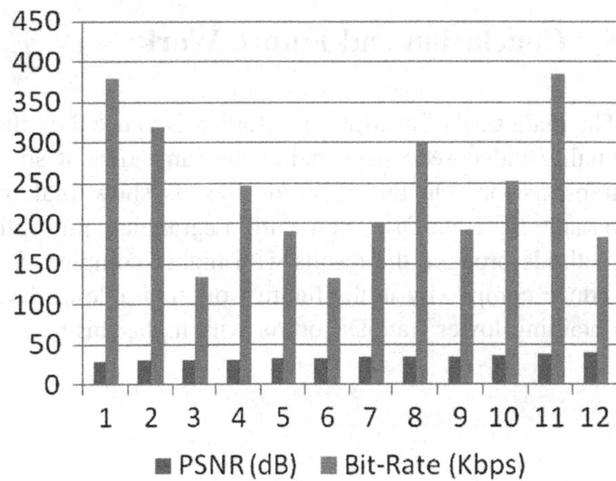

Fig. 2 Comparison of PSNR versus bit rate for different rate-distortion techniques (color figure online)

Table 2 Values of optimized PSNR versus bit rate in different techniques

PSNR (dB)	Bit rate (kbps)
29.87	379
30.45	320
31.60	246
35.00	200
35.35	192
40.40	183

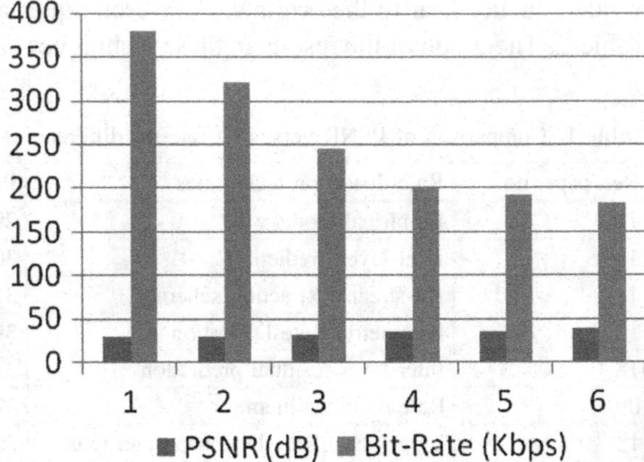

Fig. 3 Values of optimized PSNR versus bit rate in different techniques (color figure online)

in the bit rate with slight reduction in the PSNR value. In Fig. 3 the techniques with highest bit rate have reduced PSNR and vice versa. But the values of extreme, i.e., PSNR and bit rate are not the region of interest to us. The techniques which produce a better bit rate for the moderate PSNR values give us a better Rate-Distortion with highest bit rate.

5 Conclusion and Future Work

The main goal of measuring distortion is to use the different models to achieve good quality video sequences and at the same time it should yield a better bit rate of transmission. On this view, the results show that the inter-layer prediction and parametric Rate-Distortion with Lagrangian multiplier perform the best. These methods produce the results with higher complexity. This can be an objective to reduce complexity in the future work and able to derive a successful technique in obtaining lower Rate-Distortion with higher bit rates.

References

1. Sun, Jun, Wen Gao, and Debin Zhao. 2009. On rate-distortion modeling extraction of H.264/SVC fine-granular SNR scalable video. *IEEE Transactions on Circuits and Systems for Video Technology* 19 (3).
2. Mansour, Hassan, Vikram Krishnamurthy, and Panos Nasiopoulos. 2008. Rate and Distortion Modeling of Medium Grain Scalable Video Coding. *Proc ICIP08.*
3. Thomas, Nithin, David Bull, and David Redmill. 2008. *A Novel H.264 Scalable Video Coding Encryption Scheme for Secure Bit-Rate Transcoding*. UK: University of Bristol.
4. Sun, Jun, Wen Gao, and Debin Zhao. 2007. Statistical Analysis and Modeling of Rate-Distortion Function in Scalable Video Coding Fine-Granular SNR Scalable Videos. In *Proc ICME 2007*, 1135–1138.
5. Li, Xiang, Peter Amon, Andreas Hutter, and André Kaup. Lagrange Multiplier Selection for Rate-Distortion Optimization in Scalable Video Coding. *Picture Coding Symposium, (PCS).*
6. Slanina, Martin, Michal Ries, and Janne Vehkaperä. 2013. Rate Distortion Performance of H.264/SVC in Full HD with Constant Frame Rate and High Granularity. In: *Proc ICDT 2013, the Eighth International Conference on Digital Telecommunications*, 7–13.
7. Wang, Hongtao, Dong Zhang, and Houqiang Li. 2015. *Rate-Distortion Optimized Coding Method for Region of Interest in Scalable Video Coding*. *Advances in Multimedia*, vol. 2015. Hindawi Publishing Corporation, 1–11 p.
8. Xu, Long, Wen Gao, Xiangyang Ji, Debin Zhao, and Siwei Ma. 2007. *Rate Control for Spatial Scalable Coding in Scalable Video Coding*. Chinese Academy of Sciences 2007.
9. Zhu, Tao, and Xiong-wei Zhang. 2011. *Rate Control Scheme for Spatial Scalability of H.264/SVC*. PLA University of Science & Technology.
10. Cho, Yongju, Hayder Radha, Jeongju Yoo and Jinwoo Hong. 2008. A Rate-Distortion Empirical Model for Rate Adaptive Wireless Scalable Video. In *CSIS 2008*.
11. Gan, Tong, Bart Masschelein, Carolina Blanch, Antoine Dejonghe, and Kristof Denolf. 2009. Rate-Distortion Complexity Performance Analysis of the Scalable Video Coding decoder. In *Proc: ICME 2009*, 213–216.
12. Grois, Dan, and Ofer Hadar. 2011. Efficient Adaptive Bit-Rate Control for Scalable Video Coding by Using Computational Complexity-Rate-Distortion Analysis. In *Proc MM* 11–07.
13. Gupta, Mohan, Akshay Pulipaka, Patrick Seeling, Lina J. Karam, and Martin Reisslein. 2012. H.264 Coarse Grain Scalable (CGS) and Medium Grain Scalable (MGS) Encoded Video: A Trace Based Traffic and Quality Evaluation. *IEEE Transactions on Broadcasting* 58 (3): 428–439.
14. Haseeb, Abdul, Maria G. Martini, Sergio Cicalò and Velio Tralli. 2012. Rate and Distortion Modeling for Real-Time MGS Coding and Adaptation. *Wireless Advanced (WiAd)* 85–89.
15. Peng, Wen-Hsiao, John K. Zao, Hsueh-Ting Huang, Tse-Wei Wang, and Lun-Chia Kuo. 2008. Multidimensional Scalable Video Coding Bit-Stream Adaptation and Extraction for Rate-Distortion Optimized Heterogeneous Multicasting and Playback. In *Proc ICIP 2008*, 2476–2479.
16. Sun, Jun, Wen Gao, Debin Zhao, and Weiping Li. 2009. On Rate-Distortion Modeling and Extraction of H.264/SVC Fine-Granular Scalable Video. *IEEE Transactions on Circuits and Systems for Video Technology* 19 (3): 323–336.
17. Tsai, Chia-Yang, and Hsueh-Ming Hang. 2009. *Rate-Distortion Model for Motion Prediction Efficiency in Scalable Wavelet Video Coding*. National Chiao University, Hsinchu.
18. Winken, Martin, Heiko Schwarz, and Thomas Wiegand. 2008. Joint Rate-Distortion Optimization of Transform Coefficients for Spatial Scalable Video Coding Using SVC. In *Proc ICIP 2008*, 1220–1223.

A New Fibonacci Backoff Method for Congestion Control in Wireless Sensor Network

Elamathy Selvaraju and Ravi Sundaram

Abstract Congestion control is one of the predominant challenges in wireless sensor network. It has a great impact on the parameters of quality of service such as end-to-end transmission delay, packet delivery ratio (PDR), and energy consumption in wireless sensor networks. Typical congestion control schemes include binary exponential backoff (BEB) or truncated binary exponential backoff (TBEB). This involves retransmission of frames in carrier sense multiple access with collision avoidance (CSMA/CA) and carrier sense multiple access with collision detection (CSMA/CD) networks. In this work, a new backoff strategy Fibonacci Backoff Algorithm (FBA) for congestion control is proposed. Each node is allocated a wait period as an incremental period. The simulated results show its better performance by decreasing the possibility of two or more nodes choosing the same backoff period, thereby decreasing the probability of collision and energy used for retransmission.

Keywords Congestion control · Fibonacci backoff algorithm · Performance analysis · Optimal scheduling

1 Introduction

In wireless sensor networks, nodes are deployed in large quantities where sensors can organize themselves to an ad hoc multipath network environment for communication between them. Each node is aware of its neighborhood nodes. Network protocol holds a strict layered structure and implements congestion control,

E. Selvaraju (✉) · R. Sundaram
ECE Department, Dr. M.G.R. Educational and Research Institute University,
Chennai, India
e-mail: elamathy.j@gmail.com

R. Sundaram
e-mail: ravi_mls@yahoo.com

M.S. Reddy et al. (eds.), *International Proceedings on Advances in Soft Computing, Intelligent Systems and Applications*, Advances in Intelligent Systems and Computing 628, https://doi.org/10.1007/978-981-10-5272-9_23

scheduling, and routing protocols. However, the time-varying nature with different channel conditions proposes a significant challenge to accomplish the above goals. Traffic engineering, end-to-end rate adaptation, and transport layer signaling have been widely developed to prevent the network congestion.

Shashi Kiran et al. [1] proposed an algorithm to trade-off between the latency and real-time capacity to adapt the transmission schedule in accordance with the addition, removal, and changes in dynamic queries. For fixed transmission schedules, the proposed method has the ability to overcome the changes in workload more effectively than the time division multiple access. A fair end-to-end window-based congestion control protocol for a packet-switched network is demonstrated in Shashi Kiran et al. [2]. with first come-first served (FCFS) routers. Only the end host information is used by the protocol. The network considered is a multiclass fluid model. The convergence function of the protocol is proved using a Lyapunov function.

A joint optimal design of cross-layer congestion control (CCC) for wireless ad hoc network is proposed in Chen et al. [3]. The rate and scheduling constraint is formulated using multicommodity, flow variables fixed channels in networks. The resource allocation problem is addressed through natural decomposition method. Congestion control, routing, and scheduling are the subdivisions. The dual algorithm is used to handle the time-varying channels and adaptive multirate devices. The stability of the resulting system is established. The performance is characterized with respect to an ideal reference system.

The resource like time slots, frequency, power, etc., at a base station to the other network node where each flow is intended for a different receiver. A combination of queue-based scheduling at the base station and the congestion control is implemented at the base station or at the end user node leading to the fair resource allocation and provide queue-length stability [4]. A redesign framework for fluid-flow models for network congestion control is proposed [5]. An extra dynamics are introduced using augmented Lagrangian method to improve the performance of the network.

2 Congestion in Wireless Sensor Networks

Among the various reasons for the congestion in a wireless sensor network, overloading of data, coherent nature of traffic to the base station node in the network due to many-to-one topology, ingestion of sensory data into the wireless networks plays a key role for the predominant congestion. Congestion affects the continuous flow of data, loss of information, delay in the arrival of data to the destination node, unwanted consumption of energy at each node. Basically, there are two types of congestion, node level and link level.

Node level congestion
The node level congestion is found to be common in conventional network. The common reason for its occurrence is the overflow of data at node. Its major effects are packet loss and increased queuing delay.

Link level congestion
It occurs in wireless sensor network, where nodes transmit the data packets at the same time. It significantly affects the link utilization and overall throughput with an increase in delay time and energy utilization.

3 Fibonacci Backoff Algorithm

The nodes in the wireless sensor network are allowed to wait for a period which is formulated from the mathematical Fibonacci series. The wait period formulated is in incremental period. The subsequent numbers are generated directly adding the previous two numbers according to the given formulation.

$$F(x) = f(x-2) + f(x-1), \quad \text{where } x \geq 0.$$

$$F(0) = 0, \; f(1) = 1.$$

There are two methods used to prevent the congestion. The possibility of two or more nodes in the network to transmit the data at the same time increases the possibility of congestion among them.

3.1 Method I

Nodes are allowed to choose a random backoff period as the below mentioned range.

$$0 - 2^n - 1,$$

where n—number of collisions.

For the first collision, each sender node has to wait for a 0 or 1 time slot. The maximum backoff period for a node depends on the range of Fibonacci series which is chosen for the Fibonacci Backoff Algorithm (Table 1).

From this method, the retransmission attempts can be analyzed as follows,

Number of retransmission attempts	α exponential increase in the number of delay possibility

Table 1 Slot time generation

Number of collision	Slot time
1	0 or 1
2	0–3
3	0–7
4	0–256

Number of nodes considered in this work is 3, and the time slots considered is 10. The slotted time communication between the nodes and the station flow node is shown in Table 2.

3.2 *Method II*

The second method is based on the Fibonacci Backoff Algorithm (FBA), and a random Fibonacci number is generated based on the number of times the transmission failure has occurred. This method has two sections as node flow and code flow. There are three basic steps in transmission of data as follows,

1. The serial port is open,
2. Method of algorithm is selected either as 'exponential' or 'fibonacci backoff' algorithm, and
3. Data is sent.

Table 2 Transmission status for nodes 1, 2, 3

Time slot	Node1	Node2	Node3
T1	X	Txd	X
T2	X	Txd	Txd
T3	Txd	X	Txd
T4	Txd	Txd	Txd
T5	Txd	Txd	Txd
T6	Txd	Txd	Txd
T7	Txd	Txd	Txd
T8	Txd	Txd	Txd
T9	Txd	Txd	Txd
T10	Txd	X	x

Algorithm for node code flow
Algorithm for slot time communication between the nodes is as follows,

Step 1: if index < TOTAL SLOTS.
Step 2: wait for slots on serial port.
Step 3: if received slot1 = 'S', go to step1.
Step 4: data I MSG(n) is transmitted to the serial port, and node waits for acknowledgment or NACK.
Step 5: data is sent successfully, once ACK msg is received.
Step 6: if not, data is transmitted either into exponential or Fibonacci code as NACK is received.
Step 7: random number is generated, and control is transmitted to the program.
Step 8: the value returned represents the number of times, and the slot has to be skipped to avoid collision.
Step 9: if index == TOTAL SLOTS, return.
Step 10: else go to step 1.

Station flow algorithm
Algorithm for slot time communication for the station node is as follows,

Step 1: serial port is initialized for communication with node1, node2, and hyper-terminal.
Step 2: slot signal is sent to node1 and node2.
Step 3: wait period is set, until the data is received from node1 and node2.
Step 4: if node1 and node2 send data at the same time, ACK message is sent to both the nodes to signal the collision between them and discard of data.
Step 5: if anyone of the node sends data during the slot time, ACK message is send to the node and received data is displaced.
Step 6: if end of transmission is received from node1 and node2, station node is terminated else control is transferred to step2.

Flow chart for node flow

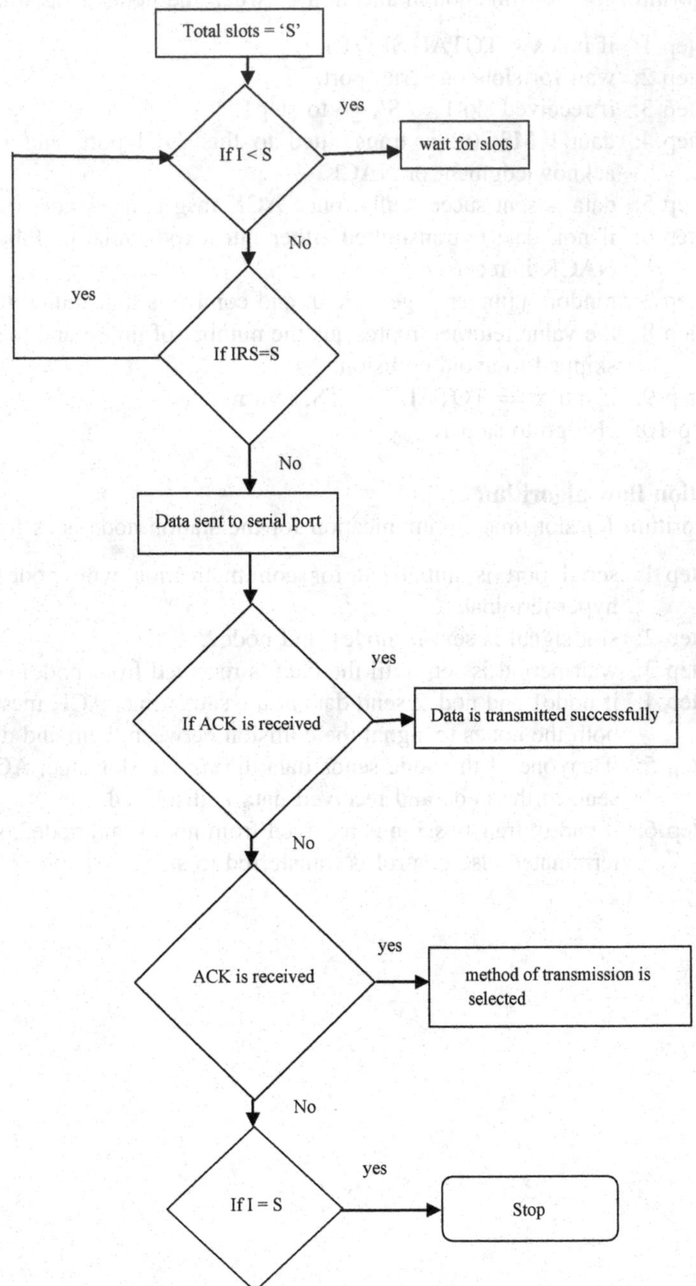

4 New Slot Allocation Table for Retransmission

4.1 *First New Slot Generation for Retransmission for Node1, Node2, Node3*

See Table 3.

4.2 *First New Slot Generation for Retransmission for Node2, Node3*

See Table 4.

4.3 *New Slot Generation for Retransmission for Station Node*

See Table 5.

Table 3 First new slot allocation

Node	Collision occurred time slot	Newly allocated time slot
Node1	Slot 1 Slot 2	Slot 6 Slot 3
Node2	Slot 1 Slot 2 Slot 3 Slot 6	Slot 4 Slot 9 Slot 8 Slot 6
Node3	Slot 3 Slot 10	Slot 5 Slot 8

Table 4 Second new slot allocation

Node	Collision occurred time slot	Newly allocated time slot
Node2	Slot 5 Slot 8	Slot 4 Slot 9
Node3	Slot 5 Slot 8	Slot 5 Slot 8

Table 5 New slot allocation for station node

Time slot	Network status	Slot	Network status
1	Tx	1	Tx
2	Tx	2	Tx
3	X	3	Tx
4	Tx	4	Tx
5	Tx	5	X
6	Tx	6	Tx
7	Tx	7	Tx
8	Tx	8	X
9	Tx	9	Tx
10	Tx	10	Tx

5 Comparative Analysis Between BEB and FBA

In a wireless sensor network, distributed coordination function (DCF) is used for this best performance. When nodes want to access the channels, the backoff algorithm is executed while its backoff time greatly depends on the random number. Hence, it is difficult to predict exactly the backoff duration. In BEB, when data is sent and if acknowledgment (ACK) message is not received represents the occurrence of collision in the transmitted channel. The nodes involved increases its contention window size, and in case of successful transmission, the slot time is reduced by one. In this work, we have proposed Fibonacci Backoff Algorithm (FBA) which makes significant changes in the channel access mechanism. Two different scenarios have been introduced for the existing protocol.

In an actual BEB, when the number of collision is 3, the backoff time slot generated is obtained from 2^n.

$$2^3 = 8(0\text{ to }7)$$
$$\text{Generated time slots} = 0, 1, 2, 3, 4, 5, 6, 7.$$

Whereas in Fibonacci Backoff Algorithm, the waiting time slot is generated using the formula,

$$F(x) = f(x-2) + f(x)$$

For the considered condition of number of collision 3,

$$F(3) = f(3-2) + f(3-1).$$
$$= 1 + 2 = 3.$$

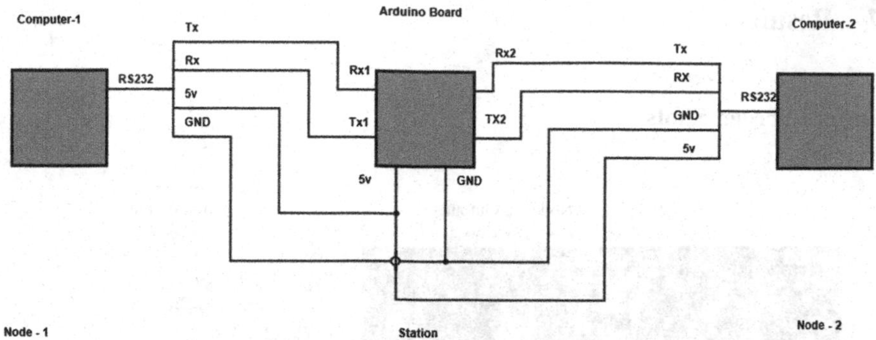

Fig. 1 Connection of nodes through Aurdino board

When $n = 3$, with the mathematical Fibonacci series generation, the waiting time slots generated are,

$$\text{Time slots} = (0, 1, 1, 2)$$

Number of time slots or the backoff/waiting period generated is 4, whereas in binary exponential backoff algorithm, the number of waiting slot generated is 8. Hence, the comparative study of BEB and FBA can be expressed as,

$$\text{Backoff period in BEB} = 2(\text{waiting time slot generated in FBA})$$

6 Hardware Setup

See Fig. 1.

7 Results

Output Screen Shots

Node - 1 output:

```
Totalslots 6
select choice
1.exponential
 2.fibonacci
1
choice= 1
index= 0
waiting for the slots, (s or n)
slots = s
data is empty

index= 1
waiting for the slots, (s or n)
slots = s
data is = hi

waiting for acknowledge,(s or n)
ACK = n
collision occured= 1 time
exponential value 1.0
index= 1
waiting for the slots, (s or n)
slots = s
data is = hi

waiting for acknowledge,(s or n)
ACK = n
collision occured= 2 time
exponential value 3.0
index= 1
waiting for the slots, (s or n)
slots = s
data is = hi

waiting for acknowledge,(s or n)
ACK = n
collision occured= 3 time
exponential value 7.0
skip= 3 slots
waiting for the slots, (s or n)
```

Node – 2 Output:

```
Totalslots 6
select choice
1.exponential
 2.fibonacci
2
choice= 2
index= 0
waiting for the slots, (s or n)
slots = s
data is empty

index= 1
waiting for the slots, (s or n)
slots = s
data is = hi

waiting for acknowledge,(s or n)
ACK = n
collision occured= 1 time
the nth number for that fibonacci sequence = 1
index= 1
waiting for the slots, (s or n)
slots = s
data is = hi

waiting for acknowledge,(s or n)
ACK = n
collision occured= 2 time
the nth number for that fibonacci sequence = 1
index= 1
waiting for the slots, (s or n)
slots = s
data is = hi

waiting for acknowledge,(s or n)
ACK = n
collision occured= 3 time
the nth number for that fibonacci sequence = 2
index= 1
waiting for the slots, (s or n)
```

Station output

```
------------------------------------
1.Press Any Key to Start Station
------------------------------------
Start of Slot
No Data Received from Node 1 or Node 2
Start of Slot
Two nodes sent data in same slot
                         Collision Occured
Start of Slot
Two nodes sent data in same slot
                         Collision Occured
Start of Slot
Two nodes sent data in same slot
                         Collision Occured
Start of Slot
Data Received from Node 2
hi

Start of Slot
No Data Received from Node 1 or Node 2
Start of Slot
Data Received from Node 2
hello how r u

Start of Slot
Data Received from Node 1
hi

Start of Slot
Data Received from Node 2
good morning

Start of Slot
```

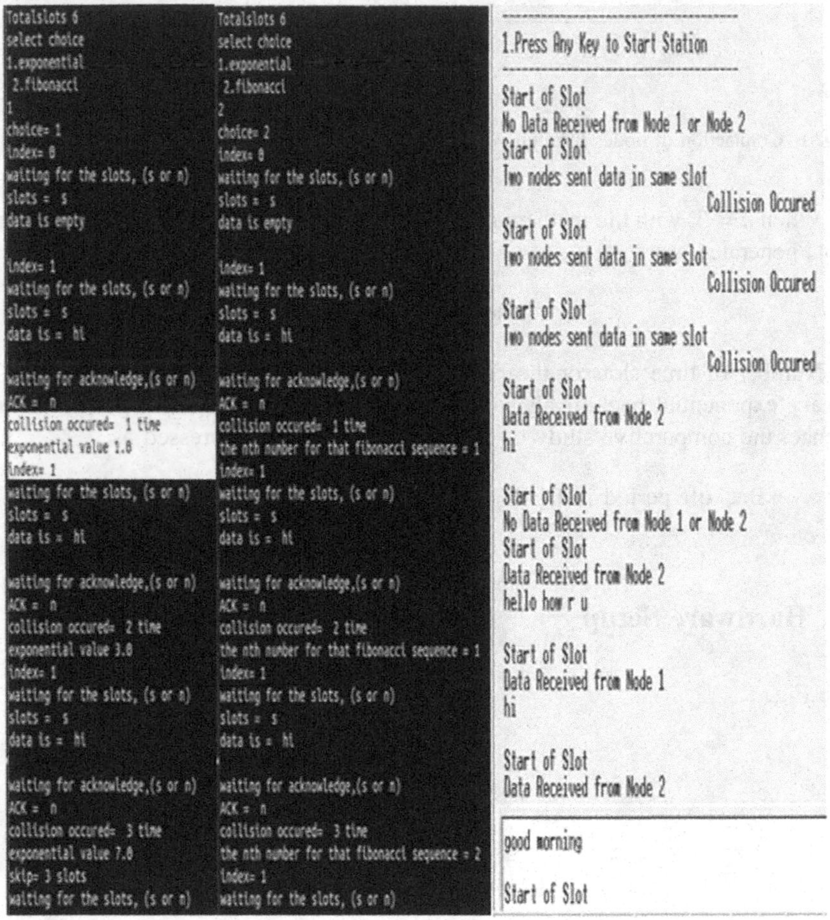

8 Conclusion

In this paper, a new Fibonacci Backoff Algorithm is proposed and implemented. A comparative analysis is done between the already existing binary exponential backoff (BEB) algorithm. The incremental nature of the Fibonacci Backoff Algorithm decreases the probability of occurrence of congestion by allocating the waiting time slots based on the number of transmission failure occurrences. The newly proposed Fibonacci Backoff Algorithm (FBA) and the results obtained prove the effectiveness of by reducing the number of waiting time slots, and the waiting period is reduced to half than the BEB algorithm. Hence, the network throughput increases by congestion prevention and decreased energy utilization.

References

1. Shashi Kiran, B.S., R.L. Pavan Shree, Sanjay K. Nagendra. 2015. Query Scheduler for Secure Collision-Less Transmission in Wireless Sensor Networks. *International Journal of Modern trends in Engineering and Research*. ISSN 2349-9745.
2. Mo, Jeonghoon, and Jean Walrand. 2000. Fair End-to-End Window-Based Congestion Control. *IEEE/ACM Transactions on Networking* 8 (5).
3. Chen, Lijun, Steven H. Low, Mung Chiang, and John C. Doyle. 2006. Cross-layer Congestion Control, Routing and Scheduling Design in Ad Hoc Wireless Networks. In *IEEE Communications Society Subject Matter Experts for Publication in the Proceedings IEEE Infocom*.
4. Zhou, Anfu, Min Liu, Zhongcheng Li, Eryk Dutkiewicz. 2015. Joint Traffic Splitting, Rate Control, Routing and Scheduling Algorithm for Maximizing Network Utility in Wireless Mesh Networks. *IEEE Transactions on Vehicular Technology*. doi:10.1109/TVT.2015.2427091.
5. Zhang, Xuan, and Antonis Papachristodoulou. 2015. Improving the Performance of Network Congestion Control Algorithms. *IEEE Transactions on Automatic Control* 60 (2).

Wireless Biometric Attendance Management System Using Raspberry Pi in IaaS Environment

Chirag Sharma, Karan Shah, Shriya Patel and Swapnil Gharat

Abstract Attendance management is one of the most important processes in an educational institute, since it is a way of evaluating the performance of the students, staff, and departments in it. Current attendance marking methods are monotonous and time-consuming. Manually recorded attendance can be easily manipulated. Maintaining attendance records integrity and security, and reducing the valuable amount of time and hassle spent in the overall process is a real challenge. We propose a system to tackle all these issues. Being one of the most successful applications of biometric verification, face recognition and fingerprint scanning have played an important role in the field of security, authorization, and authentication. Such forms of biometric verification can prove useful in case of student's attendance collection. Infrastructure as a Service (IaaS) is a form of cloud computing that can be used to provide virtualized computing resources over the Internet to thin clients like a Raspberry Pi, which is a credit card-sized computing device with respectable performance. Combining the best of all three technologies, we propose an automated, wireless, biometric attendance management system that will run in an IaaS environment that will help to implement attendance marking process in a more efficient, simple, and time-saving fashion and also provide additional services which will be discussed ahead.

Keywords Attendance management · Biometric authentication
Cloud computing · ESXi · Infrastructure as a Service (IaaS) · Wireless

---

C. Sharma (✉) · K. Shah · S. Patel
Computer Engineering, Rajiv Gandhi Institute of Technology, Mumbai, India
e-mail: chiragpsharma@gmail.com

K. Shah
e-mail: karan2010shah@gmail.com

S. Patel
e-mail: shriya894@gmail.com

S. Gharat
Rajiv Gandhi Institute of Technology, Mumbai, India
e-mail: swapnil.gharat@mctrgit.ac.in

M.S. Reddy et al. (eds.), *International Proceedings on Advances in Soft Computing, Intelligent Systems and Applications*, Advances in Intelligent Systems and Computing 628, https://doi.org/10.1007/978-981-10-5272-9_24

1 Introduction

The traditional method of taking student's attendance manually is by using attendance sheet, given by the faculty member in class. The current attendance marking methods are monotonous and time-consuming [1]. Manually recorded attendance can be easily manipulated. Moreover, it is very difficult to verify each student in a large educational institute having various branches. Biometric attendance systems used today are either manual, wired, or wireless. Such systems store attendance records locally and do not support live update of records [2]. The records are stored locally on a removable storage or on the device storage itself and need to be monitored and managed by an authorized person periodically [3]. Such systems are vulnerable to loss of data due to memory corruption, data manipulation, and damage to device or human error. Also, the installation cost of these systems is high since individual devices need to be installed for every department or classroom. Hence, this paper is proposed to tackle all these issues.

The proposed system uses a Raspberry Pi-based wireless device with a fingerprint scanner which can be used to take attendance automatically and avoids the manipulation of data as it is directly stored on a central server (based on IaaS service model) in real time.

A. Biometric Identification

As we advance in the digital age, biometric technologies are being used to fulfill the increasing security needs. Identity verification through characteristics such as fingerprints, iris [1], face [2], palm print, voice, hand-written signatures is made possible by biometric technologies. These techniques are more secure, faster, and convenient than conventional methods such as RFID cards [3], passwords, captcha verification, or physical identity-based verification.

Biometric personal authentications such as iris recognition, facial recognition, voice recognition, fingerprint matching use data taken from measurements of the corresponding features of an individual. This data is unique to the individual and remain consistent for the individual's lifetime. This technique is used at facilities requiring authorization and high level of confidentiality, but due to its evolution and easy accessibility its use in other systems such as networks, e-commerce, online banking, cellular devices is also growing rapidly. Biometric recognition becomes the most mature and popular biometrics technology used in automated personal identification. Hence, the proposed system will adopt facial recognition [2] and fingerprint scanning [4] for identification of students and automated marking of attendance (Fig. 1).

B. Infrastructure as a Service

Infrastructure as a Service (IaaS) is one of the most basic cloud computing architectures based on the client–server model which is capable of providing cloud computing services. IaaS can be used to provide physical or virtual (most often) computers and various other resources. IaaS is capable of providing various online

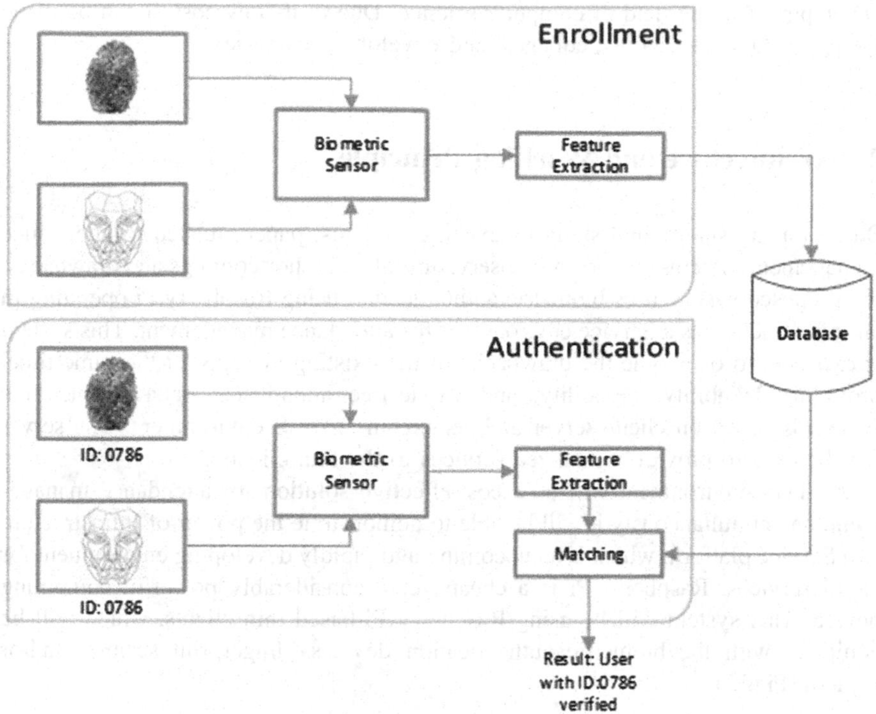

Fig. 1 General architecture of biometric authentication systems

services, abstracting the critical information of the infrastructure like physical computing resources, location, data partitioning, scaling, security, backup from the user. Hypervisors can be used to run the virtual machines as guests. Some hypervisors available in the market currently are VMware ESX/ESXi or Hyper-V. A pool of hypervisors can be created within a cloud computing system which can support a large number of virtual machines and scalability as per the customers' varying requirements. Additional resources such as a virtual-machine disk-image library, raw block storage, file or object storage, firewalls, load balancers, IP addresses, virtual local area networks (VLANs), and software bundles are offered by IaaS clouds. IaaS cloud providers supply these resources on-demand from their large pools of equipment installed in data centers. For wide-area connectivity, customers can use either the Internet or carrier clouds (dedicated virtual private networks). IaaS users pay on a per-use basis, which eliminates the capital expense of deploying in-house hardware and software.

C. Raspberry Pi

The Raspberry Pi is a small, compact, credit card-sized computer, developed in UK by the Raspberry Pi Foundation. It is a single board computer with decent specifications which makes it an excellent tool for learning and research and development in the field of computer science. Due to its low cost, it can be made easily available in schools, colleges, and developing countries.

2 Architecture and Working Principle

Based on the survey and study of existing systems, papers related to attendance management systems and general observation of their shortcomings and drawbacks, the proposed system uses biometric authentication using Raspberry Pi operating in an Infrastructure as a Service environment for attendance management. This system is expected to overcome the drawbacks of the existing systems, at the same time, providing reliability, flexibility, and hassle-free attendance management. This system is based on client–server architecture in an IaaS environment. The server will be used to provide virtualized services to the thin clients.

An IaaS environment will be a cost-effective solution for attendance management, and simultaneously it will be able to demonstrate the power of Infrastructure as a Service platform which is an upcoming and rapidly developing environment for cloud services. Raspberry Pi is a cheap yet a considerably powerful computing device. This system will be using Raspberry Pi-based thin clients, which will be equipped with the biometric authentication devices—fingerprint scanner and/or camera (Fig. 2).

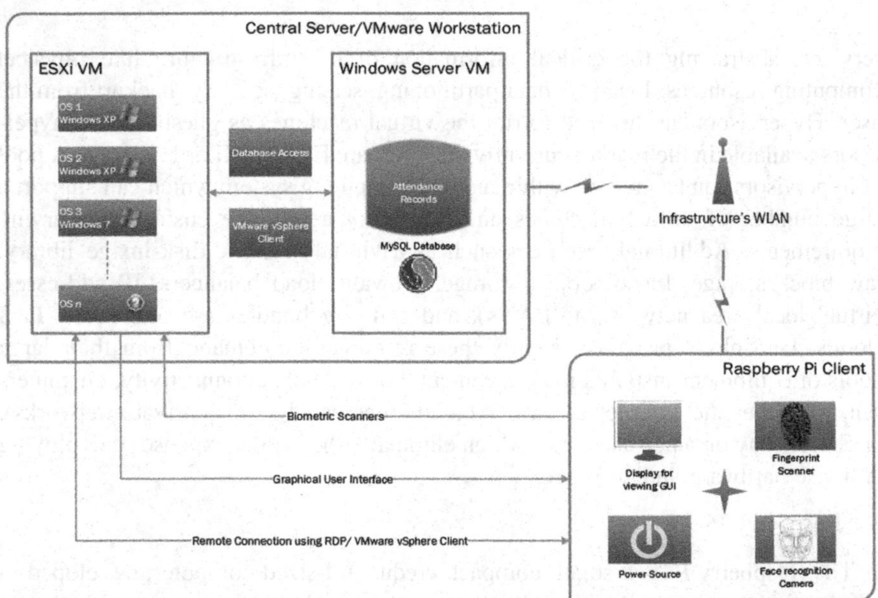

Fig. 2 Basic system architecture

These individual thin clients are expected to be in possession of every lecturer or whosoever concerned. The thin clients will be provided with an instance of an OS from the server that contains the program logic for attendance management. The device is intended to be passed to the students; they can authenticate themselves easily while the lecture is in progress, thereby saving the time and hassle involved in the conventional processes. All the updates to the attendance records will be made live on the server in the infrastructure.

On the server side, the system is expected to have been a powerful server running VMware ESXi for virtualization of operating systems. Using ESXi as the kernel, the system will have a Windows Server running on top of it. The Windows Server will be used as the centralized data store. ESXi will be used to install as many virtual machines as needed, the instances of which will be provided to the thin clients.

This demonstrates the power of IaaS. This system requires only a single licensed copy of any OS of choice, since they are technically being installed on the same machine. The cost involved in setting up such a system will also be just onetime. This system can further be enhanced for various uses in the infrastructure like report generation, notifications.

It can also be used for providing private cloud services to the entire organization or whosoever might be entitles to the same. The flowchart for attendance reporting and marking is as shown in Fig. 4.

3 Implementation

Implementation of this system requires hardware-level implementation as well as software implementation.

A. Hardware Implementation

Hardware consists of a central server with 16 GB RAM and a quad core processor (for testing purposes, and will need to be upgraded as the load and demand on the system increases) which will be used to load and run a hypervisor and database server. On the client side, we have n Raspberry Pi clients, each of which consists of a power source, a fingerprint scanner attached to it, and a display for viewing the GUI. For the communication to take place between the server and the Raspberry Pi clients, it is necessary that they should be within the same network, i.e., within the infrastructure's network.

B. Software Implementation

On the main server, we have VMware workstation loaded on which we have ESXi hypervisor and Windows Server loaded as virtual machines. Multiple operating systems are installed on ESXi which contains the biometric attendance program which deals with extraction, matching, and retrieval of records from the database. VMware vSphere client is used for managing the virtual machines on ESXi hypervisor. The Windows Server has MySQL database server installed, which is used for storing and managing the attendance records and databases. Initially, the database has to be loaded with the details of the students, professors, and lectures conducted.

The main application runs on each of the OSs running on the hypervisor, distinguished by their IP addresses.

Flow of the attendance marking procedure in the main application will be as follows (Fig. 3):

(i) Upon initialization, the main form will be loaded which will request the lecturer's fingerprint for initialization of any attendance marking procedures, as shown in Fig. 4.

(ii) Once the lecturer authenticates using his fingerprint, he will be taken to lecturer's zone, where he displayed the current lecture details and will be given the option to either automatically or manually mark up the attendance or to end the session, as shown in Fig. 5.

(iii) As soon as the lecturer initiates the auto-attendance markup procedure, the program will connect to the database and wait for a candidate to scan his fingerprint, as shown in Fig. 6.
The scanned fingerprint will be matched using 1: N matching with the records stored in the database; if match is found, that candidate's attendance will be marked and updated and a success message will be displayed, else, a failure message will be displayed asking the candidate to contact the lecturer for manual attendance markup. The success or failure message will be displayed for a few seconds, and then the program will get back to its former state of waiting for a candidate to scan his fingerprint for attendance markup.

(iv) Once all candidates have given their attendance, the lecture can go to back to the lecturer's zone after reauthenticating himself. From here, he can opt to manually mark the attendance of those candidates who might have not been able to successfully mark their attendance using biometric authentication due to any error. The lecturer will need to manually enter the candidates roll

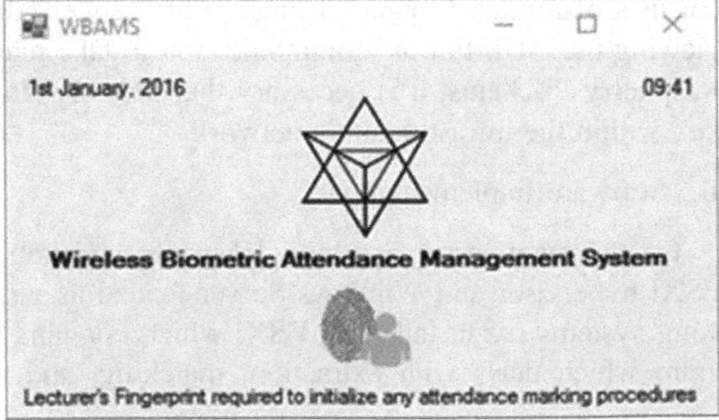

Fig. 3 Main page GUI

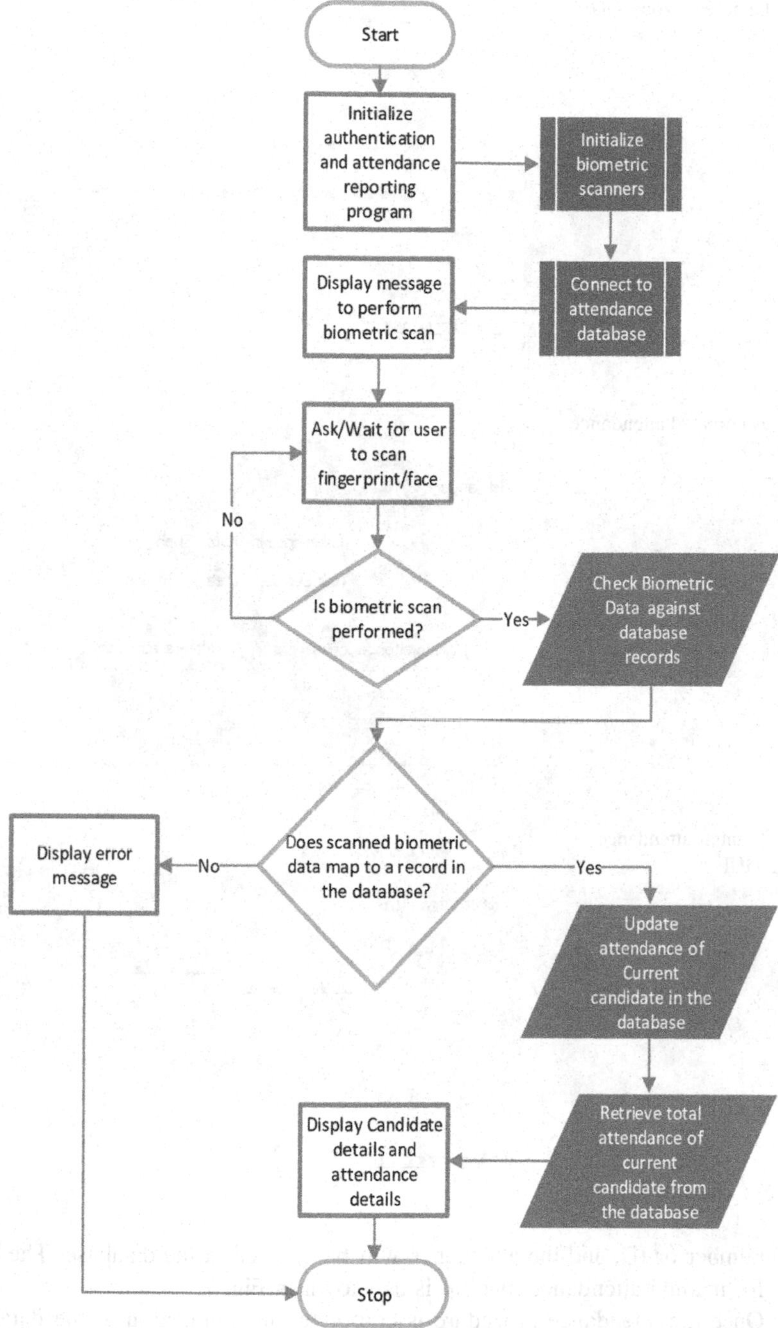

Fig. 4 Attendance reporting and marking flowchart

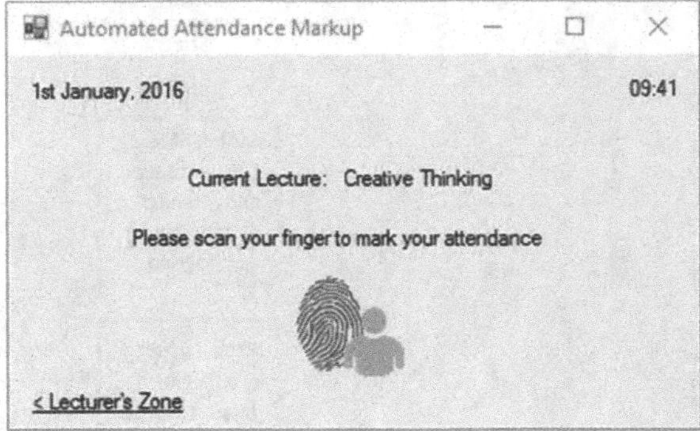

Fig. 5 Lecturer's zone GUI

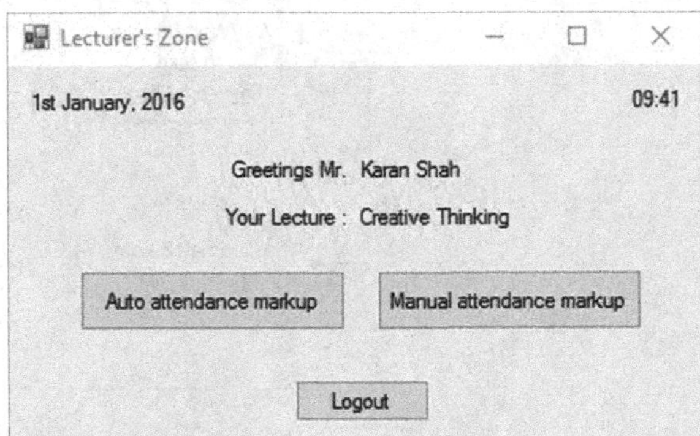

Fig. 6 Automated attendance markup GUI

Fig. 7 Manual attendance markup GUI

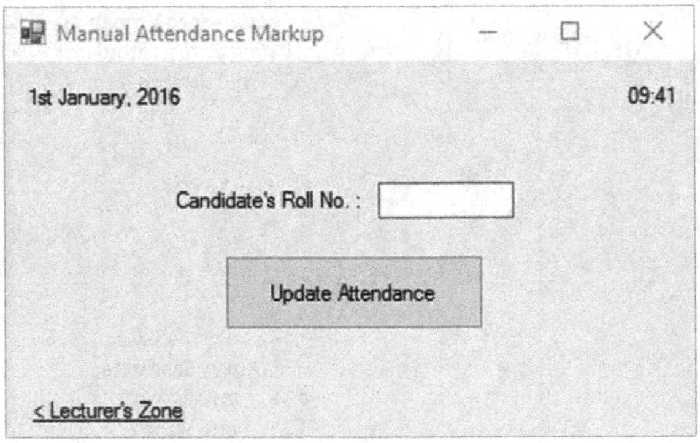

number or ID, and the attendance will be updated in the database. The GUI for manual attendance markup is as shown in Fig. 7.

Once the attendance procedure is complete, the connection to the database will be closed. In order to ensure that the same candidate cannot mark the attendance twice, authentication will be time-stamp-protected.

4 Applications and Future Scope

The main application of this system is attendance management and tracking. It can generate reports on a timely basis and can mail the reports if needed. Since it uses fingerprint for identity verification, it is a step toward ensuring security. It is far superior and reliable technology and an ultimate solution to the problem of proxy punching. Because it is flexible enough, it can be easily integrated with any other systems and scaled up if the requirement increases.

This system can be scaled for use in maintaining the records of the employees, students, or professors as per the needs of the organization. It can also be used for generating the payroll for the employers or the professors. This system can be implemented in schools, colleges, or any other organization which requires maintaining records for the same.

References

1. Seifedine, Kadry, and Khaled Smaili. 2007. A Design and Implementation of a Wireless Iris Recognition Attendance Management System. *Information Technology and Control* 36(3). ISSN 1392-124x.
2. Jomon, Joseph, and K.P. Zacharia. 2013. Automatic Attendance Management System Using Face Recognition. *International Journal of Science and Research* (*IJSR*) 2(11). ISSN 319-7064.
3. Zatin, Singhal, and Rajneesh Kumar Gujral. 2012. Anytime Anywhere-Remote Monitoring of Attendance System based on RFID using GSM Network. *International Journal of Computer Applications* (*0975–8887*) 39(3).
4. Shoewu, O., and O.A. Idowu. 2012. Development of Attendance Management System using Biometrics. *The Pacific Journal of Science and Technology* 13(1).

Ionospheric Time Delay Estimation Algorithm for GPS Applications

Bharati Bidikar, G. Sasibhushana Rao and Ganesh Laveti

Abstract The global positioning system (GPS) signal transit time delay in ionosphere comprised of ionized plasma is a major error source in GPS range measurements. As the density of the ionized plasma varies, the velocity of the radio waves differs from the velocity of light. Due to this, the GPS signals experience group delay or phase advance. Hence, the GPS signal transit time measurement is affected, and this time delay directly propagates into pseudorange measurements when scaled by the velocity of light. The delay depends on elevation angle of the satellite since the signal takes the longer propagation path when transmitted by the satellites tracked at lower elevation angle. The delay also depends on the solar activity conditions since the ionized plasma is a result of solar radiation. To achieve the precise navigation solution, the delay in ionosphere is estimated using conventional method where the total electron content (TEC) is modeled and pseudorange measurements of Link1 (L_1) and Link2 (L_2) frequencies are used. In this method, the TEC is an additional parameter to be calculated and the accurate range measurements determine the accuracy of the TEC. To overcome this, an eigenvector algorithm is proposed in this paper. The algorithm decomposes the pseudorange and carrier phase measurement coefficient matrix. The ionospheric time delay estimates of the proposed algorithm and conventional method are presented in this paper. The delays are estimated for the typical data collected on April 7, 2015, from dual frequency (DF) GPS receiver located in a typical geographic location over Bay of Bengal (Lat: 17.73° N/Long: 83.319° E). The proposed algorithm can be implemented for military and civil aircraft navigation and also in precise surveying applications.

B. Bidikar (✉) · G. Sasibhushana Rao
Department of Electronics and Communications Engineering,
Andhra University College of Engineering, Visakhapatnam, India
e-mail: bharati.bidikar@gmail.com

G. Sasibhushana Rao
e-mail: sasigps@gmail.com

G. Laveti
Department of Electronics and Communications Engineering, ANITS, Visakhapatnam, India
e-mail: ganeshlaveti2010@gmail.com

M.S. Reddy et al. (eds.), *International Proceedings on Advances in Soft Computing, Intelligent Systems and Applications*, Advances in Intelligent Systems and Computing 628, https://doi.org/10.1007/978-981-10-5272-9_25

Keywords GPS · Eigenvector · Transit time · Ionospheric time delay Pseudorange · Carrier phase · TEC

1 Introduction

The time delay of GPS Link1 (L_1) and Link2 (L_2) signals in ionosphere is one of the propagation path delays which depend on the total electron content (TEC) of the atmospheric layer [1]. This layer is part of the earth's atmosphere where ions and electrons are present in quantities sufficient to affect the propagation of radio frequency (RF) signals [2]. This plasma of free electrons will be dense during high solar activity conditions resulting in higher refraction of the GPS signal and time delay [3]. This delay constitutes a potential source of error in time measurements and tens of meters of range error [4]. Typical magnitudes of the range delay error are a few meters to tens of meters [5] and vary according to several factors such as the user's location, elevation angle, the time of the day, the time of the year, and solar cycle [5]. Hence, error estimation and correction are of prime concern in precise navigation applications.

The conventional method of calculating the ionospheric delay is by estimating the TEC along the signal propagation path since the delay is proportion to the TEC [6]. In precise positioning applications, time delay caused by an ionosphere is a major error which needs to be estimated and corrected. In this paper, an eigenvector algorithm is proposed which avoids the calculation of TEC. The paper also brings out the comparative analysis of the delay estimates of the algorithm and the conventional method involving TEC. The delays are estimated for the data obtained from the dual frequency GPS (DFGPS) receiver located in the Department of Electronics and Communications Engineering, Andhra University College of Engineering, Visakhapatnam (Lat: 17.73° N/Long: 83.319° E), for typical ephemerides collected on April 7, 2015.

2 Modeling Total Electron Content

The refraction of GPS Link1 (f_1 = 1575.42 MHz) and Link2 (f_2 = 1227.60 MHz) carrier frequencies [7] from ionospheric plasma is related to TEC and geomagnetic field. For these L_1 and L_2 electromagnetic waves, the phase index of refraction (n_p) is derived from Appleton–Hartree formula [8].

$$n_p = 1 - \frac{\omega_p^2}{2\omega^2} \pm \frac{\omega_p^2 \omega_g \cos\theta}{2\omega^3} - \frac{\omega_p^2}{4\omega^4}\left[\frac{\omega_p^2}{2} + \omega_g^2\left(1 + \cos^2\theta\right)\right] \quad (1)$$

$$\omega_p^2 = Ne^2/m\varepsilon_o; \quad \omega_g = Be/m \quad (2)$$

Here, ω is angular frequency of L_1 or L_2 signal, ε_o is free space permittivity, B is geomagnetic induction, ω_p and ω_g are electron plasma and cyclotron frequencies; these frequencies describe electron oscillations in the presence of perturbation electric field and ambient magnetic field; N, e, and m are electron density (el/m^2), charge (c), and mass (kg) respectively; the angle between geomagnetic field and direction of signal propagation is represented by θ. During high solar activity, the higher-order terms in Appleton–Hartree formula will contribute only 1–2 m of range error [9]. Hence, approximating these terms, we get

$$n_p \approx 1 - \frac{\omega_p^2}{2\omega^2} \quad n_p = 1 - \frac{1}{2}\frac{(2\pi)^2 \left(Ne^2/m\varepsilon_o\right)^2}{(2\pi)^2 f^2} \tag{3}$$

For $N = 10^{12}$ el/m^2 and $B = 32{,}000$ nT, the expression for phase refractive index can be simplified to

$$n_p = 1 - 40.3\frac{N}{f^2} \tag{4}$$

The group index of refraction (n_g) can be deduced from phase index of refraction as

$$n_g = n_p + f\frac{\mathrm{d}}{\mathrm{d}f}(n_p) \quad \Rightarrow \quad n_g = 1 + 40.3\frac{N}{f^2} \tag{5}$$

The group index of refraction given by (5) indicates it to be greater than unity, which means that the group velocity is less than the velocity of light. The integration of group refractive index given by (5) along the signal propagation path gives delay experienced by the signal (I_L) while propagating through ionosphere.

$$I_L = \frac{40.3}{f_L^2}\int_{\text{path}} n_g \quad \Rightarrow \quad I_L = \frac{40.3}{f_L^2}\,\mathrm{TEC} \tag{6}$$

where

$$\mathrm{TEC} = \frac{1}{40.3}\left(\frac{1}{f_1^2} - \frac{1}{f_2^2}\right)^{-1}(P_1 - P_2) \tag{7}$$

Subscript $L = 1$ for link 1 frequency
$= 2$ for link 2 frequency

Equations (5) and (6) show that GPS signal transit time through ionosphere is longer than the through vacuum, and the pseudorange is farther than the geomagnetic range between the satellite and the receiver [10]. Since the ionosphere is dispersive in nature, the absolute TEC is calculated using the L_1 and L_2 *GPS* carrier frequencies. In this paper, the ionospheric group delay (I_1) on L_1 carrier frequency is calculated using Eq. (6) for ephemerides data collected from DFGPS receiver located at latitude 17.73°N and longitude 83.319°E.

3 Eigenvector Algorithm

The proposed eigenvector algorithm for ionospheric delay estimation is derived from the linear combination of carrier phase and pseudorange measurements, observed on Link1 (f_1 = 1575.42 MHz) and Link2 (f_2 = 1227.60 MHz) carrier frequencies.

$$\begin{aligned} P_1 &= \rho + I_1 + \varepsilon_{p1} \\ P_2 &= \rho + \alpha \times I_1 + \varepsilon_{p2} \\ \phi_1 &= \rho - I_1 + \lambda_1 N_1 + \varepsilon_{\phi 1} \\ \phi_2 &= \rho - \alpha \times I_1 + \lambda_2 N_2 + \varepsilon_{\phi 2} \end{aligned} \tag{8}$$

where the subscripts 1 and 2 represent the measurements on the L_1 and L_2 frequencies, respectively, ε represents the effects of multipath and receiver noise [m], P_1 and P_2 are the pseudoranges on L_1 and L_2 frequencies [m], respectively, ρ is the geometric range [m], I_1 is ionospheric delay on L_1 frequency [m], α is $(f_1/f_2)^2$, ϕ_1 and ϕ_2 are carrier phase measurements [m], N_1 and N_2 are integer ambiguities, and λ_1 and λ_2 are wavelengths [m] on respective frequencies. Equation (8) is framed into a matrix form as given by (9), and here, it is assumed that multipath effect and the receiver noise are corrected.

$$Ax = B \tag{9}$$

where

$$A = \begin{bmatrix} 1 & 1 & 0 & 0 \\ 1 & -1 & \lambda_{L1} & 0 \\ 1 & \alpha & 0 & 0 \\ 1 & -\alpha & 0 & \lambda_{L2} \end{bmatrix} \quad x = \begin{bmatrix} \rho \\ I_1 \\ N_1 \\ N_2 \end{bmatrix} \quad \text{and} \quad B = \begin{bmatrix} P_1 \\ \phi_1 \\ P_2 \\ \phi_2 \end{bmatrix}$$

Equation (9) can be simplified, and the solution for '*x*' can be obtained as

$$\begin{aligned} A^T A x &= A^T B \\ x &= (A^T A)^{-1} A^T B \end{aligned} \tag{10}$$

Computation of x from (10) involves calculation of inverse coefficient matrix. This solution is numerically unstable which leads to round-off errors, or the minor errors in initial data leads to large deviation of estimated solution from the exact solution [11]. To overcome this problem, the solution for x is obtained using the eigenvector [12] decomposition. In this, the coefficient matrix A is decomposed into orthogonal matrices U and V and singular value diagonal matrix D such that [11]

$$A = UDV^T \tag{11}$$

where the diagonal matrix entries are nonnegative, and the magnitudes are in descending order, which are known as singular values or eigenvalues of matrix A. Similarly, the column vectors of U and V are left and right singular values of matrix A [12]. Entries of U and V are eigenvectors of AA^T and A^TA, respectively. In general, the eigenvector decomposition of a coefficient matrix A of the order $m \times n$ is expressed as eigenvalues and eigenvector matrices.

$$A = \left[u_1\, u_2 \ldots. u_k \,|\, u_{k+1}\, u_{k+2} \ldots. u_m\right] \left[\begin{array}{cccc|c} \sigma_1 & & & & 0 \\ & \ddots & & & 0 \\ & & & \sigma_k & 0 \\ \hline 0 & 0 & & 0 & 0 \end{array}\right] \left[\begin{array}{c} v_1^T \\ \vdots \\ v_k^T \\ \hline v_{k+1}^T \\ \vdots \\ v_n^T \end{array}\right] \tag{12}$$

where σ is singular values, and k is the rank of the matrix A. These matrices on simplification give

$$A = [u_1\, u_2 \ldots u_k] \begin{bmatrix} \sigma_1 & & \\ & \ddots & \\ & & \sigma_k \end{bmatrix} \begin{bmatrix} v_1^T \\ v_2^T \\ \vdots \\ v_k^T \end{bmatrix} + [u_{k+1}\, u_{k+2} \ldots u_m]\,[0] \begin{bmatrix} v_{k+1}^T \\ v_{k+2}^T \\ \vdots \\ v_n^T \end{bmatrix}$$

The above equation shows only k eigenvectors that contribute to matrix A. Hence, matrix A can be simplified as

$$A = [u_1 \cdots u_k] \begin{bmatrix} \sigma_1 & & \\ & \ddots & \\ & & \sigma_k \end{bmatrix} \begin{bmatrix} v_1^T \\ \vdots \\ v_k^T \end{bmatrix} \tag{13}$$

The above equation can be expressed as (11). Using these eigenvectors and the singular values, the solution for (10) is calculates as

$$x = \left(VD^2V^T\right)^{-1} VDU^T B \tag{14}$$

On further simplification the unknown vector x can be obtained as,

$$x = \left(VD^{-1}U^{T}\right)B \tag{15}$$

As seen from the above equation, the derived equation for x is much simpler in computation compared to (10). On solving the above equation, the unknown ionospheric delay on L_1 frequency is obtained from x and it is compared with delay obtained by modeling total electron content. The performance analysis of proposed eigenvector algorithm and the conventional method is presented in this paper.

4 Results and Discussion

Statistical analysis of the results shows that the ionospheric delay ranges from ≈5 m to ≈19 m. This error is estimated for geographical location (xu: 706970.90 m yu: 6035941.02 m zu: 1930009.51 m) in the Indian subcontinent over a Bay of Bengal for typical ephemerides collected on April 7, 2015, from the DFGPS receiver located in Department of Electronics and Communication Engineering, Andhra University College of Engineering, Visakhapatnam (Lat: 17.73° N/Long: 83.319° E), India.

Throughout the observation period of about 24 h, out of 32 satellites, a minimum of 9 satellites were visible in each epoch. Though the error is computed and analyzed for all the visible satellites, in this paper, the ionospheric delay for SV PRN31 is presented and the satellite was visible and was tracked for about 6 h. Table 1 details the elevation angle and the TEC for the respective time of observation, and Table 2 details the ionospheric delay estimated using eigenvector algorithm and using TEC. The change in elevation angle and the TEC is shown in Figs. 1 and 2. The delays and the electron contents are calculated for every 15 s.

Table 1 Pseudorange multipath error for satellite signal on L_1 frequency

	C/A code multipath error on L_1 frequency (m)				Error in receiver position distance (m)
	SV PRN07	SV PRN23	SV PRN28	SV PRN31	
Min	7.362	14.11	40.21	9.136	−25.8
Max	14.32	18.79	52.88	13.52	31.49
Standard deviation	1.816	1.439	1.984	1.019	10.78

Table 2 Ionospheric delay estimation of SV PRN31

	Eigenvector algorithm		Code Range technique	
	Time delay (ns)	Range error (m)	Time delay (ns)	Range error (m)
Min	16.95	5.084	16.95	5.084
Max	65.97	19.78	65.97	19.78
Mean	31.21	9.357	31.21	9.357
Standard deviation	11.03	3.309	11.03	3.309

These delays with respect to change in elevation angle are plotted in Figs. 3 and 4. From the figures, it is observed that for the elevation angle of ≈5°, the delay is as high as ≈19 m and the delay is ≈5 m when the satellite approaches zenith.

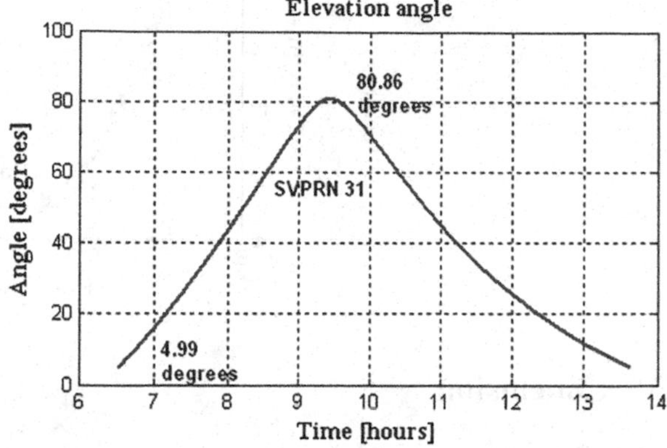

Fig. 1 Elevation angle of SV PRN31

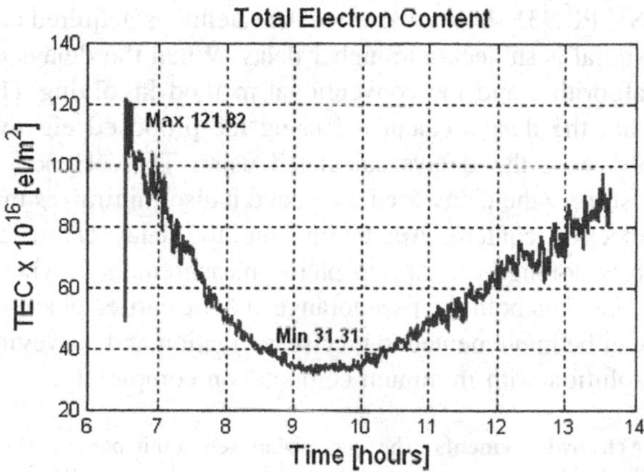

Fig. 2 TEC along the signal propagation path of SV PRN31

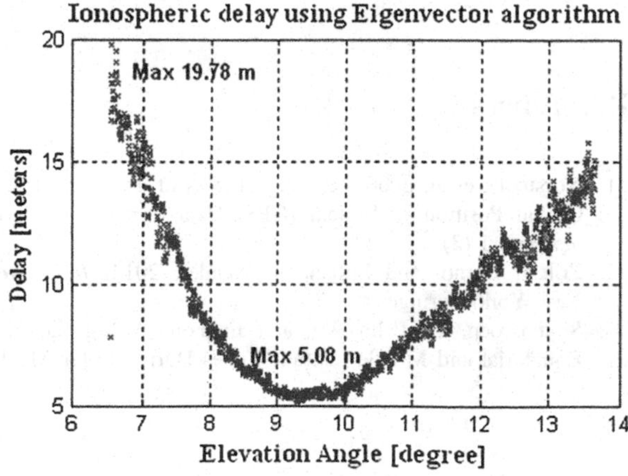

Fig. 3 Ionospheric delay of SV PRN31 using eigenvector algorithm

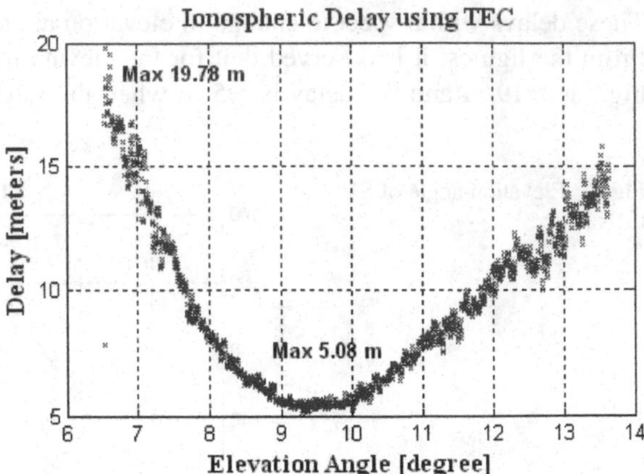

Fig. 4 Ionospheric delay of SV PRN31 using code range technique

5 Conclusion

Implementation of the proposed eigenvector algorithm to estimate the time delay of SV PRN31 shows that when the satellite is acquired at low elevation angle, the GPS signal is subjected to higher delay. When the delays calculated using the proposed algorithm and the conventional method involving TEC are compared, it is found that the delays calculated using the proposed eigenvector algorithm are same as those of the delays calculated using TEC. Hence, the proposed algorithm can estimate the delay accurately, and it also minimizes the complex calculation of total electron content. Apart from this, the delay estimates using TEC consider either pseudorange or carrier phase measurements, whereas the proposed algorithm considers both the pseudorange and the carrier phase measurements. This algorithm can be implemented in precise navigation and surveying applications to get accurate solution with minimum computation complexity.

Acknowledgements The work undertaken in this paper is supported by Ministry of Science and Technology, Department of Science and Technology (DST), New Delhi, India, under Women Scientist Scheme (WOS-A) vide, SR/WOS-A/ET-04/2013.

References

1. Norsuzila, et al. 2008. Leveling Process of Total Electron Content (TEC) Using Malaysian Global Positioning System (GPS) Data. *American Journal of Engineering and Applied Sciences* 1 (3): 223–229.
2. Zolesi, Bruno, and Ljiljana R. Cander. 2014. *Ionospheric Prediction and Forecasting*. New York: Springer.
3. Sukcharoen, T., Falin Wu, and Jingnong Weng. 2015. Characteristics of Ionosphere at Equatorial and Middle Latitude Zones During Solar Maximum. In *2015 12th International*

Conference on Electrical Engineering/Electronics, Computer, Telecommunications and Information Technology (*ECTI-CON*), 24–27 June 2015, Hua Hin.

4. Kumar, G.S. et al. 2011. GPS Satellite Transmitted Spread Spectrum Codes SNR Modeling Over the Indian Subcontinent For Precise Positioning Applications, India. In *2011 Annual IEEE Conference* (*INDICON*), 16–18 Dec 2011.
5. Ouzeau, C., F. Bastide, C. Macabiau, and B. Roturier. 2006. Ionospheric Code Delay Estimation in a Single Frequency Case for Civil Aviation. In *Proceedings of the 19th International Technical Meeting of the Satellite Division of the Institute of Navigation* (*ION GNSS 2006*), 3059–3069. Fort Worth, TX, September 2006.
6. Klobuchar. John A. 1991. Ionospheric Effects on GPS. Innovation, GPS world April 1991.
7. Teunissen, Peter J.G., and Alfred Kleusberg. 1998. *GPS for Geodesy*, 2nd ed. New York: Springer.
8. Rao, G.S. 2010. *Global Navigation Satellite Systems*, 1st ed. India: Tata McGraw-Hill.
9. Soumi Bhattacharya, et al. 2009. Ionospheric Time Delay Variations in the Equatorial Anomaly Region During Low Solar Activity Using GPS. *Indian Journal of Radio & Space Physics* 38: 266–274.
10. Lao-Sheng. 2001. Remote Sensing of Ionosphere Using GPS Measurements. In *22nd Asian Conference on Remote Sensing*, 5–9 Nov 2001, Singapore.
11. Chang, X.-W., C.C. Paige, and C.C.J.M. Tiberius. 2005. Computation of a Test Statistic in Data Quality Control. *SIAM Journal of Scientific Computing* 26: 1916–1931.
12. Melzer, T. 2004. *SVD and Its Application to Generalized Eigenvalue Problems*. Vienna: University of Technology.

A Modified Variance Kalman Filter for GPS Applications

Ganesh Laveti and G. Sasibhushana Rao

Abstract The advancements in technology have made global positioning system (GPS) part and parcel of human daily life. Apart from its domestic applications, GPS is used as a position determination system in the field of defence for guiding missiles, navigation of ships, landing aircrafts, etc. These systems require precise position estimate and is only possible with the reduced measurement uncertainty and efficient navigation solution. Due to its robustness to noisy measurements and exceptional performance in wide range of real-time applications, Kalman filter (KF) is used often in defence applications. In order to meet the increase in demands of defence systems for high precise estimates, the KF needs upgradation, and this paper proposes a new covariance update method for conventional Kalman filter that improves its performance accuracy. To evaluate the performance of this developed algorithm called modified variance Kalman filter (MVKF), real-time data collected from GPS receiver located at Andhra University College of Engineering (AUCE), Visakhapatnam (Lat/Lon: 17.72°N/83.32°E) is used. GPS statistical accuracy measures (SAM) such as distance root mean square (DRMS), circular error probability (CEP), and spherical error probability (SEP) are used for performance evaluation.

Keywords CEP · DRMS · GPS · Kalman filter · MVKF · SAM · SEP Variance

G. Laveti (✉)
Department of Electronics and Communications Engineering,
GVPCEW, Visakhapatnam, India
e-mail: ganeshlaveti2010@gmail.com

G. Sasibhushana Rao
Andhra University College of Engineering, Visakhapatnam, India
e-mail: sasigps@gmail.com

M.S. Reddy et al. (eds.), *International Proceedings on Advances in Soft Computing, Intelligent Systems and Applications*, Advances in Intelligent Systems and Computing 628, https://doi.org/10.1007/978-981-10-5272-9_26

1 Introduction

Most of today's defence operations such as rescue in dense forest, landing of aircrafts on ships, tracking of unidentified radio source depend on three main functions, navigation, tracking and Guiding, and it is unimaginable to find a system that does not have any relation with these functions. These systems require to determine the two- or three-dimensional position information of an object of interest, and they use GPS receivers for this purpose. GPS is the constellation of space bodies called satellites which revolve round the earth and provide position information based on range measurements. With a sufficient number of range measurements, GPS can provide position estimates to high degree accuracy [1]. The accuracy in turn is the function of type of equipment, geographic area, uncertainty in measurements, navigation algorithm, etc. In practice, the measurement uncertainty can never reach zero even though the system noise parameters and biases are modelled effectively and hence need a high accuracy navigation algorithm that makes out an optimal estimate from these uncertain measurements.

The various algorithms used for GPS receiver position estimate include least squares (LS), weighted least squares (WLS), evolutionary optimizers, Kalman filter (KF). The task of tracking and guiding involves estimation of objects' future course, and this could be only possible when the system dynamics are modelled into the estimator. Out of the available navigation algorithms, the only filter that makes use of the dynamics in estimation is Kalman filter [2]. In addition, KF also provides the uncertainty in its estimation whose performance varies with parameters such as process noise matrix, measurement noise matrix, geometry matrix. So this paper concentrates in improving KFE accuracy with a new geometry matrix that replaces the conventional matrix in Kalman filter's covariance update equation. The modelling of KF as GPS receiver position estimator and the details about new designed geometry matrix lead to the development of MVKF are discussed in subsequent sections.

2 Modified Variance Kalman Filter for GPS Applications

As depicted in Fig. 1, GPS uses time of arrival (TOA) measurements observed between satellite, Sat_i, and GPS receiver, R_X, to compute the receiver position. GPS $\mathrm{TOA}_{\mathrm{Sat}_i, R_x}$ measurement is the travel time of a radio signal between the receiver, R_X, and ith satellite, Sat_i, and with known signal velocity, it provides the range information. The range equation formulated in Eq. 1 is nonlinear [3] and requires minimum four such equations to be solved to get precise GPS position.

Hence, extended Kalman filter estimator (EKFE) or Kalman filter estimator (KFE) with linearized measurement equations is used to estimate the unknown receiver position. The first-order approximated linear form Taylor's series [4] of

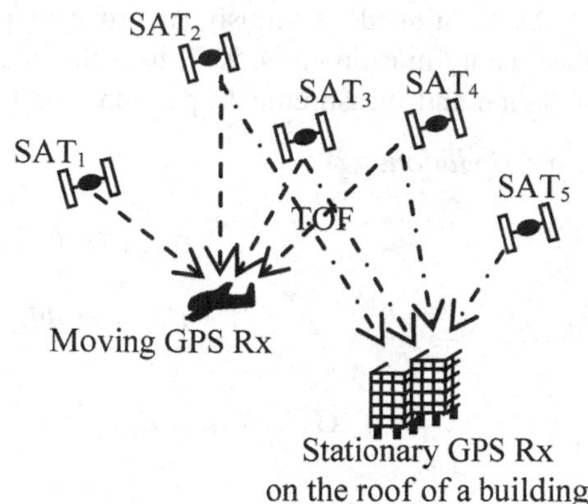

Fig. 1 Positioning with GPS TOA measurements

Eq. 1, computed at $\hat{R}_X$, is given in Eq. 2 which is used in framing the geometry matrix, ψ in KFE.

$$\mathrm{TOA}_{\mathrm{Sat}_i,R_x} = f(\mathrm{Sat}_i, R_x) = \sqrt{(x_{\mathrm{Sat}_i} - x_{R_x})^2 + (y_{\mathrm{Sat}_i} - y_{R_x})^2 + (z_{\mathrm{Sat}_i} - z_{R_x})^2} \quad (1)$$

$$i = 1, 2\ldots, \text{no. of. satellites}$$

$$f(\mathrm{Sat}, R_x) \cong f(\mathrm{Sat}, \hat{R}_x) + f'(\mathrm{Sat}, \hat{R}_x)(R_x - \hat{R}_x) \quad (2)$$

Here, $\mathrm{TOA}_{\mathrm{Sat}_i,\mathrm{R}_x}$ is the time of arrival between ith satellite, Sat_i and GPS receiver, R_x, $\mathbf{Sat} = (x_{\mathrm{Sat}_i}, y_{\mathrm{Sat}_i}, z_{\mathrm{Sat}_i})$ is the three-dimensional position coordinates of ith satellite, $\mathbf{R_x} = (x_{R_x}, y_{R_x}, z_{R_x})$, $\hat{\mathbf{R}}_x = (\hat{x}_{R_x}, \hat{y}_{R_x}, \hat{z}_{R_x})$, is the three-dimensional position coordinates of receiver, and its estimate, respectively, and f' represents the first-order derivative of nonlinear function, $f(\mathrm{Sat}, R_x)$ w.r.t, $\hat{R}_X$.

The other form Eq. 2 can be represented is

$$f(\mathrm{Sat}, R_x) - f(\mathrm{Sat}, \hat{R}_x) \cong f'(\mathrm{Sat}, \hat{R}_x)(R_x - \hat{R}_x) \Rightarrow \delta\mathrm{TOA}_{\mathrm{Sat},R_x} \cong \frac{\partial f(\mathrm{Sat}, R_x)}{\partial \hat{R}_x}\delta R_x$$

$$\Rightarrow \delta\mathrm{TOA}_{\mathrm{Sat}}, R_x \cong \psi(\mathrm{Sat}, \hat{R}_x)\delta R_x$$

where

$$\frac{\partial f(\mathrm{Sat}, R_x)}{\partial \hat{R}_x} = \frac{(\hat{X}_{R_x} - X_{\mathrm{Sat}})}{\mathrm{TOA}_{\mathrm{Sat},\hat{R}_x}}\delta\hat{X}_{R_x} + \frac{(\hat{y}_{R_x} - Y_{\mathrm{Sat}})}{\mathrm{TOA}_{\mathrm{Sat},\hat{R}_x}}\delta\hat{Y}_{R_x} + \frac{(\hat{Z}_{R_x} - Z_{\mathrm{Sat}})}{\mathrm{TOA}_{\mathrm{Sat},\hat{R}_x}}\delta\hat{Z}_{R_x} \quad (3)$$

Here, $\delta\mathrm{TOA}_{\mathrm{Sat},\mathrm{R}_x}$ is the error or change in range measurements, δR_x is the error or change in receiver position and the term $\partial f(\mathrm{Sat}, \hat{R}_x)/\partial \hat{R}_x$ represents geometry matrix, ψ or Jacobian matrix, J, given in Eq. 3.

As mentioned previously in order to provide the position estimates and the uncertainty in estimation, KFE uses the geometry matrix, ψ in two stages [5], time updation and measurement updation which are formulated as below.

Time Updation:

$$R_{x_{t/t-1}} = \Phi R_{x_{t-1/t-1}} + P_{R_x} \tag{4}$$

$$C_{t/t-1} = \Phi R_{x_{t/t-1}} \Phi^T + \xi \tag{5}$$

$$G_{\mathrm{K}} = C_{t/t-1} \psi^T_{t/t-1} \left(\psi_{t/t-1} C_{t/t-1} \psi^T_{t/t-1} + \beta \right)^{-1} \tag{6}$$

Measurement Updation:

$$R_{x_{t/t}} = R_{x_{t/t-1}} + G_{\mathrm{K}} (\mathrm{TOA}_t - \psi_{t/t-1} R_{x_{t/t-1}}) \tag{7}$$

$$C_{t/t} = \left(I - G_{\mathrm{K}} \psi_{t/t-1} \right) C_{t/t-1} \tag{8}$$

where $R_{x_{t/t-1}}$ is the receiver position estimate at time (t) prior collection of measurements, $R_{x_{t/t}}$ is the estimate of receiver position at time (t) on post reception of measurements; TOA at time (t), $C_{t/t\text{-}1}$ represents the uncertainty in the estimated receiver position at time (t) prior measurements, $C_{t/t\text{-}1}$ represents the uncertainty in estimated receiver position at time (t) post the measurements, G_{K} is the Kalman Gain, Φ represents receiver position state transition matrix, ψ is geometry matrix, I is the identity matrix, β the measurement error covariance matrix, P_{Rx} and ξ are the process noise and its uncertainty, respectively.

It is observed from Eq. 3 that the geometry matrix, ψ is the resultant of first-order approximation from Taylor's series and as defined in [6] a nonlinear function like f(Sat, R_x) is modifiable if there exist a linear structure which is the function of the predicted state, $\hat{R}_x$ and the actual measurement, $\mathrm{TOA}_{\mathrm{Sat},R_x}$, i.e. if f (Sat, R_x) is modifiable than Eq. 3 can be rewritten as Eq. 9.

$$\delta\mathrm{TOA}_{\mathrm{Sat},R_x} = \Gamma\left(\mathrm{TOA}_{\mathrm{Sat},R_x}, \hat{R}_x\right) \times \delta R_x \tag{9}$$

In practice, the range measurements, TOA, are noisy, and hence, $\mathrm{TOA}^*_{\mathrm{Sat},R_x}$ is used in Eq. 9 instead of $\mathrm{TOA}_{\mathrm{Sat},R_x}$. The developed MVKF makes use of the new observation matrix, Γ instead of ψ in Eq. 8, while retaining the other KFE equations as same [7]. Kalman gain, G_{K} in Eq. 6, is the function of ψ, $C_{t/t-1}$ and β, which decides the amount of weight to be imposed on current measurements while updating the receiver position and hence making this equation the function of current measurements, $\mathrm{TOA}^*_{\mathrm{Sat},R_x}$ results in poor performance [6]. This is the reason the new observation matrix Γ is used in Eq. 8, and the resultant new updated covariance matrix for MVKE is given in Eq. 10

$$C_{t/t-1} = \left(I - G_{\mathrm{K}}\Gamma_{t/t-1}\right)C_{t/t-1} \quad (10)$$

3 Results and Discussion

The GPS receiver located at AUCE, Visakhapatnam, is used for the collection of real-time data, which is used in the performance evaluation of KFE and the developed MVKF estimator. The three-dimensional position of the receiver is estimated over a period of 23 h 56 min (2872 epochs) with a randomly chosen initial position estimate of X: 785 m, Y: 746 m and Z: 3459 m. The data is collected at a sampling interval of 30 s, and the position error of epochs the estimator took to reach convergence (i.e. position error in three dimensions <100 m) is plotted in Figs. 2 and 3. The details pertaining to the convergence epochs are also given in Table 1. It is observed from the figures and Table 1 that the developed algorithm, MVKF, converges at faster rate compared to conventional KFE.

Also the GPS statistical accuracy measures (SAM) for both the algorithms are calculated for the entire range of data and tabulated in Table 2. Various SAM [8] such as $1(\sigma)$ sigma error, circular error probability (CEP), spherical error probability (SEP), spherical accuracy standard (SAS), mean radial spherical error (MRSE), distance root mean square (DRMS) are used in the evaluation of the algorithms

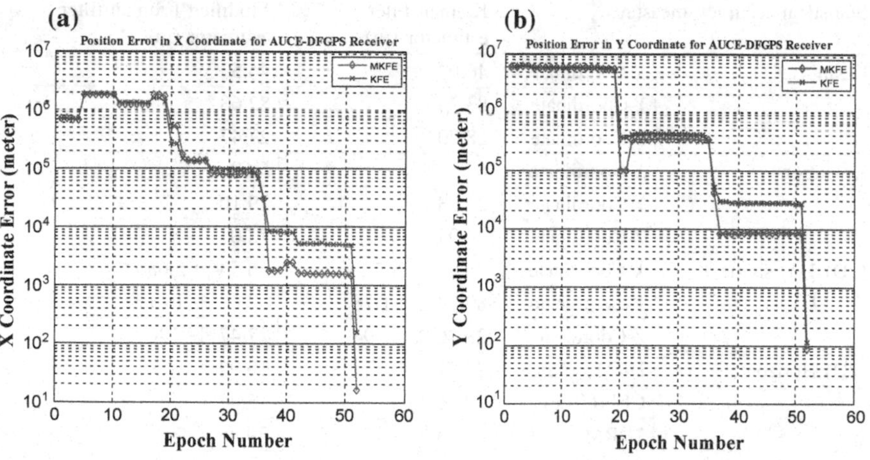

Fig. 2 Position error versus epochs. **a** Position error in X-coordinate with KFE and MVKF for AUCE, Visakhapatnam GPS receiver. **b** Position error in Y-coordinate with KFE and MVKF for AUCE, Visakhapatnam GPS receiver

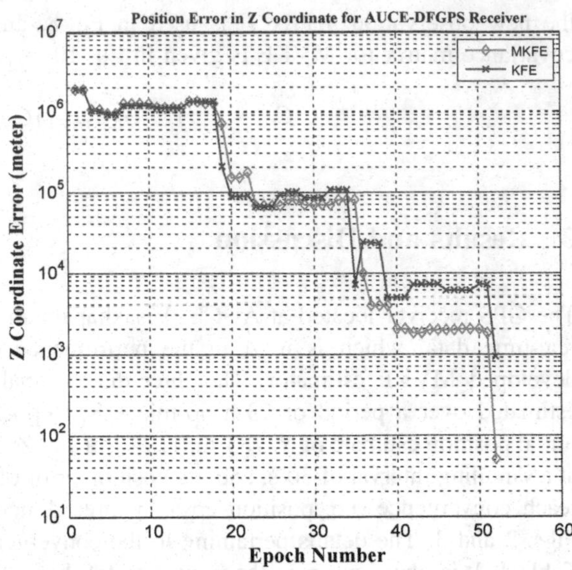

Fig. 3 Position error in *Z*-Coordinate with KFE and MVKF for AUCE, Visakhapatnam GPS receiver

Table 1 Estimator convergence performance

GPS receiver	KFE	MVKF
AUCE, Visakhapatnam	92	52

Table 2 SAM of KFE and the developed MVKF for AUCE, Visakhapatnam GPS receiver

AUCE, GPS receiver			
Statistical accuracy measures		Kalman filter estimator (m)	Modified Kalman filter estimator (m)
Mean	*X*-Coordinate	46.99	46.53
	Y-Coordinate	83.73	82.68
	Z-Coordinate	28.20	27.47
Deviation	*X*-Coordinate	6.7	5.68
	Y-Coordinate	21.78	19.10
	Z-Coordinate	24.94	7.20
(1D) 1(σ) sigma (68%)	*X*-Coordinate	46.99 ± 6.7	46.53 ± 5.68
	Y-Coordinate	83.73 ± 21.78	82.68 ± 19.10
	Z-Coordinate	28.20 ± 24.94	27.47 ± 7.20
(2D) Horizontal	DRMS (65%)	22.81	19.92
	CEP (50%)	17.30	15.02
	2DRMS (95%)	45.63	39.85
(3D) Horizontal and vertical	SEP (50%)	27.29	16.31
	MRSE (61%)	33.80	21.19
	SAS (99%)	44.57	26.64

accuracy performance. Table 2 depicts the SAM of both the algorithms on AUCE and Visakhapatnam GPS receiver and is given below.

It is obvious from the accuracy measures that for AUCE, Visakhapatnam receiver the position estimated by the MVKF will be within 26.64 m from its true position with a probability of 0.99, where KFE estimates the position within 44.57 m. This shows the efficiency of developed algorithm over conventional KFE.

4 Conclusion

A new algorithm for the GPS receiver position estimation was developed based on predict and update concept of KFE. The geometry matrix designed out of first-order Taylor's approximation of nonlinear measurement function is modified. The designed new geometry matrix is then used for modification of covariance update equation in KFE. The developed algorithm, MVKF performance, is evaluated with the collected real-time GPS data over a period of 23 h 56 min. The GPS receiver position located at AUCE, Visakhapatnam, is estimated with the collected data, and the algorithms' performance is evaluated with various SAM. Results demonstrated that MVKF converges in less time (with an epoch difference of 40 for AUCE, Visakhapatnam) and has an accuracy difference of 2 m CEP when compared to conventional KFE. This also showed that the MVKF has faster convergence rate with high accuracy and is suitable for real-time defence applications such as navigation of ships, landing of CAT I and II aircrafts.

Acknowledgements Part of the work undertaken in this paper is supported by University Grants Commission (UGC), Govt. of India, New Delhi, India. Vide Ltr. No.42-126/2013 (SR), Dated 25.03.2013.

References

1. Ganesh Laveti, G. Sasibhushana Rao. 2014. *GPS Receiver SPS Accuracy Assessment Using LS and LQ Estimators for Precise Navigation*. IEEE INDICON 1–5. ISBN 978-1-4799-5364-6.
2. Ramsey, Faragher. 2012. *Understanding the Basics of the Kalman Filter Via a Simple and Intuitive Derivation*. IEEE Signal Processing Magazine 128–132, September 2012. ISSN 1053-5888.
3. Rao, G.S. 2010. *Global Navigation Satellite Systems*. India: McGraw Hill Education Pvt. Ltd.
4. Guanrong, Chen. 1993. *Approximate Kalman Filtering*. USA: World Scientific Publishing Co. Pte. Ltd.
5. Nagamani, Modalavalasa, and G. Sasibhushana Rao. 2015. A New Method of Target Tracking by EKF Using Bearing and Elevation Measurements for Underwater Environment. *Elsevier Journal of Robotics and Autonomous Systems* 74: 221–228.
6. Song, T.L., and J.L. Speyer. 1985. A Stochastic Analysis Of A Modified Extended Kalman Filter with Application to Estimation with Bearings Only Measurements. *IEEE Transactions on Automatic Control* AC-30 (10): 940–949.

7. Rao, S.K. 1998. Modified Gain Extended Kalman Filter with Application to Angles Only Underwater Passive Target Tracking. *IEEE Signal Processing Proceedings* 2: 1439–1442.
8. *Statistics and Its Relationship to Accuracy Measure in GPS*. NovAtel, APN-029, Rev-1, 1–6, December 2003.

Genetic Algorithm Designed for Solving Linear or Nonlinear Mixed-Integer Constrained Optimization Problems

Hemant Jalota and Manoj Thakur

Abstract Genetic algorithms (GA) initially were not developed to handle the integer restriction or discrete values to the design variables. In the recent years, researchers has focused their work on developing/modifying GAs for handling integer/discrete variables. We have modified the BEX-PM algorithm developed by Thakur et al. [1] to solve the nonlinear constrained mixed-integer optimization problems. Twenty test problems have been used to conduct a comparative study to test the effectiveness of proposed algorithm (MI-BEXPM) with other similar algorithms (viz. MILXPM, RST2ANU, and AXNUM) in this class available in the literature. The efficacy of the results acquired through MI-BEXPM is compared with other algorithms on two well-known criteria. The performance of MI-BEXPM is also analysed and compared for solving real-life mixed-integer optimization problems with other methods available in the literature. It is found that MI-BEXPM is significantly superior to the algorithms considered in this work.

Keywords Real-coded genetic algorithms · Bounded exponential crossover Power mutation · Mixed-integer optimization · Constraint optimization

1 Introduction

Mixed-integer nonlinear programming problem (MINLP) is the important class of nonlinear optimization problems. A MINLP is an optimization problem where the objective functions and constraints are nonlinear functions of the decision variables with some of the decision variables having integer restriction. If objective functions as well as constraints are linear function, then the corresponding problem is called a

H. Jalota (✉) · M. Thakur
Indian Institute of Technology Mandi, Mandi 175001, Himachal Pradesh, India
e-mail: hemant.jalota4@gmail.com

M. Thakur
e-mail: manojpma@gmail.com

M.S. Reddy et al. (eds.), *International Proceedings on Advances in Soft Computing, Intelligent Systems and Applications*, Advances in Intelligent Systems and Computing 628, https://doi.org/10.1007/978-981-10-5272-9_27

mixed-integer linear programming problem (MILP). A MINLP having the all the variable as integer is called integer nonlinear programming problem (INLP).

In last few decades, several population-based heuristic algorithms have been designed. These algorithms try to search the global optimal solution of general nonconvex optimization problems. Several variants of these algorithms have been suggested to handle constraints and for solving mixed-integer optimization problems. Two of the major classes of algorithms where the research had been focused are evolutionary algorithms and swarm-based algorithms. Evolutionary algorithms try to mimic the process of natural evolution. Some of the algorithms which belong to the class of evolutionary algorithms are genetic algorithms, genetic programming, evolutionary strategies, evolutionary programming.

Swarm intelligence is based on group behaviour of simple individuals (swarms) in which individuals independently may not show intelligence but as group they show intelligent behaviour. Swarm intelligence is intelligent behaviour shown by a system which emerges due to cooperative interaction of components of the system to achieve a goal which may not be achievable by individual efforts. Ant colony optimization, particle swarm optimization, and artificial bee colony algorithm are among some of the most popular swarm intelligence-based techniques used to solve optimization problems.

Evolutionary algorithms and swarm intelligence have been quite successful in solving many engineering applications having highly nonlinear, nonconvex, non-differentiable, and multimodal models, and a variety of modifications have been proposed to tackle MINLPs. Many algorithms based on EAs have been effectively used to find the solution of MINLP problem [2, 3].

Lin [4] designed a mimetic algorithm combined with an evolutionary Lagrange method for solving MINLPs. Lin et al. [5] proposed a hybrid differential evolution method to solve MINLPs. The method works with two phases called accelerated phase and migrating phase used to maintain exploration and exploitation. Later a modified coevolutionary hybrid differential evolution for MINLPs was suggested by Lin et al. [6]. Yan et al. [7] introduce a memory-based lineup competition algorithm having cooperation and bi-level competition mechanism for exploration and exploitation. Xiong et al. [8] introduced a hybrid genetic algorithm for finding a globally compromise solution of a mixed-discrete fuzzy nonlinear programming. Cheung et al. [9] developed a hybrid algorithm which combines genetic algorithm and grid search to solve MINLP.

Cardoso et al. [10] presented a modified simulated annealing (M-SIMPSA). This method uses combination of simulate annealing and Nelder and Mead simplex method in the inner loop and Metropolis algorithm ([11–13]) in the outer loop. Some of the other population-based algorithms used to solve MINLP are simulated annealing technique [10, 14], tabu search method [15], multistart scatter search [16], and particle swarm optimization [17].

Among the above-discussed methods, genetic algorithms (GAs) [3, 9, 18, 19] are the most successful. GA is an iterative process that works with a set of the solutions (population) which are modified by genetic operators to guide the solution towards the optimum solution in the search space. Crossover and mutation are one of the essential operators of GA. Crossover helps in exploring the promising zones of the search space, using the information from chromosomes (solutions), and mutation assists in avoiding premature convergence by preserving sufficient disparity within the population. This work extends the recently developed RCGA, BEXPM by Thakur et al. [1], named as "MI-BEXPM". The performance of MI-BEXPM is analysed and compared with other algorithms on the basis of twenty test problems as well as real-life problems.

The rest of paper is organized as the following: Sect. 2 proposed the algorithm to find the solution of the mixed-integer optimization problem. The experimental setup used in current study is detailed in Sect. 3. Analysis of the results for twenty test problems and discussion is given in Sect. 4. The efficiency of MI-BEXPM for solving real-life mixed-integer optimization problems is analysed and compared with other algorithms in Sect. 5. Finally, conclusions from the comparative study are drawn in Sect. 6.

2 Proposed GA

GA belongs to a class of population-based iterative algorithms, which tries to find the near global optimal solution of an optimization problem. The search in GA is governed by three main genetic operators, viz. selection, crossover, and mutation. These operators are applied iteratively to direct the search during the evolution of the population during the search process. We will discuss about the operators used in this study in the following subsections.

2.1 Selection

Selection operator works on the principle of the survival of fittest. It is used to discard the inferior individuals from the population and filter relatively better fit individuals to participate in the biological evolution process. After applying selection operator, an intermediate pool of the population (mating pool) is constructed. Lot of selection techniques have been proposed in the literature. Some of the popular selection operators are ranking [20], roulette wheel [21], stochastic uniform sampling (SUS) [22], and tournament [20] which are widely used selection operators. We have employed tournament selection operator in this work. Tournament selection selects a subset of the population randomly and conducts a fitness-based competition among the chosen solutions. The cardinality of this subset is called tournament size. The winner of the tournament becomes a part of mating pool. The process is repeated until the cardinality of this mating pool becomes equal to the population size.

2.2 Crossover

Crossover operator in GA mimics the process of chromosomal crossover in biology to produce the recombinant chromosomes. Individuals from mating pool are randomly chosen to participate in the crossover process with certain probability called crossover probability (p_c). Here, we used BEX crossover [1] to produce a pair of offspring solutions within the variable bounds from a pair of parent solutions lying within the variable bounds. BEX crossover is a parent-centric operator and has one scale parameter λ. For small (large) values of λ, offspring produced are spread near (away) from the parents. Also for a fixed value of λ, the spread of child chromosome is proportional to the that of the parent solutions. Steps to generate child chromosomes C_1 and C_2 using parent chromosomes P_1 and P_2 via BEX crossover operator are as follows:

1. Randomly choose two parent chromosomes P_1 and P_2 from the mating pool (population after employing GA operator).
2. Randomly generate a uniform number $r_c \in (0, 1)$. If r_c is less than prescribed crossover rate p_c, then crossover is applied to P_1 and P_2, otherwise it will pass as such for mutation.
3. If $r_c < p_c$, then C_1 and C_2 is produced using Eqs. (1) and (2)

$$C_1 = P_1 + \gamma_1 |P_2 - P_1| \tag{1}$$

$$C_2 = P_2 + \gamma_2 |P_2 - P_1| \tag{2}$$

where

$$\gamma_j = \begin{cases} \lambda \, ln\left\{\exp\left(\frac{B_l - P_j}{\lambda|P_2 - P_1|}\right) + u\left(1 - \exp\left(\frac{B_l - P_j}{\lambda|P_2 - P_1|}\right)\right)\right\} & \text{if } p \leq 0.5 \\ -\lambda \, ln\left\{1 - u\left(1 - \exp\left(\frac{B_u - P_j}{\lambda|P_2 - P_1|}\right)\right)\right\} & \text{if } p > 0.5 \end{cases}$$

for $j \in \{1, 2\}$, $u, p \in (0, 1)$ are random numbers following uniform distribution, $\lambda > 0$ is a scaling parameter, $B_l = \{B_l^1, B_l^2, \ldots, B_l^n\}$ and $B_u = \{B_u^1, B_u^2, \ldots, B_u^n\}$ are lower and upper bound of the decision variable.

2.3 Mutation

It is inspired from the biological mutation in which changes in a DNA sequence occurs by altering one or more gene values of the chromosome. Mutation operator in GA is applied for minimizing the possibility to stick into the local or suboptimal solution. It gives a small random perturbation to explore the neighbourhood of the current solution. Not all the individuals go through the mutation phase. It is applied with a relatively small probability (p_m) called probability of mutation and tries to give a random drift to solution to be in a promising zones of the search space. Here, power mutation [23] is being used for this purpose. The search power of mutation is controlled by index parameter (p). Larger (smaller) the value of p has higher (smaller) possibility to introduce perturbation in the muted solution. The probability of generating mutated solution on either side is proportional to its relative position from the variable bounds of the decision variable. The muted solution (x_i^{k+1}) from the current solution (x_i^k) is produced as follows:

$$x_j^{k+1} = \begin{cases} x_j^k - t_j\left(x_j^k - L_j\right) & \text{if } \frac{x_j^k - L_j}{U_j - L_i} \leq s \\ x_j^k + t_j\left(U_j - x_j^k\right) & \text{Otherwise} \end{cases}$$

Here, k refers to the current generation, $s \in (0, 1)$, t_j ($j \in \{1, 2, \ldots, n_v\}$; $n_v = \#$ of decision variables) are random numbers which follow uniform distribution and power distribution, respectively.

2.4 *Truncation Technique*

In this work, truncation technique based on floor and ceiling function is applied to each $x_i \in I$ (here, I refer to the set of variables having integer restrictions). It helps in maintaining the randomness within the newly generated population and reduces the chance of producing similar integer value for same real values that lies within two similar successive integer values [18, 24].

$$x_i^{k+1} = \begin{cases} \lfloor x_i^k \rfloor & \text{if } p \leq 0.5 \\ \lceil x_i^k \rceil & \text{Otherwise} \end{cases}$$

where $p \in (0, 1)$ is a uniform random number.

2.5 *Constraint Handling Technique*

Due to non-requirement of penalty parameter unlike other penalty constraint handling techniques, parameter-free penalty (PFP) method suggested by Deb [25] is applied to handling constraints. This approach can be easily embedded and executed while evaluating the fitness function during the search process. The fitness function using PFP is evaluated as follows

$$\Theta(z) = \begin{cases} G(z); & \text{if } z \text{ is feasible} \\ G_{\max}(z) + \sum_{k=1}^{r_1} (\Gamma_k(z)) + \sum_{k=1}^{r_2} |\zeta_j|; & \text{otherwise} \end{cases}$$

where $G_{\max}(z)$, Γ_k, r_1, ζ_j, and r_2 are the worst feasible value, inequality constraint, # of inequality constraints, equality constraint, and # of equality constraints, respectively.

3 Experimental Setup

The test bed selected for the comparative study consists of a number of real and/or integer decision variables having linear and/or nonlinear inequality constraints. The best/optimum solutions reported in the literature are summarized in Table 1.

Table 1 Problems considered for study

Problem	Variables	Objective function	Inequality constant	Global optimum value	References
1	2(1+1+0)	Linear	2(1+1)	2	[3, 10, 18, 26]
2	2(1+1+0)	Nonlinear	1(0+1)	2.124	[3, 10, 18]
3	3(2+1+0)	Quadratic	3(2+1)	1.07654	[3, 10, 18, 26]
4	2(2+0+0)	Cubic	2(0+2)	−6961.81	[18]
5	3(0+3+0)	Quadratic	2(1+1)	−68	[18]
6	4(0+4+0)	Quadratic	1(1+0)	−6	[10, 18]
7	3(2+1+0)	Nonlinear	4(2+2)	99.24521	[3, 10, 18]
8	7(3+4+0)	Nonlinear	9(5+4)	3.557463	[3, 10, 18, 26]
9	5(3+2+0)	Quadratic	3(0+3)	32217.4	[3, 10, 18]
10	8(0+8+0)	Nonlinear	3(3+0)	0.94347	[10, 18]
11	5(0+5+0)	Quadratic	6(6+0)	8	[18]
12	7(0+7+0)	Nonlinear	6(3+3)	14	[18]
13	2(0+2+0)	Nonlinear	2(2+0)	−42.632	[18]
14	3(1+2+0)	Nonlinear	0(0+0)	0	[18, 24]
15	5(0+5+0)	Quadratic	8(8+0)	807	[18, 24]
16	40(0+40+0)	Linear	3(3+0)	1030361	[18, 24]
17	40(20+20+0)	Linear	3(3+0)	1030361	[18, 24]
18	100(0+100+0)	Nonlinear	2(2+0)	3.03E+08	[18, 24]
19	100(50+50+0)	Nonlinear	2(2+0)	3.03E+08	[18, 24]
20	8(4+4+0)	Nonlinear	3(0+3)	0.999955	[18]
Gear train	4(0+4+0)	Nonlinear	0	–	[27]
Reinforced concrete beam	3(1+1+1)	Nonlinear	0	–	[28]
Speed reducer	7(6+1)	Nonlinear	0	–	[29]
Welded beam (A)	4(3+0+1)	Nonlinear	0	–	[30]
Welded beam (B)	6(1+1+4)	Nonlinear	0	–	[31]

Table 2 Parameter settings of MI-BEXPM

Parameter	Values
Crossover rate	0.9
Mutation rate	0.009
Crossover index for real variables	0.35
Mutation index for real variables	10.0
Crossover index for integer variables	0.65
Mutation index for integer variables	4.0
Elitism size	1
Tournament size	2

As discussed in [18], each problem is run 100 times with distinct initial population. Here, successful run is that run whose objective function value lies within 1% range of the reported best/optimal solution. For each problem percentage of successful runs (ps), average function evaluations of successful runs (avg) are measured.

$$ps = \frac{T_s * 100}{T_r} \quad (3)$$

$$\text{avg} = \frac{T_f}{T_s} \quad (4)$$

Here, T_s = total # of successful runs, T_r = total runs, and T_f = sum of function evaluations of successful runs.

The parameter settings of MI-BEXPM algorithm used to conduct this experiment are shown in Table 2. Population size = 10 times the total number of decision variable for each considered problem except 16, 17, and 18 problems (chosen to be three times).

4 Results and Discussions

Results observed using MI-BEXPM are compared with MILXPM, RST2ANU, and AXNUM on the basis of ps and avg for twenty problems. Table 3 demonstrates successful runs for each problem corresponding to each algorithm for twenty test problems. MI-BEXPM shows 100% success rate in twelve problems, whereas MILXPM, RST2ANU, and AXNUM show 100% success rate only in ten, eleven, and eight problems, respectively. Moreover, the minimum success rate of MI-BEXPM to solve problem is greater than 50% for each problem, while MILXPM, RST2ANU (unsuccessful to obtain optimal solution within 100 runs for 4th problem), and AXNUM have two, seven, and six problems, respectively, which have less than 50% success rate.

Table 3 Percentage of successful runs and average function evaluations of successful runs—twenty problems

Problem No.	Percentage of successful runs				Average function evaluations of successful runs			
	MI-BEXPM	MILXPM	RST2ANU	AXNUM	MI-BEXPM	MILXPM	RST2ANU	AXNUM
1	85	84	47	86	135	172	173	1728
2	89	85	57	67	100	64	657	82
3	68	43	4	35	813	18608	221129	65303
4	80	95	2	82	221	10933	1489713	45228
5	100	100	75	95	84	671	2673	13820
6	100	100	100	100	80	84	108	432
7	57	59	0	45	916	7447	–	16077
8	71	41	15	3	2064	3571	180859	1950
9	100	100	100	100	100	100	189	4946
10	100	93	100	33	160	258	545	700
11	95	100	100	97	275	171	2500	863
12	87	71	29	19	1159	299979	6445	380115
13	100	99	100	91	40	77	35	456
14	100	100	100	100	70	78	214	1444
15	100	92	19	9	11067	2437	3337	267177
16	100	100	100	100	240	1075	1114	2950
17	100	100	100	100	240	1073	1189	3016
18	100	100	100	100	600	600	2804	600
19	100	100	100	100	600	600	1011	600
20	100	100	100	100	160	250	697	256

MI-BEXPM, MILXPM, and AXNUM show equivalent success rate for Problem 1, but MI-BEXPM has lesser average number of function execution (shown in Table 3), while in problem-4 and problem-7 MILXPM outperforms better in terms of success rate but MI-BEXPM performs better in average number of function evaluation than other algorithms. And for problem-11, MILXPM performs better than other algorithms in terms of both *ps* and avg.

For overall performance comparison of MI-BEXPM for the chosen test suite, performance index (PI) is applied, suggested by Bharti [32] and used by several to compare the performance of the algorithm [18, 23, 24, 33]. It is based on successful run, average number of function evaluations, and average time of execution of successful runs. In the current study, all the algorithms considered to be compared are not run on same machine so it is absurd to consider time. Consequently

$$\mathrm{PI} = \frac{1}{P_n}\sum_{i=1}^{P_n}(k_1 s_1^i + k_2 s_2^i) \tag{5}$$

where s_1^i and s_2^i is the ratio of T_s to T_r and minimum of average function evaluation among algorithm to the average function evaluation of each algorithm, respectively, for the *i*th problem. k_j is the weight correspond to each $s_j, j = \{1, 2\}$ such that $\sum_{j=1}^{2} k_j = 1$. Assume $\{k_1 = k\}$, then $k_2 = \{1 - k\}$. From Fig. 1, it can be observed that MI-BEXPM is superior than MILXPM, RST2ANU, and AXNUM in terms of *PI*.

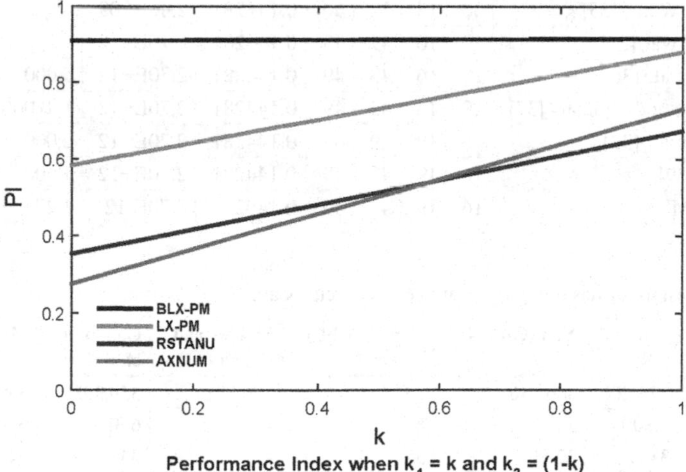

Fig. 1 Performance index

5 Application of MI-BEXPM

In previous section, we have demonstrated the effectiveness of MI-BEXPM for solving benchmark test problems. In this section, we further analyse the performance of MI-BEXPM on a set of mixed-integer real-life problems. The problems considered for this purpose are some of the popular problems available in the literature (stated in Table 1). The parameter setting of MI-BEXPM while solving these problems is kept same as discussed in Sect. 3.

The best results obtained by MI-BEXPM and those reported in the literature are shown in Table 4. From these results, it is observed that result obtained by MI-BEXPM is better than that reported in [27, 34, 35]. Cases where the results are similar to [31, 36–39], it is uses lesser function calls.

Table 5 refers to the results obtained by MI-BEXPM and reported in the literature. From these results, it can be easily observed that solution obtained by MI-BEXPM and reported by Gandomi et al. [30] are superior to rest of the algorithms considered. It is to be noted that the best solution obtained by MI-BEXPM and stated in Gandomi et al. [30] are same, but MI-BEXPM is able to solve this problem with lesser number of function evaluation than Gandomi et al. [30].

Table 4 Results of designing of gear train

Author(s)	β_1	β_2	β_3	β_4	Gear	Minimum	Function evaluation
Sandgren [27]	18	22	45	60	0.146667	5.70E−06	–
Kannan and Kramer [34]	13	15	33	41	0.144124	2.20E−08	–
Zhang and Wang [35]	30	15	52	60	0.14423	2.40E−09	–
Deb and Goyal [31]	19	16	49	43	0.144281	2.70E−12	–
Gandomi et al. [36]	19	16	43	49	0.144281	2.70E−12	5,000
Parsopoulos and Vrahatis [37]	19	16	43	49	0.144281	2.70E−12	100,000
Gandomi et al. [38]	16	19	49	43	0.144281	2.70E−12	2,000
Ali et al. [39]	16	19	43	49	0.144281	2.70E−12	1,500
MI-BEXPM	16	19	43	49	0.144281	2.70E-12	1,273

Table 5 Results of designing of reinforced concrete beam

Attributes	Amir [28]	Yun (GA) [40]	Yun (GA-FL) [40]	Montes and Ocaña [41]	Gandomi et al. [30]	MI-BEXPM
Best	374.200	366.146	364.854	376.298	359.208	359.208
$A(\beta_1)$	7.800	7.200	6.160	–	6.320	6.320
β_2	31	32	35	–	34	34
α_1	7.790	8.045	8.750	–	8.500	8.500
Γ_1	−4.201	−2.878	−3.617	–	−0.224	−0.224
Γ_2	−0.021	−0.022	0	–	0	0
Function evaluation	396	100,000	100,000	30,000	25,000	9,457

Tables 6, 7 and 8 demonstrates the best results obtained by MI-BEXPM and available in the literature. The solution obtained by MI-BEXPM for these problems is better among the feasible solutions considered in this study.

Table 6 Results of designing of speed reducer

Attributes	Ray and Saini [42]	Kuang et al. [29]	Mollinetti et al. [43]	MI-BEXPM	Gandomi et al. [36]	Akhtar et al. [44]	Montes et al. [45]
Best	2732.901†	2876.118†	2894.901†	2996.36	3000.981	3008.08	3025.005
β	17	17	17	17	17	17	17
α_1	3.514185	3.6	3.5	3.50001	3.5015	3.506122	3.506163
α_2	0.700005	0.7	0.7	0.7	0.7	0.700006	0.700831
α_3	7.497343	7.3	7.3	7.30006	7.605	7.549126	7.460181
α_4	7.8346	7.8	7.8	7.8	7.8181	7.85933	7.962143
α_5	2.9018	3.4	2.9	3.35024	3.352	3.365576	3.3629
α_6	5.0022	5	5.286683	5.28669	5.2875	5.289773	5.309
Γ_1	−0.0777	−0.0996	−0.07392	−0.07392	−0.0743	−0.0755	−0.0777
Γ_2	−0.2012	−0.2203	−0.198	−0.198	−0.1983	−0.1994	−0.2013
Γ_3	−0.036	−0.5279	−0.10795	−0.49918	−0.4349	−0.4562	−0.4741
Γ_4	−0.8754	−0.8769	−0.90147	−0.90147	−0.9008	−0.8994	−0.8971
Γ_5	0.5395	−0.0433	0.541785	−2.26E−05	−0.0011	−0.0132	−0.011
Γ_6	0.1805	0.1821	1.30E−07	−3.84E−06	−0.0004	−0.0017	−0.0125
Γ_7	−0.7025	−0.7025	−0.7025	−0.7025	−0.7025	−0.7025	−0.7022
Γ_8	−0.004	−0.0278	0	−2.86E−06	−0.0004	−0.0017	−0.0006
Γ_9	−0.5816	−0.5714	−0.79583	−0.79583	−0.5832	−0.5826	−0.5831
Γ_{10}	−0.166	−0.0411	−0.14384	−0.05133	−0.089	−0.0796	−0.0691
Γ_{11}	−0.0552	−0.0513	−0.198	−0.01085	−0.013	−0.0179	−0.0279
Mean	2758.888	–	2894.901	2996.38	3007.2	3012.12	3088.778
Worst	2780.307	–	–	2996.43	3009.0	3028.28	3078.592
SD	–	–	0.00E+00	0.01586	4.9634	–	–

†Values refer to the infeasible constraints

Table 7 Results of designing of welded beam (A)

Method	α_1	α_2	α_3	α_4	Γ_1	Γ_1	Γ_3	Γ_4	Γ_5	Best
MI-BEXPM	0.240	6.356	8.294	0.2443	0.2070	16.351	0.0038	0.0312	0.2342	2.390
Gandomi et al. [30]	0.201	3.562	9.041	0.2057	9800	−27.368	−0.0042	2210	−0.2355	1.73†

†Values refer to the infeasible constraints

Table 8 Results of designing of welded beam (B)

Method	α_1	α_2	α_3	α_4	α_5	α_6	Γ_1	Γ_2	Γ_3	Γ_4	Best
MI-BEXPM	0.187	1.6848	8.25	0.25	4	Steel	−3.80E +02	−4.02E +02	−0.2343	−0.0159	1.9418
Deb et al. [31]	0.187	1.6849	8.25	0.25	4	Steel	−3.80E +02	−4.02E +02	−0.2346	−0.1621	1.9422
Wang and Yin [46]	0.187	1.6842	8.25	0.25	4	Steel	−3.80E +02	−4.02E +02	−0.2343	2.4199	1.9421[†]

[†]Values refer to the infeasible constraints

6 Conclusions

In the present study, BEX-PM GA developed for continuous variable constrained optimization problems is modified to solve mixed-integer variables constraint optimization problems. The new variant of BEX-PM GA is named as MI-BEXPM. The results obtained are compared on a set of twenty mixed-integer constrained optimization benchmark problems mentioned in the literature. It is observed that MI-BEXPM outperforms other algorithms MILXPM, RST2ANU, and AXNUM individually in several problems. The overall performance is found to be better than these algorithms on the basis of performance index used extensively in the literature for such type of comparative studies.

Moreover, the performance of MI-BEXPM has been analysed for solving real-life mixed-integer optimization problems in comparison with other existed methods from the literature. Analysing the obtained results, MI-BEXPM not only performs well for test benchmark problems, also performs well for real-life problems.

From the above study, it can be concluded that MI-BEXPM is a promising algorithm among the class of algorithms considered in this study for solving test problems as well as real-life problems.

References

1. Thakur, M., S.S. Meghwani, and H. Jalota. 2014. A modified real coded genetic algorithm for constrained optimization. *Applied Mathematics and Computation* 235: 292–317.
2. Yokota, T., M. Gen, Y. Li, and C.E. Kim. 1996. A genetic algorithm for interval nonlinear integer programming problem. *Computers and industrial engineering* 31 (3): 913–917.
3. Costa, L., and P. Oliveira. 2001. Evolutionary algorithms approach to the solution of mixed integer non-linear programming problems. *Computers and Chemical Engineering* 25 (2): 257–266.
4. Lin, Y. 2013. Mixed-integer constrained optimization based on memetic algorithm. *Journal of applied research and technology* 11 (2): 242–250.
5. Lin, Y.C., K.S. Hwang, and F.S. Wang. 2004. A mixed-coding scheme of evolutionary algorithms to solve mixed-integer nonlinear programming problems. *Computers and Mathematics with Applications* 47 (8): 1295–1307.
6. Lin, Y.C., K.S. Hwang, and F.S. Wang. 2001. Co-evolutionary hybrid differential evolution for mixed-integer optimization problems. *Engineering Optimization* 33 (6): 663–682.

7. Yan, L., K. Shen, and S. Hu. 2004. Solving mixed integer nonlinear programming problems with line-up competition algorithm. *Computers and Chemical Engineering* 28 (12): 2647–2657.
8. Xiong, Y., and S.S. Rao. 2004. Fuzzy nonlinear programming for mixed-discrete design optimization through hybrid genetic algorithm. *Fuzzy Sets and Systems* 146 (2): 167–186.
9. Cheung, B.S., A. Langevin, and H. Delmaire. 1997. Coupling genetic algorithm with a grid search method to solve mixed integer nonlinear programming problems. *Computers and Mathematics with Applications* 34 (12): 13–23.
10. Cardoso, M.F., R. Salcedo, S.F. de Azevedo, and D. Barbosa. 1997. A simulated annealing approach to the solution of MINLP problems. *Computers and Chemical Engineering* 21 (12): 1349–1364.
11. Metropolis, N., A.W. Rosenbluth, M.N. Rosenbluth, A.H. Teller, and E. Teller. 1953. Equation of state calculations by fast computing machines. *The Journal of Chemical Physics* 21 (6): 1087–1092.
12. Kirkpatrick, S., M. Vecchi, et al. 1983. *Optimization by simulated annealing. science* 220 (4598): 671–680.
13. Kirkpatrick, S. 1984. Optimization by simulated annealing: quantitative studies. *Journal of Statistical Physics* 34 (5–6): 975–986.
14. Wah, B.W., Y. Chen, and T. Wang. 2007. Simulated annealing with asymptotic convergence for nonlinear constrained optimization. *Journal of Global Optimization* 39 (1): 1–37.
15. Exler, O., L.T. Antelo, J.A. Egea, A.A. Alonso, and J.R. Banga. 2008. A tabu search-based algorithm for mixed-integer nonlinear problems and its application to integrated process and control system design. *Computers and Chemical Engineering* 32 (8): 1877–1891.
16. Ugray, Z., L. Lasdon, J.C Plummer, F. Glover, J. Kelly, J., and R. Marti. 2005. A multistart scatter search heuristic for smooth NLP and MINLP problems. In *Metaheuristic optimization via memory and evolution*, 25–57. Springer.
17. dos Santos Coelho, L. 2009. An efficient particle swarm approach for mixed-integer programming in reliability redundancy optimization applications. *Reliability Engineering and System Safety* 94 (4): 830–837.
18. Deep, K., K.P. Singh, M. Kansal, and C. Mohan. 2009. A real coded genetic algorithm for solving integer and mixed integer optimization problems. *Applied Mathematics and Computation* 212 (2): 505–518.
19. Hua, Z., and F. Huang. 2006. An effective genetic algorithm approach to large scale mixed integer programming problems. *Applied Mathematics and Computation* 174 (2): 897–909.
20. Goldberg, D.E., and K. Deb. 1991. A comparative analysis of selection schemes used in genetic algorithms. *Urbana* 51: 61801–2996.
21. da Silva, E.L., H.A. Gil, and J.M. Areiza. 1999. Transmission network expansion planning under an improved genetic algorithm. In *Proceedings of the 21st 1999 IEEE international conference on power industry computer applications*, PICA'99, 315–321. IEEE.
22. Blickle, T., and L. Thiele. 1996. A comparison of selection schemes used in evolutionary algorithms. *Evolutionary Computation* 4 (4): 361–394.
23. Deep, K., and M. Thakur. 2007. A new mutation operator for real coded genetic algorithms. *Applied Mathematics and Computation* 193 (1): 211–230.
24. Mohan, C., and H. Nguyen. 1999. A controlled random search technique incorporating the simulated annealing concept for solving integer and mixed integer global optimization problems. *Computational Optimization and Applications* 14 (1): 103–132.
25. Deb, K. 2000. An efficient constraint handling method for genetic algorithms. *Computer Methods in applied Mechanics and Engineering* 186 (2): 311–338.
26. Floudas, C.A. 1995. *Nonlinear and mixed-integer optimization: fundamentals and applications*. Oxford University Press.
27. Sandgren, E. 1990. Nonlinear integer and discrete programming in mechanical design optimization. *Journal of Mechanical Design* 112 (2): 223–229.
28. Amir, H.M., and T. Hasegawa. 1989. Nonlinear mixed-discrete structural optimization. *Journal of Structural Engineering* 115 (3): 626–646.

29. Ku, K.J., S. Rao, and L. Chen. 1998. Taguchi-aided search method for design optimization of engineering systems. *Engineering Optimization* 30 (1): 1–23.
30. Gandomi, A.H., X.S. Yang, and A.H. Alavi. 2011. Mixed variable structural optimization using firefly algorithm. *Computers and Structures* 89 (23): 2325–2336.
31. Deb, K., and M. Goyal. 1996. A combined genetic adaptive search (GeneAS) for engineering design. *Computer Science and Informatics* 26: 30–45.
32. Bharti. 1994. Controlled random search technique and their applications. PhD thesis, Department of Mathematics, University of Roorkee, Roorkee, India.
33. Deep, K., and M. Thakur. 2007. A new crossover operator for real coded genetic algorithms. *Applied Mathematics and Computation* 188 (1): 895–911.
34. Kannan, B., and S.N. Kramer. 1994. An augmented lagrange multiplier based method for mixed integer discrete continuous optimization and its applications to mechanical design. *Journal of Mechanical Design* 116 (2): 405–411.
35. Zhang, C., and H.P. Wang. 1993. Mixed-discrete nonlinear optimization with simulated annealing. *Engineering Optimization* 21 (4): 277–291.
36. Gandomi, A.H., X.S. Yang, and A.H. Alavi. 2013. Cuckoo search algorithm: a metaheuristic approach to solve structural optimization problems. *Engineering with Computers* 29 (1): 17–35.
37. Parsopoulos, K.E., and M.N. Vrahatis. 2005. United particle swarm optimization for solving constrained engineering optimization problems. In *Advances in natural computation*, 582–591. Springer.
38. Gandomi, A.H., G.J. Yun, X.S. Yang, and S. Talatahari. 2013. Chaos-enhanced accelerated particle swarm optimization. *Communications in Nonlinear Science and Numerical Simulation* 18 (2): 327–340.
39. Ali, M.Z., N.H. Awad, P.N. Suganthan, R.M. Duwairi, and R.G. Reynolds. 2016. A novel hybrid cultural algorithms framework with trajectory-based search for global numerical optimization. *Information Sciences* 334: 219–249.
40. Yun, Y. 2005. Study on adaptive hybrid genetic algorithm and its applications to engineering design problems. Master's thesis, Waseda University.
41. Mezura-Montes, E., B. Hernández-Ocaña. 2009. Modified bacterial foraging optimization for engineering design. In *Intelligent engineering systems through artificial neural networks*. ASME Press.
42. Ray, T., and P. Saini. 2001. Engineering design optimization using a swarm with an intelligent information sharing among individuals. *Engineering Optimization* 33 (6): 735–748.
43. Mollinetti, M.A.F., D.L. Souza, R.L., Pereira, E.K.K. Yasojima, O.N. Teixeira, O.N. 2016. ABC+ES: Combining artificial bee colony algorithm and evolution strategies on engineering design problems and benchmark functions. In *Hybrid intelligent systems*, 53–66. Springer.
44. Akhtar, S., K. Tai, and T. Ray. 2002. A socio-behavioural simulation model for engineering design optimization. *Engineering Optimization* 34 (4): 341–354.
45. Mezura-Montes, E., C.A.C. Coello, and R. Landa-Becerra. 2003. Engineering optimization using simple evolutionary algorithm. In *Proceedings of 15th IEEE international conference on tools with artificial intelligence*, 149–156. IEEE.
46. Wang, J., and Z. Yin. 2008. A ranking selection-based particle swarm optimizer for engineering design optimization problems. *Structural and Multidisciplinary Optimization* 37 (2): 131–147.

Smart Healthcare Management Framework

D. Vidyadhara Reddy and Divakar Harekal

Abstract Healthcare IT has emerged as one of the basic necessities of life. Information technology could provide the solutions for various health ailments. It is done by processing numerous health parameters of the individual. These parameters could be measured using different electronics devices to keep up the health record and notify the individual and the doctor about any aberrations. Due to growing population and urbanization, various life style health problems are on the rise. Majority of the healthcare devices work in silos. Hence health vitals recorded remain locally stored and enhance the chances of data getting loss. By interconnecting these devices, we could maintain the health records centrally and derive insights. This could be incorporated with the advanced wireless technology. Smartphones could be leveraged for smart functionalities like voice recognition and Google services available. In this project, we are integrating the glucose meter monitors with the wireless communication to take advantage of the IoT technology and collect the blood sugar readings from the different individual, store them and provide the insights based on the health of the individual.

Keywords Healthcare · Communication system · Insights

1 Introduction

Healthcare systems, that measure, monitor and provide solutions for various types of health issues. With the help of these healthcare systems there has been a greater improvement in the quality of our life of each individual. Due to the advancement in electronics from past several years, various healthcare devices that help in

D. Vidyadhara Reddy (✉) · D. Harekal
Department of Computer Science, MS Ramaiah Institute of Technology, Bangalore, India
e-mail: vidhureddy02@gmail.com

D. Harekal
e-mail: divakar.h@msrit.edu

M.S. Reddy et al. (eds.), *International Proceedings on Advances in Soft Computing, Intelligent Systems and Applications*, Advances in Intelligent Systems and Computing 628, https://doi.org/10.1007/978-981-10-5272-9_28

monitoring and tracking our health parameters are developed, used, and have found considerable success.

Even though these devices help up in maintaining our health vitals they are independent in providing their support. Due to the aging of population, more and more of such devices are being used and they generate the huge health data that is stored individually and possibility of loss of this health data is high, which is critical.

The idea is to primarily integrate and maintain the health data obtained from the various heterogeneous healthcare devices centrally which are inter dependent to each other. Second, by integrating these devices and collecting all the health data we can provide the valuable insights to make it useful and easy for the doctors, caretakers or any health personals to monitor and manage the health of each individual.

With the advances in the networking technologies, the primary task can be easily fulfilled by the new and rigorous growing technology called Internet of Things. Internet of Things (IoT), as the name suggests, is the idea to connect various things (technically devices) to form a network where they can share their individual information. This Internet of Things reflecting its use cases in various fields of the technology can also step into the health care to provide the significant support in the integrating the healthcare devices. As the architecture of the Internet of Things varies with the scenario of the application in which it is implemented, health care can have its own framework to maintain the health data from the different health care devices.

As the data from the various sources grew to be called as a "big" data, also the various analytical methods were developed and implemented to process this structured and unstructured, unused data. The analytical tools integrate, select, and apply the various techniques such as predictive analysis, text analysis, and so on, to process the data into certain form which aid us in proper decision making. This data analysis can be applied on the big health data, which reflects the various health parameters of the individual, to process the health parameters at different times and provide the insights on the health vitals that support the doctors, caretakers, and other family persons to decide on future steps to be taken to maintain the individual's health.

2 Literature Survey

Nowadays people prefer to stay in homely comfort and expect their medications in their homely environment. This healthcare service provided by the organization maybe in the form of prescription, medications, or insights. Maintaining long-term healthcare treatment for the growing elderly population and escalating healthcare expenditure pose a challenge to the current healthcare system.

Due to advancement in the healthcare systems there are different ways to provide healthcare services to the people. In the healthcare information systems using agent technology [1], they have proposed the architecture that has many agents and these agents are used for obtaining the expertise results. The interface agent will filter the unwanted information provided in the report and will communicate the essential information the user. Doctor agent will receive test result report, receive unread test results alerts, informs about the available test results, query, and receive diagnosis suggestions. Prescription agent will electronically generate the prescription and will search for the availability of medications and will monitor the patient's medication and will eliminate patient's waiting time. Mobile agent will search, retrieve, and deliver data to the other agents or to the user. Lab agent will interface with the equipments and will provide real-time report on the examined. Diagnostic agent will select the method of treating the patient according to the data collected. Mobile device agent will provide an application where the patient's essential data can be entered. Home patient agent is a mediator who will do routine medications and will send the abnormal situations and receive emergency medications to the patient. Schedule agent will schedule the doctor's appointment for the patient and also will reschedule the time according to the emergencies. Electronic Health Record (HER) agent is an integrated database holding patient's relevant digital information.

The advancement in communication, data analysis, and information storage has decreased the mortality rate of the people. The health care is the challenging section for every country, in its social and economic state. The health care is the section where there will be more spiraling of cost for medications. Using IoT there are many devices that are connected, where the devices that connected to internet will exceed the number of people present on earth. The healthcare devices that are connected to the internet can be used by the people who are willing to require the homely comfort and also the doctor's assistance [2].

For the regular monitoring of the patient, these healthcare devices need to be connected to the internet. The IoT is revolutionizing networking in exponential rate over applications, including sensing, enhanced learning, e-health, and automatic applications. The architecture of H3 IoT is used because it is more convenient for e-health, where the homely environment for monitoring health status of the patients is easier [3, 4].

The H3 IoT architecture is confined to very small area where all the healthcare devices are connected to the centrally microcontroller, using Bluetooth or zigbee technology which has the communication range of only few meters and the devices out of this range cannot communicate.

Nowadays we find many patients who are bedridden, paralyzed, suffering from Alzheimer's disease, and they don't have anyone to look after them. They require assistance of any other people and even if they have someone to take care of them they need assistance for home health care. When the patient is in critical condition the people who are with the patient may not know what is actually happening with the patient, for example, high BP, low BP, extremes of sugar level, etc. These cannot be known by the assisting people. The healthcare devices need to be connected to the internet.

One of the approaches to monitor patient's health is by devices connected to android smartphone where there is an application which will monitor the health of the patients and that application will provide the caretaker the daily status of the patient through android smartphone. It has few liabilities like the user should have the android smartphone and the application to which the devices communicate with, and also the android smartphone or the healthcare devices may go out of range of communication [5].

There is much of the raw data related to the health care in today's world. This healthcare data originates from the various sources such as healthcare devices, institutions, and various agencies. They need to be processed with various analytical tools such as Mahout, Pig, and Hive, which can help in the better analysis. The health data should be carried through the analytical phases such as data standardization, data integration, data selection, data analysis, and data visualization. Standardization of the big health data is achieved with the help of standard terminology system such as SONMED, MESH, which yield the consistent interpretable data. Integration and selection can be done by achieved by the various approaches such as the data warehousing approach, virtual data integration, so on. Finally, the system needs to handle various algorithms for data analysis and provide various kinds of graphical data visualization to users [6].

3 Smart Healthcare Management Framework

Healthcare devices that measure and monitor various health vitals are helping the doctors, caretakers, and the individual himself to keep up health. But these devices are in silo and can manage the health vitals locally. On the other hand, the advancement in the wireless communication systems is making the significant progress with a goal to connect each and every thing or person through the evolving IoT. Healthcare can make use of these advances and make the healthcare devices remotely maintain the health data. Various architectures are proposed to connect the healthcare devices which are either dependent on the fixed control system to transmit data or data feeding is done manually. This framework makes the healthcare devices smart by enhancing their independent functionality to communicate with the central system. The central system may be specific to the health vital or to all of the health vitals of the individual. This overcomes the dependency of the healthcare devices on the external device for communication across the internet. The block diagram of the framework is shown in Fig. 2. The overview of the framework explains the different platforms on which the framework works and services carried out on the client/server model (Fig. 1).

The framework aims at making all the healthcare devices individually remote by equipping the devices with the separate communication system, which communicates with the central server that maintains all the health parameters of the individual person which help in maintaining their health in the far better way.

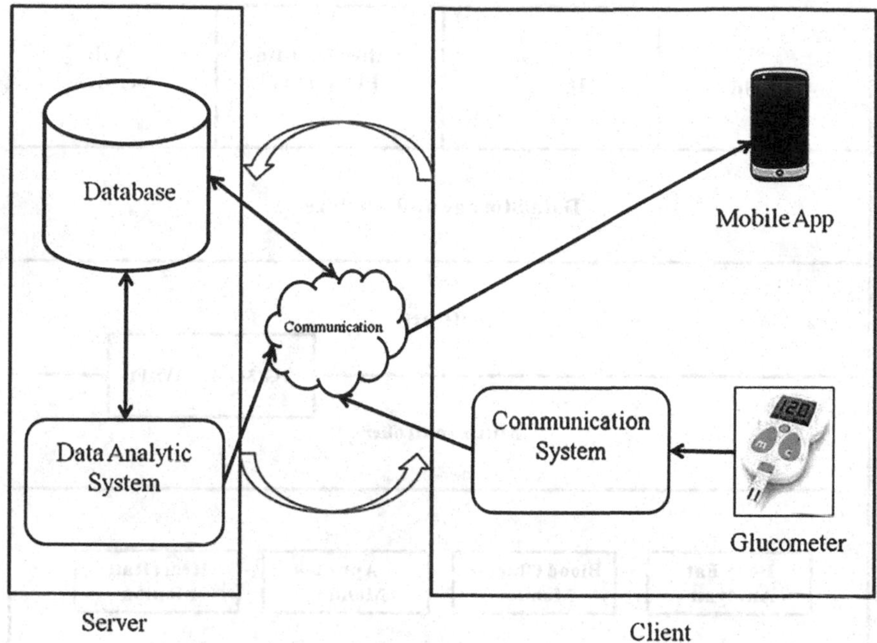

Fig. 1 Overview of the framework

This central system provides the easy access to the doctors and the caretakers with the person's health information and makes their task easy. The architecture proposed in this framework is simple and has five layers of operations. Each of the layers is independent of their platform on which they operate.

3.1 Physical Device Layer

All the healthcare devices that measure various health vitals are part of this layer. Each healthcare device such as blood glucose meter, digital thermometer, weighing scale, and so on, sense and measure different health parameters based on their functionality. This health data is sent to the above layers.

3.2 Physical Processing Layer

This layer is responsible for processing the data that is retrieved from the lower layer. Various microcontrollers with functionality to retrieve the data from the healthcare devices via serial communication or any possible communication

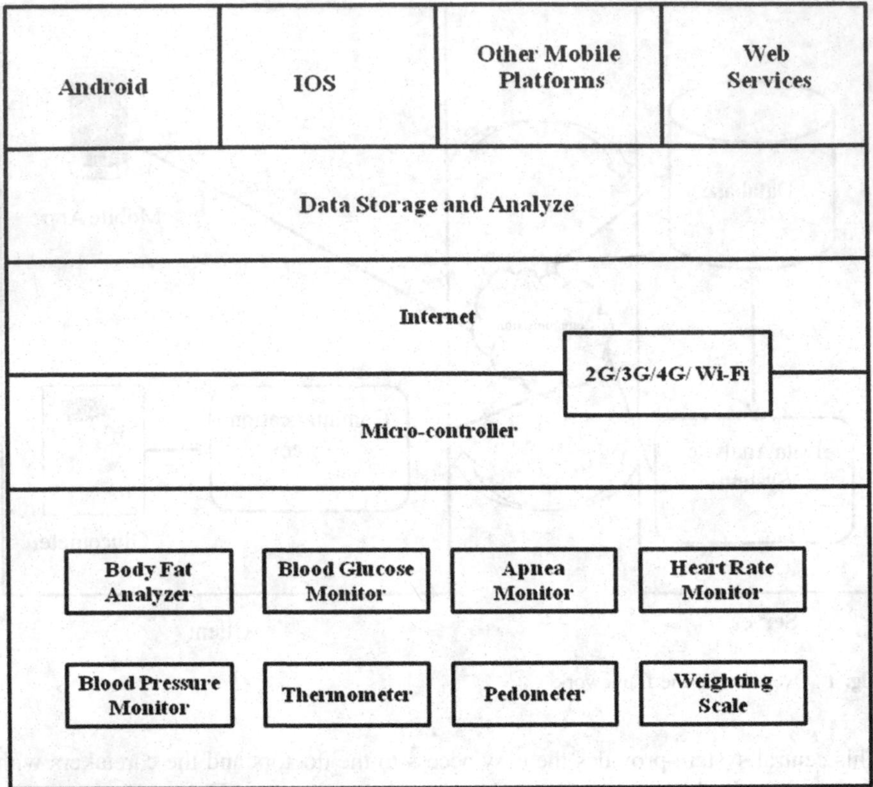

Fig. 2 Layered architecture of the framework

paradigm are being interfaced with devices in lower layer. These microcontrollers should also have the functionality to send the data to the central server through the 2G/3G using the GSM standard. This is the important layer in the framework which acts as the interface between the hardware and the software.

3.3 *Internet Layer*

Internet of Things, as the name implies, uses the internet as the network to exchange the data between the devices that are in various remote areas. This layer describes the services that act as carrier of the health data obtained from the different healthcare devices that are in the different networks. The health data obtained is transferred to the central server to provide the services of the above layers.

3.4 *Data Storage and Analyze*

Health data obtained to the server is to be refined and stored in a manner which can be helpful for analyzing the data in the future. Various RDBMS systems such as MySQL, Oracle database, and various cloud service providers such as Amazon S3, Google Drive can be used to store the data. Data that is stored is analyzed by using the various data analytical techniques, based on the requirement, and the insights on the data are being generated. These insights convey the status of the person's health parameters and provide suggestions on further diagnosis to gone through so as to keep up his health.

3.5 *Mobile Application Layer*

Various mobile application development platforms such as the Android, iOS, Windows, so on can be used to develop the mobile applications to provide the user with fast and easy access to his health status. These applications interact with the user to collect certain basic information from the user and provide the insights on his health. Web services are provided where the user can access his health status, in case if the user is not exposed the mobile applications.

4 Conclusion

Integration of the healthcare devices can be achieved using the various trending technologies like Internet of Things. By integrating, we can collect the health data from all the devices, store, and make the data available to various heath personals who are interested. Also we can provide the insights on the data using the analytics. This framework provides the layered approach of integrating the healthcare devices by making each of the devices to remotely communicate with the central server, which provides the functionality of structured data storage and data analysis to generate the insights on the health data.

References

1. Omar Al-Sakran, Hasan. 2015. *Framework Architecture for Improving Healthcare Information Systems Using Agent Technology*, 17. doi: 10.5121/ijmit.2015.7102.
2. Chiuchisan, Iuliana, Hariton-Nicolae Costin, and Oana Geman. 2010. *Adopting the Internet of Things Technologies in Health Care Systems*. 978-1-4244-8217-7110/$26.00 ©201O IEEE.
3. Ray, Partha P. 2014. *Home Health Hub Internet of Things (H3 IoT): An Architectural Framework for Monitoring Health of Elderly People*. 978-1-4799-7613-3/14/$31.00 ©2014 IEEE.

4. Memon, M., S.R. Wagner, C.F. Pedersen, F. Hassan AyshaBeevi, and F. Overgaard Hansen. 2014. Ambient Assisted Living Healthcare Frameworks, Platforms, Standards, and Quality Attributes. *Sensors* 14 (4312–4341): 2014.
5. Hennessy, Mark, Chris Oentojo, and Steven Ray. 2013. *A Framework and Ontology for Mobile Sensor Platforms in Home Health Management*. 978-1-4673-6333-4/13/$31.00 c 2013 IEEE.
6. Cha, Sangwhan, Ashraf Abusharekh, and Syed S.R., Abidi. 2015. *Towards a 'Big' Health Data Analytics Platform*. 978-1-4799-8128-1/15 $31.00 © 2015 IEEE.
7. Huang, Yinghui, and Guanyu Li. 2010. Descriptive Models for Internet of Things. In *International Conference on Intelligent Control and Information Processing* 13–15 Aug, 2010—Dalian, China, 978-1-4244-7050-1/10/$26.00 c2010 IEEE.
8. Pedro Diogo, Luís Paulo Reis, and Nuno Vasco Lopes. *Internet of Things: A System's Architecture Proposal*.
9. Syaifuddin, Muhammad, Kalaiarasi Sonai Muthu Anbananthen. 2013. *Framework: Diabetes Management System*. IMPACT-2013, 978-1-4799-1205-6/13/$31.00 ©2013 IEEE.
10. Chen, Jian-xun, Shih-Li Su, Che-Ha Chang. 2010. Diabetes Care Decision Support System. In *2010 2nd International Conference on Industrial and Information Systems*. 978-1-4244-8217-7110/$26.00 ©201O IEEE.
11. Claudio Cobelli, Chiara Dalla Man, Giovanni Sparacino, and Lalo Magni, Giuseppe De Nicolao, and Boris P. Kovatchev. 2009. Diabetes: Models, Signals, and Control. *IEEE Reviews in Biomedical Engineering* 2, 1937-3333/$26.00 © 2009 IEEE.
12. Kustanowitz, Jack. *Personal Medical Monitoring Devices*. CMSC 828.
13. Park, Chan-Yong, Joon-Ho Lim, and Soojun Park. 2011. *ISO/IEEE 11073 PHD Standardiza tion of Legacy Healthcare Devices for Home Healthcare Services*. 978-1-4244-8712-7/11/ $26.00©2011 IEEE.
14. Jung, Sang-Joong, Risto Myllylä, and Wan-Young Chung. 2013. Member, IEEE, Wireless Machine to-Machine Healthcare Solution Using Android Mobile Devices in Global Networks. *IEEE Sensors Journal* 13(5).
15. Bassi, Alessandro, Martin Bauer, Martin Fiedler, Thorsten Kramp, Rob van Kranenburg, Sebastian Lange, and Stefan Meissner. *Enabling Things to Talk: Designing IoT solutions with the IoT Architectural Reference Model*. Heidelberg, New York, Dordrecht, London: Springer. ISBN 978-3-642-40402-3.
16. Amit Sehgal, and Rajeev Agrawal. 2014. *Integrated Network Selection Scheme for Remote Healthcare Systems*. 978-1-4799-2900-9/14/$31.00 ©2014 IEEE.

Hybrid Nature-Inspired Algorithms: Methodologies, Architecture, and Reviews

Abhishek Dixit, Sushil Kumar, Millie Pant and Rohit Bansal

Abstract Evolutionary computation has turned into a significant problem-solving approach among several researchers. As compared to other existing techniques of global optimization, the population-based combined learning procedure, robustness, and self-adaptation are some of the vital topographies of evolutionary algorithms. In spite of evolutionary algorithms has been broadly acknowledged for resolving numerous significant real applications in various areas; however in practice, occasionally they carry only fringe performance. There is slight motivation to assume that one can discover an unvaryingly finest optimization algorithm for resolving all optimization problems. Evolutionary algorithm depiction is resolute by the manipulation and survey liaison retained during the course. All this evidently elucidates the necessity for fusion of evolutionary methodologies, and the aim is to enhance the performance of direct evolutionary approach. Fusion of evolutionary algorithms in recent times is gaining popularity owing to their proficiencies to resolve numerous legitimate problems such as, boisterous environment, fuzziness, vagueness, complexity, and uncertainty. In this paper, first we highlight the necessity for fusion of evolutionary algorithms and then we explain the several potentials of an evolutionary algorithm hybridization and also discuss the general architecture of evolutionary algorithm's fusion that has progressed all through the recent years.

A. Dixit (✉) · S. Kumar
Amity University, Noida, India
e-mail: abhishekdixitg@gmail.com

S. Kumar
e-mail: kumarsushiliitr@gmail.com; skumar21@amity.edu

M. Pant
Indian Institute of Technology, Roorkee, Uttarakhand, India
e-mail: millifpt@iitr.ac.in

R. Bansal
Rajiv Gandhi Institute of Petroleum Technology, Noida, India
e-mail: rohitbansaliitr@gmail.com

M.S. Reddy et al. (eds.), *International Proceedings on Advances in Soft Computing, Intelligent Systems and Applications*, Advances in Intelligent Systems and Computing 628, https://doi.org/10.1007/978-981-10-5272-9_29

Keywords Nature inspired computation (NIC) · Evolutionary computation (EC) Genetic algorithm (GA) · Differential evolution (DE)

1 Introduction

A study in computation science to find the procedures for the 'best' solutions is optimization. Optimization has been used widely in a various fields such as transportation, manufacturing, physics, and medicine [1, 2]. Problems associated with real-world optimization [3] are:

- Problems in differentiating between global optimal and local optimal solutions.
- Noise presence in evaluating the solutions.
- With the problem aspect, the 'jinx of dimensionality' roots the scope of the search space to propagate exponentially.
- Issues related with specified limitations and compulsion.

Various orthodox patterns of optimization have been suggested and established, and implementation/applications is also been observed, going from designing new drug or prediction of protein structure or to scheduling power system. These techniques are also facing problems in achieving the new emergent needs of current industry, where the prevailing optimization problems lean toward to be constrained, vigorous, multivariate, multi-objective and modalities [4, 5]. The existing and conventional optimization methods when attempt to highly nonlinear optimization task have been delimited by a frail global search capability, uncertainty, and inadequacy. Additionally with practical large-scale systems, various conventional optimization methodologies are not proficient to be implemented.

On the basis of qualities, we can divide optima into global or local optima. Figure 1 explains a belittlement issue $F \subseteq S$ in the viable search space. For minimization problems, we can define global and local optima as:

- Global Minima: objective function $f(x)$, solution $x^* \in F$ is global optima, if

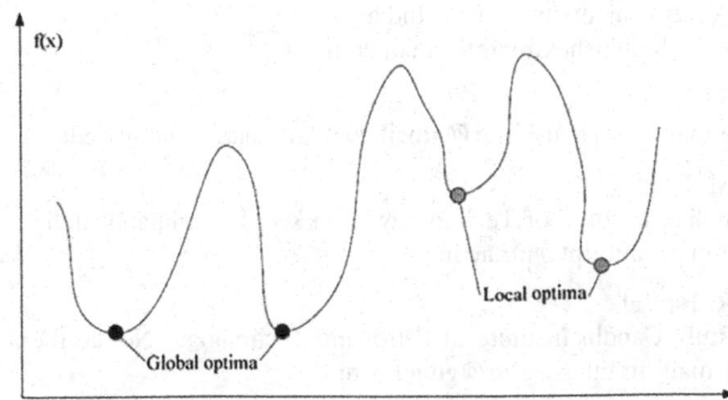

Fig. 1 Types of optima

$$f(x^*) < f(x) \because \forall x \in F \tag{1}$$

where $F \subseteq S$

- Local Optima: for the objective function $f(x)$ the solution $x_N^* \in \mathrm{N} \subseteq F$

$$f(x_N^*) < f(x) \forall x \in N \tag{2}$$

where $N \subseteq F$.

2 Evolutionary Algorithm

The development of evolutionary algorithms in last few years is very important amid all other search and optimization procedures. EAs [6] are a division of evolutionary computation, are a set of meta heuristics, population based used effectively in various applications having prodigious intricacy [7]. It success on solving difficult problems has been the engine of a field known as evolutionary computation (EC).

Evolutionary algorithms are based on the principles of adaptation and natural evolution of Charles Darwin [8] in 'On the Origin of Species'. Evolutionary algorithms (EAs) are pertinent to handle a varied kind of complex problems because of their ease, adequate tractability, and overall usability. There are few important characteristics which this evolutionary algorithm owns, and those can support them in belonging to the group of generated and test approaches:

- EAs are based on population, viz. these algorithms are a simultaneous collection of candidate solutions.
- These algorithms typically use amalgamation of one of more candidate solutions info into a current one.
- These algorithms are stochastic in nature.

A population to resolve for the optimization task is to be primed. Mutation and/or crossover are applied to generate new and different solutions. On the basis of resultant's fitness, appropriate selection approach is enforced to decide whether the selections of solutions are to be sustained into the subsequent generation. This process is then repeated to get the best fitness solution.

Most of the existent implementation of EA originates from any one of these 3 basic types: genetic algorithm (GA) [9], evolutionary programming (EP), and evolutionary strategies [10, 11]. In general, EAs is categorized by three facts:

- A set of solution candidate is sustained, which
- Goes through a selection procedure and
- Is operated by genetic operators typically recombination and mutation.

3 Hybrid Modal

It may not be adequate to find the preferred resolution of the problem from a simple and direct evolutionary algorithm for various problems. Previous studies suggest that a straight evolutionary algorithm might be unsuccessful to obtain a desired optimum solution for numerous types of problems which clearly demand the necessity for fusion of evolutionary algorithms with other probing optimization algorithms. Hybridization is usually done to mend:

- Performance of evolutionary algorithm
- Quality of the optimal results
- And as a part of higher structure, evolutionary algorithm can be integrated.

A number of studies in the past are being done on hybrid evolutionary optimization, and it has been observed the prodigious attainment of these techniques that can commendably conflict with their distinct shortcomings profiting from individual's strong points. The aims of developing fusion tactics are to handle very specific types of optimization problems. For instance, to resolve one of the most significant power system optimization glitches known as the unit commitment (UC) scheduling, a fusion of genetic algorithm (GA) and differential evolution (DE), termed hGADE has been proposed [12]. Fusion of the ACO, PSO, and 3-Opt algorithms can result in a hybrid evolutionary algorithm for solving traveling salesman problem [13]. Hybrid linkage crossover (HLX) is incorporated into differential algorithm (DE) to alleviate the disadvantages of DE and improve its performance [14]. To resolve multi-robot path planning, meta heuristic algorithms such as ACO-GA [15] and tree structure encoding-based hybrid EA [16] are used. Fusion of improved particle swarm optimization (IPSO) with an improved gravitational search algorithm (IGSA) has been proposed to determine the optimum route of the path for multi-robot in a muddle environment [17]. To precis; the hybridization inspiration is to improve reliability, conjunction hastening, and heftiness. Hybridization of evolutionary algorithm in general can be categorized into various categories as per the techniques or procedures, for example, hybridization motivation and hybridization design. To clarify, this can be divided into 'pre-processors and post-processors,' 'co-operators,' and 'implanted operators' built on the affiliation between all the nature-inspired computation methods involved. Essentially, a cautious and broad exploration of the taxonomy of hybridization would not only benefit us in gaining a profound considerate for the nature-inspired computation techniques but also elect the preeminent amalgamations for the optimization problems targeted.

Evolutionary techniques stake many resemblances, like adaptation, learning, and evolution. Alternatively, these techniques also have some distinctive dissimilarity, and individually have their individual benefits and drawbacks [18]. Advantages and disadvantages of evolutionary algorithms are summarized in Table 1. For example, differential algorithm (DE) requires high computational effort but the search results are effective.

Table 1 Advantages and disadvantages of the evolutionary algorithm

Evolutionary algorithm	Advantages	Disadvantages
DE	Active search	High computational work
PSO	Distribution of information	Hasty
MEC	Timely	Sluggish
SA	Heftiness	Long computation time
ACO	Pheromone-based exclusiveness	Over similarity
CSA	Multiplicity	Sluggish convergence speed
HS	Algorithm simplicity	Obsolete information

A. Motivation of Hybridization

Hybridization techniques are superior to standalone algorithm as these techniques have the competency of overpowering the shortcomings of standalone algorithms without losing their benefits. Hybridization of evolutionary algorithm can be done by several ways, and few of them are summarized below:

- The solutions can be created by problem-specific heuristics for the initial population.
- Obtained solutions from EAs can be enhanced by local improvement search procedure.
- Genotype solutions are mapped to phenotype solution by the algorithm and present the solution in indirect way.
- Problem knowledge is exploited by variation operator.

B. Hybridization Architecture

As represented in below diagram (Fig. 2), as per the nature of evolutionary algorithms we can divide hybrid nature-inspired algorithms into three groups.

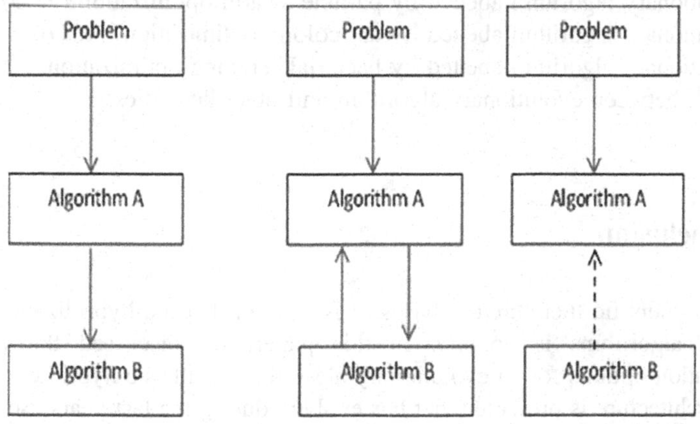

Fig. 2 Architectures of the hybrid NIC algorithms

- Pre-processors and post-processors: This type of hybridization NIC classification is the utmost prevalent and common, and in this type the optimization procedures are executed in sequence, i.e., results obtained by Algorithm 1 (pre-processor) are tweaked by Algorithm 2 (post-processor). For instance, fusion of algorithm based on particle swarm optimization (PSO) algorithm and ant colony optimization (ACO) algorithm called PSO-ACO is proposed to inherit their advantages and overcome their drawbacks [19].
- Co-operators (Fig. 2): This is the type of hybrid association where two algorithms are involved in simultaneous to adjust each other output as shown in Fig. 2. Both the algorithms share the common information during the process of search. As an example, fusion of fuzzy inference and the evolutionary computation technique is done for the approximation of nonlinear function. Parameter of both the algorithms can be enhanced, tweaked, and controlled by them [20, 21].
- Embedded operators (Fig. 2): These types of technique are categorized by their design and architectures, where one algorithm is entrenched inside another algorithm. Usually in this type of approach from different optimization techniques, the local search and global search are combined together so as to enhance the hybridization convergence. As an example, in [22–24], to detect premature detection of high quality solutions, the heuristic local search strategies are engaged in the algorithm and then to further refine the pheromone concentration or to generate the new solution, the given solution by local search strategy can be applied.

To enhance the generic efficiency of evolutionary algorithms, several different heuristics/meta heuristic procedures have been used. Following are the few of the common hybrid design architectures used:

- One evolutionary algorithm is hybridized with another.
- Evolutionary algorithm assisted by neural network.
- Evolutionary algorithm assisted by fuzzy logic.
- Evolutionary algorithm abetted by particle swarm optimization (PSO).
- Evolutionary algorithm abetted by ant colony optimization (ACO).
- Evolutionary algorithm abetted by bacterial foraging optimization.
- Fusion between evolutionary algorithm and other heuristics.

4 Conclusion

From the scientific literature/databases, it is evident that the hybridization of evolutionary algorithms is popular. In this paper, we discussed the numerous hybridization options for an evolutionary algorithm, and basic hybrid evolutionary design/architecture is presented that has evolved during the last years. Some of the popular and important hybrid architectures are also reviewed as reported in literature. In future the above discussed framework and architecture can be efficiently applied to develop hybrid NIC algorithm that will provide a various other ways of resolving the real-world problems more efficiently and quickly with accuracy.

References

1. Chan, K.Y., T.S. Dillon, and C.K. Kwong. 2011. Modeling of a Liquid Epoxy Molding Process Using a Particle Swarm Optimization-Base Fuzzy Regression Approach. *IEEE Transactions on Industrial Informatics* 7 (1): 148–158.
2. Tang, K.S., R.J. Yin, S. Kwong, and K.T. Ng. 2011. A Theoretical Development and Analysis of Jumping Gene Genetic Algorithm. *IEEE Transactions on Industrial Informatics* 7 (3): 408–418.
3. Spall, J.C. 2003. *Introduction to Stochastic Search and Optimization*. Hoboken, NJ: Wiley.
4. Deb, K. 2001. *Multiobjective Optimization Using Evolutionary Algorithms*. Chichester, UK: Wiley.
5. Coello, C.A.C., D.A.V. Veldhuizen, and G.B. Lamont. 2002. *Evolutionary Algorithms for Solving Multiobjective Problems*. Norwell, MA: Kluwer.
6. Eiben, A.E., and J.E. Smith. 2003. *Introduction to Evolutionary Computation*, 175–188. Berlin: Springer.
7. Fonseca, C., and P. Fleming. 1995. An Overview of Evolutionary Algorithms in Multiobjective Optimization. *IEEE Transactions on Evolutionary Computation* 3 (1): 1–16.
8. Jones, G., A. Robertson, C. Santimetvirul, and P. Willett. 1995. Non-hierarchic Document Clustering Using a Genetic Algorithm. *Information Research* 1 (1): 1.
9. Ahn, C. 2006. *Advances in Evolutionary Algorithms: Theory, Design and Practice*. New York: Springer.
10. Bäck, T., and H. Schwefel. 1993. An Overview of Evolutionary Algorithms for Parameter Optimization. *IEEE Transactions on Evolutionary Computation* 1 (1): 1–23.
11. Beyer, H.G., and B. Send Hoff. 2008. Covariance Matrix Adaptation Revisited—The CMSA Evolution Strategy. In *Lecture Notes in Computer Science*, 5199, 123–132.
12. Anupam, Trivedia, Dipti, Srinivasana, Subhodip, Biswasc, Thomas, Reindlb. 2015. Hybridizing Genetic Algorithm with Differential Evolution for Solving the Unit Commitment Scheduling Problem. *Science Direct Swarm and Evolutionary Computation* 23: 50–64.
13. Mahi, Mostafa, Ömer Kaan Baykan, and Halife Kodaz. 2015. A New Hybrid Method Based on Particle Swarm Optimization, Antcolony Optimization and 3-opt Algorithms for Traveling Salesman Problem. *Applied Soft Computing* 30: 484–490.
14. Cai, Yiqiao, and Jiahai Wang. 2015. Differential Evolution with Hybrid Linkage Crossover. *Information Sciences* 320: 244–287.
15. Oleiwi, B.K., H. Roth, and B.I. Kazem. 2014. A Hybrid Approach Based on ACO and GA for Multi Objective Mobile Robot Path Planning. *Applied Mechanics and Materials* 527: 203–212.
16. Ming-Yi Ju, Siao-En Wang, and Jian-Horn Guo. 2014. Path Planning Using a Hybrid Evolutionary Algorithm Based on Tree Structure Encoding. *The Scientific World Journal*.
17. Das, P.K., H.S. Behera, and B.K. Panigrahi. 2016. A Hybridization of an Improved Particle Swarm Optimization and Gravitational Search Algorithm for Multi-Robot Path Planning. *Swarm and Evolutionary Computation*.
18. Wolpert, D.H., and W.G. Macready. 1997. No Free Lunch Theorems for Optimization. *IEEE Transactions on Evolutionary Computation* 1 (1): 67–82.

19. Jing, Yang, Wen-Tao, Li, Xiao-Wei, Shi, Li, Xin, Jian-Feng, Yu. 2013. A hybrid ABC-DE Algorithm and Its Application for Time-Modulated Arrays Pattern Synthesis. *IEEE Transactions on Antennas and Propagation* 61 (11).
20. Yu-Jun, Zheng, Hai-Feng, Ling, Sheng-Yong, Chen, Jin-Yun, Xue. 2015. A Hybrid Neuro-Fuzzy Network Based on Differential Biogeography-Based Optimization for Online Population Classification in Earthquakes. *IEEE Transactions on Fuzzy Systems* 23 (4).
21. Jie, Li, Junqi, Zhang, ChangJun, Jiang, and MengChu, Zhou. 2015. Composite Particle Swarm Optimizer with Historical Memory for Function Optimization. *IEEE Transactions on Cybernetics* 45 (10).
22. Lixin, Tang, and Xianpeng, Wang. 2013. A Hybrid Multi-Objective Evolutionary Algorithm for Multi-Objective Optimization Problems. *IEEE Transactions on Evolutionary Computation* 17 (1).
23. Liangjun, Ke, Qingfu, Zhang, and R. Battiti. 2014. Hybridization of Decomposition and Local Search for Multiobjective Optimization. *IEEE Transactions on Cybernetics* 44 (10).
24. Sindhya, K., K. Miettinen, and K. Deb. 2013. A Hybrid Framework for Evolutionary Multi-Objective Optimization. *IEEE Transactions on Evolutionary Computation* 17 (4): 495–511.

Hybrid Computational Intelligent Attribute Reduction System, Based on Fuzzy Entropy and Ant Colony Optimization

P. Ravi Kiran Varma, V. Valli Kumari and S. Srinivas Kumar

Abstract Attribute reduction plays a crucial role in reducing the computational complexity and therefore the resource consumptions in the area of artificial intelligence, machine learning and computing applications. Rough sets are a very promising technique in attribute reduction or feature selection. Fuzzy and rough set hybrids have been proven to be more effective in selecting important features from the available data, particularly in the case of real-time data. There is a need for global searching strategies to find the best possible, minimal combination of features, and at the same time to maintain the originality of information. This paper proposes a hybrid computational intelligent attribute reduction system based on fuzzy entropy, fuzzy rough sets, and ant colony optimization, which do not depend on fuzzy dependency degree. Experimentation conducted on several UCI universal benchmark data sets proves this method to be feasible in obtaining minimal feature set with undisturbed or improved classification accuracy when compared to fuzzy entropy and dependency degree-based fuzzy rough quick reduct.

Keywords Rough sets · Fuzzy rough sets · Fuzzy entropy · Ant colony optimization · ACO · Feature selection · Attribute reduction

P. Ravi Kiran Varma (✉)
MVGR College of Engineering, Vizianagaram, AP, India
e-mail: ravikiranvarmap@gmail.com

V. Valli Kumari
Andhra University College of Engineering, Visakhapatnam, AP, India
e-mail: vallikumari@gmail.com

S. Srinivas Kumar
University College of Engineering Kakinada, JNTU Kakinada, Kakinada, AP, India
e-mail: samay_ssk2@yahoo.com

M.S. Reddy et al. (eds.), *International Proceedings on Advances in Soft Computing, Intelligent Systems and Applications*, Advances in Intelligent Systems and Computing 628, https://doi.org/10.1007/978-981-10-5272-9_30

1 Introduction

Computational intelligent systems often require a preprocessing stage called feature selection or attribute reduction. Feature reduction will help to optimize the computational complexity of the knowledge processing task; sometimes it also comes with additional benefit of improved accuracy, and also the volume of data features to be collected reduces. There is a continuous research in the area of feature selection [1] or attribute reduction; the main goal of which is to optimize the features at the same time retain the quality and originality of data objects. Rough sets [2, 3] are introduced by Pawlak and became very popular as a tool which mathematically deals with vagueness, lack of preciseness, and uncertainness in the knowledge extraction, and analysis of data. The advantage of rough sets is that it does not depend on any external inputs and also do not transform the existing data. It is best suitable for dimensionality reduction. Other applications of rough sets include rule generation and prediction [4]. Traditional rough sets [5] can be applied directly for discrete data, where there is a need for additional step of discretization for real-valued data. To deal with real-time data a combination of fuzzy systems and rough sets were utilized in several works [6–9] for the purpose of attribute reduction. While performing rough set reduction for real-time data, the process of discretization may incur loss of information for some applications [10]. To avoid such a loss, fuzzy–rough approach can be employed. The fuzzy–rough technique mandates fuzzification of data, which does not require any further information except the number of membership functions to be used for each attribute.

Fuzzy approximation is the basis of most of the literature in fuzzy–rough reduction [11]. Dependence degree-based reduct algorithm was employed in [6] applied to Web categorization. Fuzzy entropy was used in the fuzzy–rough reduction process in [10, 12] using a greedy approach. The work of Eric et al. [8] has used discernibility matrix-based approach for fuzzy–rough reduction. Jensen and Shen proposed fuzzy similarity-based approaches, which use fuzzy boundary region, fuzzy lower approximation, and fuzzy discernibility matrix-based approaches [7]. In [13], the authors have utilized principal component analysis as an algorithm for attribute reduction. Fuzzy–rough reduction aims at finding optimal set of attributes from the available list; however, finding the best combination is always a NP-hard problem. There is a need for some metaheuristic search techniques in order to find the best possible combination among the given attribute list [14]. Few examples of metaheuristic searching techniques are genetic algorithms [15], simulated annealing [16], cuckoo search [17], tabu search [18], etc. Ant colony optimization (ACO), bee colony optimization [19], particle swarm optimization [20], social cognitive optimization [21] fall under the category of swarm intelligence-based techniques [22]. ACO is a metaheuristic global search algorithm acquainted by Dorigo [23]. Jensen and Shen [24] have proposed ACO search for attribute reduction using fuzzy dependency degree, which in turn was derived from fuzzy-positive region, fuzzy lower approximation. Liangjun et al. [25] proposed ACOAR which is a rough set-based ACO and makes use of rough set dependency degree in the algorithm; however, it is mostly suitable for discrete data sets. In [14], Ravi Kiran Varma et al. proposed mutual information-based attribute reduction using ACO. ACO was proven to be fruitful in attribute reduction in these papers [26–28]. Hulian and Zong have done a similar work of attribute reduction using rough sets with swarm optimization [29]. Ant colony optimization was also used in the area of vehicular traffic control systems [30].

This paper aims at demonstrating the ability of fuzzy entropy-based fuzzy rough attribute selection using ant colony optimization as the global minima search technique, and the results were compared with similar works in this area [7, 10, 12]. A quick reduct algorithm was developed using fuzzy entropy without the need for fuzzy dependency degree, and also to avoid the disadvantage of local search, ACO was further employed to obtain a best reduct. A similar work is proposed by Varma et al. [31] where the authors applied fuzzy rough sets and ACO for eliminating unimportant features of real-time intrusion detection system (IDS) data set.

2 Back Ground

2.1 Fuzzy Rough Sets and Attribute Selection

In rough sets [2, 12], the approximations done to the original object set is crisp. In the case of fuzzy rough sets, the lower and upper approximations are fuzzy. For real-time data, discretization has to be performed on the original data before applying rough set reduction, and in this process, there may be loss of information in some cases. In fuzzy rough sets, the attributes are converted to fuzzy data where each attribute is divided into membership groups, thereby avoiding the information loss for real-time data [32]. Rough sets can be treated as a special case of fuzzy rough sets where the elements of the set generated by lower approximation have a membership of 1. However, in fuzzy rough sets, the elements of lower approximation contain a membership of [0, 1] and therefore can handle uncertainty in a flexible way than the crisp counterpart [10].

An information system $\mathbb{Z}$ can be represented as a set of finite objects $\mathbb{O}$ and a set of finite attributes $\mathbb{T}$, $\mathbb{Z} = (\mathbb{O}, \mathbb{T})$. The actual attributes $\mathbb{T}$ of the table consist of conditional as well as decision attributes, $\mathbb{T} = \{T_C \cup T_D\}$ [33]. Like the crisp equivalence class for rough set, fuzzy equivalence class is the important thing in the fuzzy rough set. In fuzzy rough set, the conditional and the decision features are fuzzy in nature. The definitions of fuzzy equivalence and fuzzy similarity relation, fuzzy lower and upper approximations can be found in [10].

Typically in fuzzy rough attribute selection procedures, the lower approximation or the positive region is widely used; the upper approximation is not employed. For any attribute $\mathscr{b}$, the fuzzy equivalence is denoted by $\mathbb{O}/IND(\mathscr{b})$. Say, an attribute has two fuzzy sets $\mathscr{b}_1$ *and* $\mathscr{b}_2$ based on two membership functions, then the partition $\mathbb{O}/IND(\mathscr{b}) = \{\mathscr{b}_1, \mathscr{b}_2\}$. In calculating the dependency degree for fuzzy rough reduction,

$\mathbb{O}/IND(\mathcal{W})$ or in short form $\mathbb{O}/\mathcal{W}$ has to be obtained. $\mathbb{O}/\mathcal{W}$ in the case of rough set is nothing but the set which contains objects that are indiscernible between the attribute t_1, t_2 given $\mathcal{W} = \{t_1, t_2\}$. In the case of fuzzy rough reduction, the set $\mathbb{O}/\mathcal{W}$ is based on the Cartesian product as shown below:

$$\mathbb{O}/IND(\mathcal{W}) = \mathbb{O}/\mathcal{W} = \otimes\{t \in \mathcal{W} \mid \mathbb{O}/IND(\{t\})\} \tag{1}$$

If $= \{l, m\}$ and $\mathbb{O}/IND(l) = \{l_1, l_2\}$ and $\mathbb{O}/IND(m) = \{m_1, m_2\}$ then

$$\mathbb{O}/\mathcal{W} = \{l_1 \cap m_1, l_1 \cap m_2, l_2 \cap m_1, l_2 \cap m_2\} \tag{2}$$

$$\mu_{E_1} \cap \mu_{E_2} \cap \cdots \cap \mu_{E_n}(o) = \min(\mu_{E_1}, \mu_{E_2}, \ldots, \mu_{E_n}). \tag{3}$$

3 Fuzzy Entropy-Based Quick Reduct

The information entropy, for an event E which consists of n outcomes, was proposed by Shannon [34] and is given as:

$$\mathbb{H}(E) = -\sum_{j=0}^{n} p_j \log_2 p_j. \tag{4}$$

Information system $\mathbb{Z}$ can be represented as a set of finite objects $\mathbb{O}$ and a set of finite attributes $\mathbb{T}$, $\mathbb{Z} = (\mathbb{O}, \mathbb{T})$. The actual attributes $\mathbb{T}$ of the table consist of conditional as well as decision attributes, $\mathbb{T} = \{T_C \cup T_D\}$. Let $\mathbb{F}_1, \mathbb{F}_2, \ldots, \mathbb{F}_n$ be the fuzzy membership subsets of an attribute, then the fuzzy entropy of a given membership function $\mathbb{F}_j$ is given by:

$$\mathbb{H}(\mathbb{F}_j) = -\sum_{Q \in \mathbb{O}/T_D}^{n} p(Q|\mathbb{F}_j) \log_2 p(Q|\mathbb{F}_j), \tag{5}$$

where the decision relative probability is given by

$$p(Q|\mathbb{F}_j) = \frac{|Q \cap \mathbb{F}_j|}{|\mathbb{F}_j|} \tag{6}$$

The fuzzy entropy given an attribute subset $\mathcal{W}$ is given by Parthalain et al. [12]:

$$\xi(\mathcal{W}) = \sum_{\mathbb{F}_j \in \mathbb{O}/\mathcal{W}} \frac{|\mathbb{F}_j|}{\sum_{X_j \in \mathbb{O}/\mathcal{W}} |X_j|} \mathbb{H}(\mathbb{F}_j). \tag{7}$$

Input: Tc & Td Conditional and Decision Attributes.

```
β ←{ }
Let k = a1 be the first attribute of Tc
∀ s ∈( Tc − a1)
    If ξ(s) < ξ(k)
      k = s
β ← {k}
do
{
    R ← β
    flag = False
    min = ξ(β)
    ∀ s ∈ (Tc − R)
    {
        if ξ( R ∪ { s } ) < min
        {
            min = ξ( R ∪ { s } )
            a = s
            flag = True
        }
    }
    If(flag = =True)
        β ← R ∪ { a }
} while(flag == True)
Output: β
```

Fig. 1 Fuzzy entropy-based quick reduct algorithm

This fuzzy entropy can be used to bubble out the best contributing attributes, similar to the fuzzy dependency degree. But, in the case of fuzzy entropy, the smaller the better, whereas in the case of fuzzy dependency degree, the larger the better. The earlier fuzzy entropy-based algorithms [10] used both dependency degree as well as fuzzy entropy; however, we propose a fuzzy entropy-based quick attribute reduction which does not need to calculate the fuzzy dependency degree and hence takes shorter time to compute the reduct and also with comparable or improved classification accuracy. The hill climbing approach-based fuzzy entropy quick reduction algorithm is as shown in Fig. 1.

3.1 *Fuzzy Entropy-Based Quick Reduct Algorithm (FEQR)*

See Fig. 1.

4 Attribute Reduction Using Fuzzy Entropy and Ant Colony Optimization

To overcome the disadvantage of hill climbing-based attribute reduction where there is no guarantee of global best solution, we propose a hybrid intelligent attribute reduction system which makes use of ACO to search the global best attribute reduction set. ACO is an evolutionary swarm-based computational intelligent algorithm proposed by Dorigo et al. [35, 36]. Here a colony of ants work together for a common goal of searching the global best solution. 'n' no of ants will be released for 'm' no of iterations, and for each iteration one ants solution will be marked as best solution and the path searched by the iteration best ant will be updated with pheromone trail similar to the biological ants. There is another parameter called the heuristic information, which will guide the ants to take a decision of selecting next node from the current node in the path toward a solution construction, with a probability calculation [37]. Therefore, there are two important values that are considered in the ACO solution construction, the pheromone and the heuristic information. In this work, we propose a relative fuzzy entropy-based heuristic information, in ant colony optimization in search of global best attribute set. Relative fuzzy entropy is used as heuristic information. An ant has to select next attribute y from present attribute x using the formula:

$$\eta(x, y) = \xi(S_{\bar{a}}) - \xi(S_{\bar{a}} \cup y) \tag{8}$$

$p^{\bar{a}}_{x,y}(\mathrm{ITR}_n)$ is the probability of selecting the next attribute y by the ant ā from current attribute position x, at nth iteration. $\tau_{x,y}$ is the pheromone concentration and $\eta_{x,y}$ is the heuristic information at the branch x, y. α and β are the tuning factors for pheromone and heuristic information, respectively.

$$p^{\bar{a}}_{x,y}(\mathrm{ITR}_n) = \frac{\tau^{\alpha}_{x,y}\eta^{\beta}_{x,y}(\mathrm{ITR}_n)}{\sum_{z \in (T_C - S_{\bar{a}})} \tau^{\alpha}_{x,z}\eta^{\beta}_{x,z}(\mathrm{ITR}_n)}, y \in (T_C - S_{\bar{a}}) \tag{9}$$

The proposed algorithm of ant colony optimization of fuzzy entropy-based fuzzy rough feature reduction is shown in Fig. 2.

```
Input: D_T = (U, T_C ∪ T_D, v, f), a Decision Table
Output: The minimum Feature Reduct, F_red its length L_red.
Fuzzify the Information Decision Table D_T = (U, T_C ∪ T_D,
v, f)
Let F_C ∪ F_D be the Fuzzified conditional and decision
attributes.
F_red ← { T_C } and L_red ← | T_C |
ITR = 0, be the iteration counter.
while( ITR < ITR_max)
{
    For every ant ā
    {
        F_ā ← { 0 }, L_ā = 0;
        if(ITR == 0)
            select attribute t_ā ∈ { T_C } randomly;
        else
            select attribute t_ā by eq.(9)
        F_ā ← { t_ā }, L_ā = 1;
        do
        {
            flag = FALSE , ⓓ = 0
            Let v_ā be the last attribute of F_ā
            for every attribute x_ā ∈ { T_C - F_ā }
            {
                calculate η(v_ā, x_ā) by eq.(8)
                if(η(v_ā, x_ā) > 0)
                ⓓ = ⓓ + [ η^β_{v_ā,x_ā} τ^α_{v_ā,x_ā} ]
            }
            if(ⓓ != 0)
            {
                select next attribute x_ā by eq.(9) using
                ⓓ as denominator.
            flag= TRUE;
            F_ā ← { F_ā ∪ x_ā }
            L_ā = L_ā + 1
            }
        }while(flag == TRUE);
        Pheromone Update:
        for every l,m ∈ F_ā
            τ(l,m) = τ(l,m) + ( q / |F_ā| )
        if (L_ā < L_red)
        {
            F_red ← F_ā;
            L_red = L_ā;
        }
    }
    ITR= ITR+1
    Pheromone Evaporation:
        for every l,m ∈ T_C
            τ(l,m) = τ(l,m) ρ
}
return F_red its length L_red
```

Fig. 2 Ant colony optimization of fuzzy entropy-based fuzzy rough reduction algorithm (ACOFEFR)

4.1 Ant Colony Optimization of Fuzzy Entropy-Based Fuzzy Rough Reduction Algorithm (ACOFEFR)

In this algorithm, a group of ants are released whose duty is to traverse the nodes (attributes) in search of best possible attribute set based on the pheromone and heuristic values. The same process is repeated for few iterations, which can be decided on trial and error basis. In each iteration, all the ants will find a solution, but the global best solution will be considered. In the algorithm, it can be observed that an ant will start with a node randomly in the first iteration, and select the next attribute by calculating the selection probability which in turn depends on heuristic and pheromone values, but there is no randomness from the second iteration onwards, instead they depend on selection probability function even for the first node. It was also observed from trial and error basis that the updating of pheromone for every ants solution produced better result than updating of pheromone per iteration. The evaporation of pheromone, however, is done only per iteration. Relative fuzzy entropy is used as heuristic information in this algorithm, which was shown in Eq. (8). The probability selection formula is shown in Eq. (9). The fuzzy entropy value plays a crucial role in the convergence of the solution by the ants.

5 Result Analysis and Discussion

Several experiments were conducted to provide support for the proposed algorithm, the UCI machine learning data sets are used as the standard benchmark data. All the experiments were conducted over a Windows 7-based i3 machine with 4 GB RAM, and Java program environment. C4.5 decision tree classifier [38] is used for evaluating the performance of proposed algorithms. The parameters used for the ACO algorithm are as shown in Table 1. Triangular membership functions were used here in fuzzification process. The solution provided by the algorithm depends on number of ants released and number of iterations the process of releasing ants will be carried. Initially the program was started with 10 ants and 10 iterations, later on trial and error basis, it was observed that optimal consistent solution was attained for three ants and five iterations for all the above data sets. The advantage of fuzzy entropy-based reduction is with real-valued data sets, and hence, the experiments were conducted on real valued or combination of real and discrete valued data sets are taken. Table 2 presents the experimentation results over 15 UCI data sets [39], column 2, is the name of the data set, the third column is the number of attributes of that data set, the fourth column contains the number of samples of that data set, the fifth column is the reduced attribute set obtained by the greedy hill climbing-based fuzzy entropy quick reduct (FEQR) algorithm, the sixth column gives the size of

Table 1 ACO parameters

Parameter	Value
α (heuristic tuning parameter)	1
β (pheromone tuning parameter)	0.01
ρ (evaporation constant)	0.9
q (updating constant)	0.1
ω (heuristic limiting constant)	0.001
Initial pheromone	0.5

reduced attributes obtained by the proposed ACOFEFR algorithm, the seventh and eighth columns are the list of attributes obtained by the FEQR and ACOFEFR algorithms, respectively, the ninth column is the classification accuracy obtained by J48 decision tree algorithm for the full attribute data sets, the tenth and eleventh columns are the classification accuracies obtained with the reduced attribute sets of the FEQR and ACOFEFR algorithms, respectively. It was very clear from the table that the ACO-based global searching strategy has a great deal of advantage over the greedy hill climbing fuzzy entropy-based quick reduct method (FEQR), which can be seen from the reduct sizes found by both the algorithms. Except for the case of Iris, Wine, yeast, and diabetes, the ACOFEFR has produced reduced attribute set compared to FEQR. The ACOFEFR for the case of Olitos data set has produced five attributes, whereas the FEQR has produced 15 attributes out of the total 25 attributes, at the same time keeping the classification accuracy unchanged. It can be observed that for most of the datasets, the classification accuracies with the reduced attribute sets generated by ACOFEFR are very close or even better when compared to the full attribute accuracies. The process of solution construction by one ant is shown in Fig. 3 as a sample, for the data set *diabetes*. The ant has started with attribute number three and is selecting the next attribute based on the selection probability, which indirectly depends on fuzzy entropy, and it can be seen that the ant has constructed a solution 3, 2, 6, which is nothing but the reduced attribute set.

Table 3 shows a comparison with similar works in this area, which have used fuzzy entropy-based algorithms for reduction. The proposed approach has produced better results in the case of Cleveland, Iris, Olitos, and Wine data sets with shorter attribute size and near classification accuracies. The '—' in the table 3 for Iris data set indicates that the result is not available. In the case of Iris data set our algorithm selected the best two attributes and yet maintaining the accuracy mark. In the case of wine data also, our approach has outperformed by selecting only four attributes out of 13. In the case of Olitos data, comparable accuracy was retained and also the attributes are reduced to five which is the best among others. For Cleveland data also the best attribute set was attained.

Table 2 Results over 15 UCI data sets

S. No	Data set	Full attr. size	No of objects	FEQR reduct size	Proposed algorithm reduct size (ACOFEFR)	List of attr. selected by FEQR	List of attr. selected by proposed algorithm (ACOFEFR)	Full attr. acc. (%)	FEQR acc. (%)	Proposed algorithm acc. (ACOFEFR) (%)
1	Ecoli (real)	7	336	6	5	1, 3, 4, 5, 6, 7	1, 3, 4, 5, 7	84.23	82.73	80.35
2	Diabetic_retinopathy (real)	19	1151	17	15	15, 3, 16, 14, 12, 13, 11, 9, 10, 5, 6, 17, 19, 2, 4, 1, 7	1, 2, 3, 4, 5, 6, 7, 8, 9, 10, 11, 16, 17, 18, 19	64.38	63.16	63.51
3	Heart (real)	13	294	10	08	9, 2, 3, 5, 6, 7, 1, 4, 13, 12	1, 2, 3, 4, 5, 6, 8, 10	77.89	80.95	78.23
4	Cleveland (real)	13	303	7	6	9, 12, 2, 7, 6, 3, 4	2, 3, 6, 7, 9, 13	52.14	57.75	53.47
5	Glass (real)	9	214	9	9	1, 2, 3, 4, 5, 6, 7, 8, 9	1, 2, 3, 4, 5, 6, 7, 8, 9	66.82	66.82	66.82
6	Iris (real)	4	150	2	2	3, 4	3, 4	96.00	96.00	96.00
7	Vehicle (real)	18	846	16	6	8, 18, 17, 16, 15, 14, 13, 12, 11, 10, 9, 7, 6, 5, 4, 3	8, 18, 17, 16, 15, 14	72.46	72.10	69.50
8	Olitos (real)	25	120	15	5	3, 2, 7, 20, 5, 4, 13, 23, 16, 19, 14, 9, 18, 11, 10	6, 11, 19, 20, 22	67.5	66.67	66.67
9	Wine (real)	13	178	4	4	7, 11, 12, 13	1, 7, 10, 13	93.82	91.57	94.94
10	Yeast (real)	8	1484	5	5	1, 2, 3, 5, 6	1, 2, 3, 5, 6	56.00	46.56	46.56
11	teachingAssistant (semi real)	5	151	4	3	1, 3, 4, 5	1, 3, 4	52.98	49.00	49.00
12	Badges2 (semi real)	10	294	10	1	1, 2, 3, 4, 5, 6, 7, 8, 9, 10	4	100	100	100
13	Abalone (real)	8	4177	7	6	5, 6, 3, 7, 8, 2, 1	5, 6, 7, 8, 2, 1	21.16	21.23	21.85
14	Diabetes (real)	8	768	3	3	2, 3, 6	2, 3, 6	73.82	74.35	74.35
15	Bupa (real)	6	344	5	4	1, 2, 4, 5, 6	1, 2, 5, 6	68.69	61.73	62.32

Ant : 1
Step : 1
Ant started: 3; Entropy: 0.9328212930736577
Attr: 1; Entropy: 0.9406430545480182
Attr: 2; Entropy: 0.8945825980306055
Attr: 4; Entropy: 0.9351936954092879
Attr: 5; Entropy: 0.9387060443941186
Attr: 6; Entropy: 0.9281874981290594
Attr: 7; Entropy: 0.9401106271565158
Attr: 8; Entropy: 0.9465781826234946

Step: 2
Ant covered: 3, 2; Entropy:
0.8945825980306055
Attr: 1; Entropy: 0.9120014019111715
Attr: 4; Entropy: 0.9013872870939019
Attr: 5; Entropy: 0.9050104912034126
Attr: 6; Entropy: 0.894109659097509
Attr: 7; Entropy: 0.9105647507853382
Attr: 8; Entropy: 0.9199695462125276

Step: 3
Ant covered: 3, 2, 6; Entropy:
0.894109659097509
Attr: 1; Entropy: 0.914138532221694
Attr: 4; Entropy: 0.9016891090478448
Attr: 5; Entropy: 0.9057505257650875
Attr: 7; Entropy: 0.9120336120975364
Attr: 8; Entropy: 0.9223383696914151
: Ant Solution: 3, 2, 6.

Fig. 3 Process of solution construction by one ant

Table 3 Comparison of proposed algorithm

S. No	Data set (real)	Original attr. size and full attr. acc. (%)	Proposed alg, reduct and acc. ACOFEFR (%)	[10] (%)	[7] (%)	[12] (%)
1	Heart	13, 77.89	8, 78.23	9, 78.52	8, 78.52	9, 80.37
2	Cleveland	13, 52.14	6, 53.47	10, 52.53	9, 54.55	10, 53.53
3	Glass	9, 66.82	9, 66.82	9, 69.63	9, 65.89	9, 65.89
4	Iris	4, 96.00	2, 96.00	3, 96.00	–	–
5	Olitos	25, 67.5	5, 66.67	8, 68.33	6, 63.33	6, 67.50
6	Wine	13, 93.82	4, 94.94	9, 93.26	6, 88.20	6, 94.94

6 Conclusion

This paper proposed an ant colony optimization-based fuzzy entropy attribute reduction algorithm, which produces global best solution when compared to greedy hill climbing-based approaches. The proposed algorithm was proven to be feasible and attaining comparable and sometimes better results for real-valued data sets in terms of both, number of attributes reduced and classification accuracies. This approach is suitable for real-valued data sets, since the fuzzification process preserves the originality of data in a much better way compared to discretization. Another feature of our algorithm is that it does not need to calculate fuzzy dependency degree. However, fuzzy-based attribute reduction suffers from high computational complexity when compared to non-fuzzy techniques like rough set reduction using discretized data.

References

1. Dash, M., and H. Liu. 1997. Feature Selection for Classification. *Intelligent Data Analysis* 1 (3): 131–156.
2. Pawlak, Z. 1982. Rough Sets. *International Journal of Computer and Information Sciences* 11 (5): 341–356.

3. Pawlak, Z. 2002. Rough Set Theory and Its Applications. *Journal of Telecommunications and Information Technology* 3 (2): 7–10.
4. Duntsch, I., and G. Gediga. 2000. Rough Set Data Analysis. In *Encyclopedia of Computer Science and Technology*, vol. 43(28), 281–301. ed. A. Kent and J. G. Williams.
5. Nguyen, H.S. 1998. Discretization Problem for Rough Set Methods. RSCTS98, LNAI 1424: 545–552.
6. Richard, J., and Q. Shen. 2004. Fuzzy Rough Attribute Reduction with Application to Web Categorization. *Fuzzy Sets and Systems* 141: 469–485.
7. Jensen, R., and Q. Shen. 2009. New Approaches to Fuzzy-Rough Feature Selection. *IEEE Transactions on Fuzzy Systems* 17 (4): 824–838.
8. Tsang, E.C., D. Chen, D.S. Yeung, X.Z. Wang, and J.W. Lee. 2008. Attributes Reduction Using Fuzzy Rough Sets. *IEEE Transactions of Fuzzy Systems* 16 (5): 1130–1141.
9. Chris, C., J. Richard, H. Germán, and S. Dominik. 2010. Attribute Selection with Fuzzy Decision Reducts. *Information Sciences* 180: 209–224.
10. Parthalain, N.M., R. Jensen, and Q. Shen. 2006. Fuzzy Entropy Assisted Fuzzy Rough Feature Selection. IEEE International Conference on Fuzzy Systems, 423–430.
11. Yueng, D.S., D. Chen, E.C.C. Tsang, J.W.T. Lee, and W. Xizhao. 2005. On the Generalization of Fuzzy Rough Sets. *IEEE Transactions on Fuzzy Systems* 13 (3): 343–361.
12. N.M. Parthalain, R. Jensen, and Q. Shen. 2008. Finding Fuzzy Rough Reducts with Fuzzy Entropy. IEEE International Conference on Fuzzy Systems, 1282–1288.
13. Ravi Kiran Varma, P., and V. Valli Kumari. 2012. Feature Optimization and Performance Improvement of a Multiclass Intrusion Detection System using PCA and ANN. *International Journal of Computer Applications* 44 (13): 4–9.
14. Ravi Kiran Varma P, V. Valli Kumari, and S. Srinivas Kumar. 2015. A Novel Rough Set Attribute Reduction Based on Ant Colony Optimization. *International Journal of Intelligent Systems Technologies and Applications* 14 (3/4), 330–353.
15. Holland, H.J. 1975. Adaptation in Natural and Artificial Systems. University of Michigan Press.
16. Kirkpatrick, S., D.C. Gelatt, Jr, M.P. Vecchi. 1983. Optimization by Simulated Annealing. *Science* 220 (4598): 671–680.
17. Yang, X.S., S. Deb. 2009. Cuckoo Search Via Levy Flights. IEEE World Congress on Natural and Biologically Inspired Computing, 210–214.
18. Glover, F. 1986. Future Paths for Integer Programming and Links to Artificial Intelligence. *Computers and Operations Research* 13 (5): 533–549.
19. Pham, D.T., A. Ghanbarzadeh, E. Koc, S. Otri, S. Rahim, and M. Zaidi. 2005. The Bees Algorithm. Technical Note, Manufacturing Engineering Centre, Cardiff University, UK.
20. J. Kennedy. 1997. The Particle Swarm: Social Adaptation of Knowledge. IEEE International Conference on Evolutionary Computing, 303–308.
21. Xie, X.-F., Z. Wen-Jun, Y. Zhi-Lian. 2002. Social Cognitive Optimization for Non Linear Programming Problems. International Conference on Machine Learning and Cybernetics (ICMLC), 779–783.
22. Blum, C., and A. Roli. 2003. Metaheuristics in Combinatorial Optimization. *ACM Computing Surveys* 35 (3): 268–308.
23. Dorigo M. 1992. Optimization Learning and Natural Algorithms. Italy: PhD thesis, Politecnico di Milano.
24. Richard, J., and S. Qiang. 2005. Fuzzy-Rough Data Reduction with Ant Colony Optimization. *Fuzzy Sets and Systems* 149: 5–20.
25. Liangjun, K., F. Zuren, and R. Zhigang. 2008. An Efficient Ant Colony Optimization Approach to Attribute Reduction in Rough Set Theory. *Pattern Recognition Letters* 29: 1351–1357.
26. Tamara, Q., A.A. Qasem, and A.S. Sawsan. 2012. A Reduct Computation approach based on Ant Colony Optimizatoin. *Abhath Al Yarmouk Basic Science and Engineering* 21 (1): 29–40.

27. Majdi, M., and E. Derar. 2013. Ant Colony Optimization Based Feature Selection in Rough Set Theory. *International Journal of Computer Science and Electonics Engineering* 1 (2): 244–247.
28. Jensen, R., and Q. Shen. 2003. Finding Rough Set Reducts with Ant Colony Optimization. In *Proceedings of the 2003 UK Workshop on Computational Intelligence*, 15–22.
29. Huilian, F., and Y. Zong. 2012. A Rough Set Approach to Feature Selection Based on Wasp Swarm Optimization. *Journal of Computational Information Systems* 8 (3): 1037–1045.
30. Jabbarpour, M., H. Malakooti, R. Noor, N. Anuar, and N. Khamis. 2014. Ant Colony Optimisation for Vehicle Traffic Systems: Applications and Challenges. *International Journal of Bio-Inspired Computation* 6 (1): 32–56.
31. Ravi Kiran Varma, P., V. Valli Kumari, and S. Srinivas Kumar. 2016. Feature Selection Using Relative Fuzzy Entropy and Ant Colony Optimization Applied to Real-Time Intrusion Detection System. *Procedia Computer Science* 85: 503–510.
32. Jensen, R., and Q. Shen. 2004. Semantics Preserving Dimensionality Reduciton, Rough and Fuzzy Rough Based Approaches. *IEEE Transactions on Knowledge Data Engineering* 16: 1457–1471.
33. Dubios, D., and H. Prade. 1992. *Putting Rough Sets and Fuzzy Sets Together, in Intelligent Decision Support*, 203–232. Dordrecht: Kluwer Academic Publishers.
34. Shannon, C.E. 1948. A Mathematical Theory of Communication. *Bell System Technical Journal* 27 (3): 379–423.
35. Dorigo, M., V. Maniezzo, and A. Colorni. 1996. The Ant System, Optimization by a Colony of Cooperating Agents. *IEEE Transactions on Systems, Man, and Cybernetics—Part B* 26 (1): 29–41.
36. Dorigo, M., and Caro, D.G. 1997. Ant Colony Algorithm for the Travelling Salesman Problem. *BioSystems*, 73–81.
37. Dorigo, M., and D.G. Caro. 1999. *Ant Colony Optimization Meta Heuristic, in New Ideas in Optimization*, 11–32. Mc Graw Hill.
38. Quinlan, J.R. 1993. *C4.5: Programs for Machine Learning*. Morgan Kaufmann Publishers.
39. Lichman, M. 2013. {UCI} Machine Learning Repository, University of California, Irvine, School of Information and Computer Sciences, 2013. [Online]. Available: http://archive.ics.uci.edu/ml. Accessed 2015.

Design and Development of ASL Recognition by Kinect Using Bag of Features

B.P. Pradeep Kumar and M.B. Manjunatha

Abstract In this paper, we proposed a methodology to recognize communication using gestures with Kinect sensor. The real issues are the tracker, and it is extremely touchy to light intensity variations since the measurement model is based on a color-based point search. The present system design faces the problems like a non-homogeneous background; the cloth color should not match with the skin color and having poor segmentation, single user. Proposed systems use aligning of depth map and RGB image and extract the features using bag of features. The system utilizes strong features using surf and classified by *k* means clustering for creating bag of features. To evaluate our system, we collected an American Sign Language (ASL) untrained data set which included approximately 1950 samples, while each samples were taken by Kinect sensor, and hence, included color, depth, and skeleton information. Analysis for various scenarios on our samples shows effectiveness of the proposed method for recognition at different distance using SVM.

Keywords American Sign Language · Bag of features · Surf · Kinect sensor

1 Introduction

As of late, the hand signal acknowledgment has turned into a noteworthy exploration challenge because of its huge use in human–computer interaction (HCI). As opposed to signals, a common segment of talked dialects and the gesture-based communications show the characteristic route for correspondence with hard of hearing individuals. In communication through signing, every motion as of now has relegated meaning, and solid guidelines of connection and language structure might

B.P. Pradeep Kumar (✉)
Jain University, Bangalore, India
e-mail: pradi14cta@gmail.com

M.B. Manjunatha
A.I.T, Tumkur, India
e-mail: manju.kari29@gmail.com

M.S. Reddy et al. (eds.), *International Proceedings on Advances in Soft Computing, Intelligent Systems and Applications*, Advances in Intelligent Systems and Computing 628, https://doi.org/10.1007/978-981-10-5272-9_31

be connected to make acknowledgment tractable. American Sign Language (ASL) is the dialect of decision for most hard of hearing in the United States. ASL utilizes roughly 6,000 motions for normal words and finger spelling for imparting dark words or formal people, places or things. Correspondence in American Sign Language (ASL) regularly depends on available shapes put in or moved crosswise over specific areas in respect to the endorser's body, notwithstanding developments of the head and arm, and outward appearance. In any case, appropriate names and words with no unitary sign are spelled, letter by letter, in English, and understudies of ASL regularly start their studies by taking in the 26 hand shapes that constitute the manual letter set. Gesture-based communications, i.e., dialects that basically pass on data by means of visual examples, regularly serve as an option or reciprocal method of human correspondence. Visual examples, rather than the sound ones utilized as a part of the oral dialects, are framed by hand shapes and manual or general body movement, lip developments, and outward appearances. Their expressiveness encourages human association and trade of data not just in the presence of listening to hindered individuals additionally in circumstances where discourse is unrealistic, e.g., in boisterous workspaces. Three issues ought to be settled to perceive gesture-based communication. The principal test is the dependable following of the hands trailed by powerful component extraction as the second issue. At long last, the third undertaking concerns the understanding of the fleeting component succession. The execution of the gesture-based communication can be partitioned into manual (hand introduction, area, and direction) and non-manual (head, mouth, and outward appearance) parameters. Once in a while, the utilization of manual parameters is sufficient to recognize a few signs; however, there are ambiguities in different signs which require non-manual data to distinguish them. Nianjun et al. [1] proposed a technique to observe every one of the 26 letters from start to finish by utilizing distinct HMMs topologies with various states. Nguyen et al. [2] proposed a continuous structure to perceive 36 hand vocabularies like American Sign Language (ASL) and digits in unrestricted situations. Their frameworks are utilized to contemplate and dissect hand stances, not the hand movement direction as in our framework. Yang et al. [3] presented an ASL acknowledgment framework in view of a period delay neural system. The greater part of the introduced works is exceptionally moving and has diverse fascinating ways to deal with overcome distinctive issues of gesture-based communication acknowledgment. A large portion of the presented frameworks is running in disconnected from the net mode, i.e., they gather the element arrangement and begin acknowledgment when the motion has as of now been performed.

The main contribution of this paper is a hand gesture recognition system to recognize American Sign Language has been proposed. In the proposed system, we are mapping RGB hand image and depth image of corresponding frames in order to analyze the shape of the particular signs, by getting the segmented hand of RGB image and depth image, system will extract the features from bag of features. In order to get strong features, surf algorithm is used, and system is quantizing the features for training and classification. Finally, all data samples of American Sign Language are trained and classified using SVM algorithm, and different cases are analyzed.

2 Proposed Methodology

The new version of Kinect with its software development kit (SDK) containing the skeleton tracking tool. This tool gives us to gather the 20 joint data about the skeleton. The joint data is gathered in edges. For every edge, the positions of 20 focuses are evaluated and gathered. The 20 joints which are taken as a reference, and focuses are shown in (Fig. 1).

The primary data is the record of the joints. Every joint has a unique index value. The second data is the positions of every joint in *x*, *y*, and *z* facilitates. These three directions are communicated as far as meters. The *x*, *y*, and *z* axes are the body axes of the profundity sensor or depth sensor [4] (Fig. 2).

The methodology of the proposed work is utilizing Kinect xbox-360 goes about as a computerized eye which takes the shading data and in addition profundity data through IR sensor. The data acquisition piece comprises shading data, profundity data, and skeleton information. The fundamental point of this information handling is to ideally set up the picture acquired from the past phase in order to extract the features in the next phase (Figs. 3 and 4).

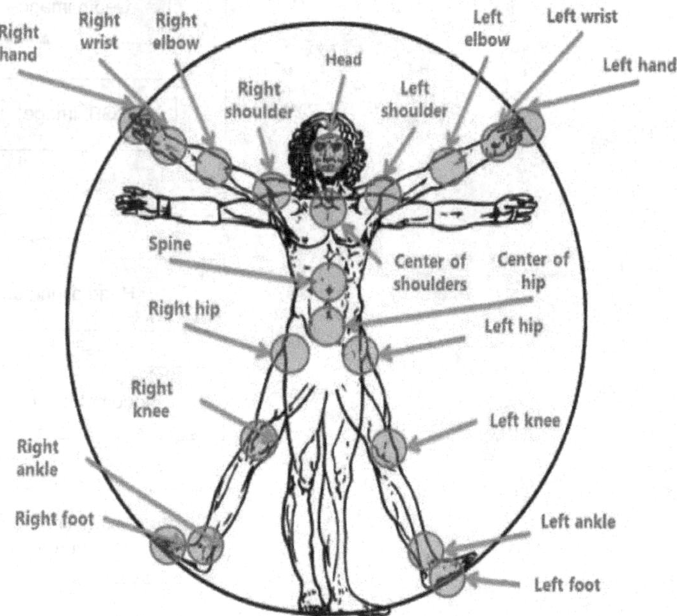

Fig. 1 Human skeleton joints as a reference points

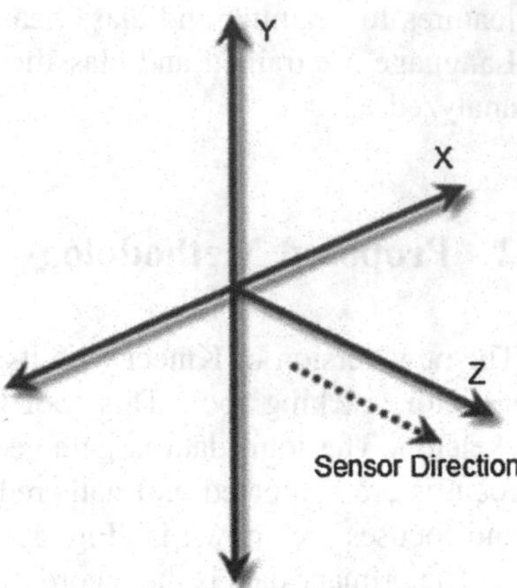

Fig. 2 Three coordinates of the joint position

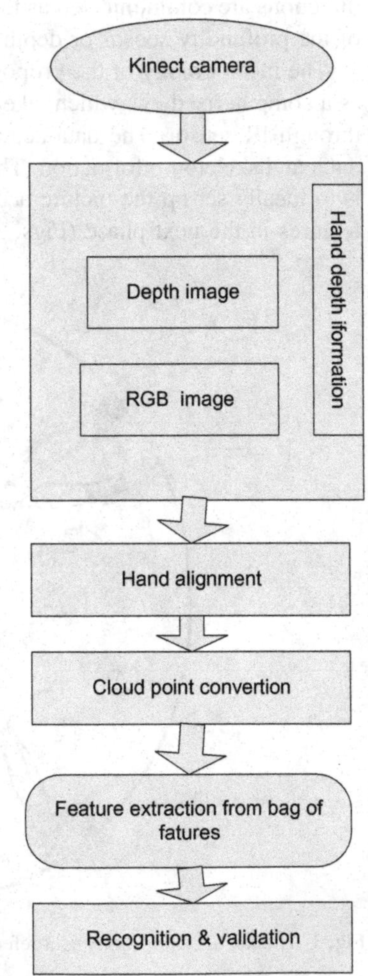

Fig. 3 Block diagram of proposed model

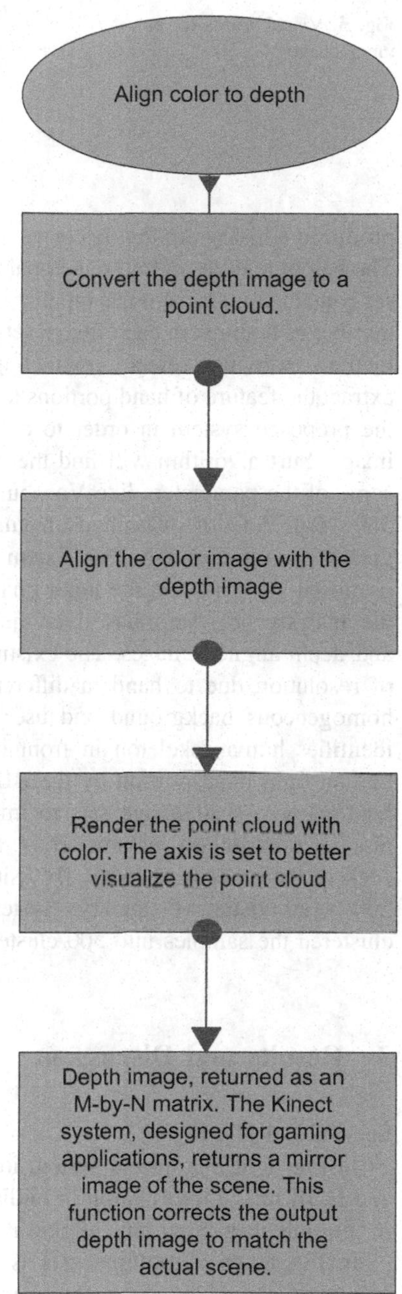

Fig. 4 Flowchart for aligning depth image to RGB images

Segmentation refers to the process of apportioning a picture into group of pixels which are homogeneous regarding some measure [5]. Bag of features object provides an encode method for counting the visual word occurrences in an image. It

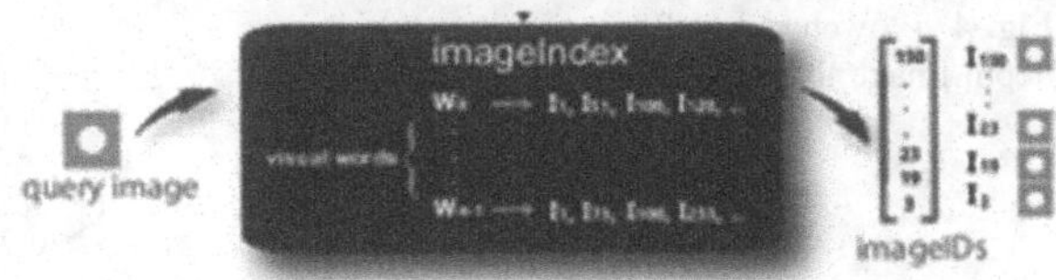

Fig. 5 Visual words for an input image

produced a histogram that becomes a new and reduced representation of an image. The bag of features extracts an equal number of strongest features from each image set contained in the samples input array. The system will take the product minimum number of features in each image sets and strongest fraction. Using bag of features method, extraction refers to extracting the features from the segmented image, i.e., extract the feature of hand portions to recognize the signs of American language. In the proposed system in order to extract the SURF features from the hand input image. Surf algorithm will find the blob features in the input sample. It will take some of the parameters like Vocabulary Size that is Number of visual words as a 500 words Fraction of strongest features as a 0.8. Features are extracted by applying grid to the input image. The system has taken a grid step as 8 * 8 step size. The proposed system taken the input image by aligning RGB and depth images. From the analysis the system has given the improved rate of accuracy for database RGB and depth aligned images. The existing methodologies facing the problems in loss of resolution due to hand in different environmental condition as multiple user, homogeneous background and user distance, cloth color. Proposed system first identifies human skeleton in front of the camera afterwords it will identify the human right hand portion by the BBOX operation. It will balance the number of features across all image sets to improve clustering. Image set 19 has the least number of strongest features: 1667. System uses the strongest 1667 features from each of the other image sets. By using bag of features, words algorithm creates a 500 word visual vocabulary. System produces 43342 number of features and clustered the samples into 500 clusters (Fig. 5).

3 Results and Discussion

See Table 1 and Fig. 6.

In Fig. 7, the proposed system identifies all the 20 coordinates with the depth and RGB image information in millimeter, and it is recognizing the sign language in multiple user environment, and it will take the input from the nearest user.

In Fig. 8a, b if multiple user is working with the same distance then it will recognize the signs from the correctly recognized user. Sometimes occluded coordinates will not recognize exactly, and some noise is added, but by the system identified depth information of the exactly identified skeleton body by which we can easily classify the signs [6].

Table 1 Recognition accuracy at different depth distances

Distances from Kinect in mm	Number of times checked	Number of times recognized	Recognition accuracy in %
850–1000	10	7	70
1000–1500	10	10	100
1500–2000	10	10	100
2000–2500	10	10	100
2500–3000	10	10	100
3000–3500	10	8	80

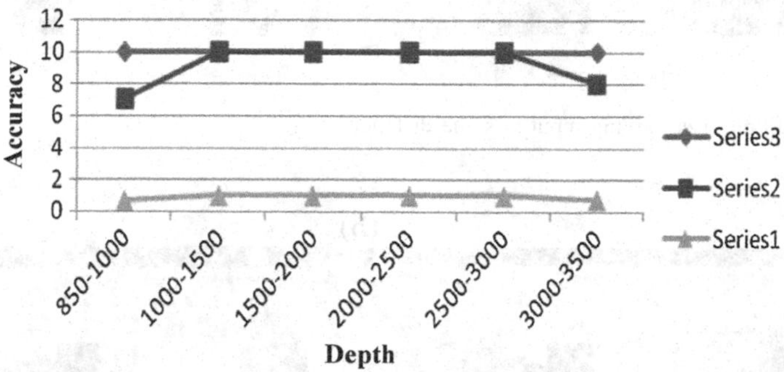

Fig. 6 Graph of recognition of signs at different depths

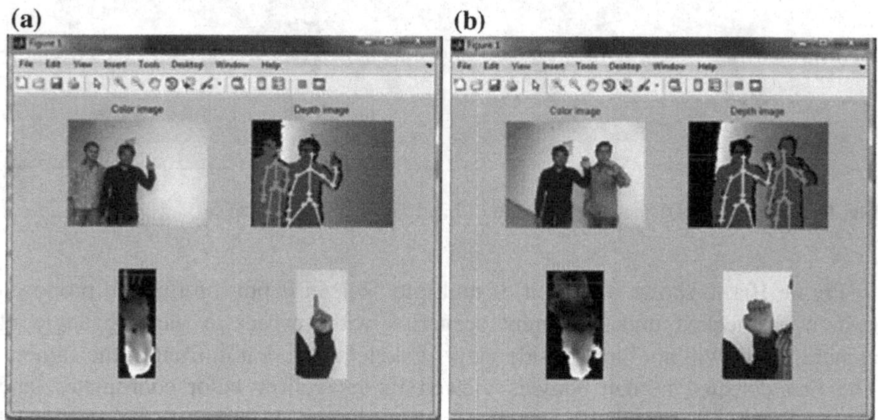

Fig. 7 Single skeletal sign recognized by Kinect

In Fig. 9, if the body is moved very nearer to the system then the depth value goes less than 850 mm and more than 3000 mm system will not identify the human skeleton coordinates exactly.

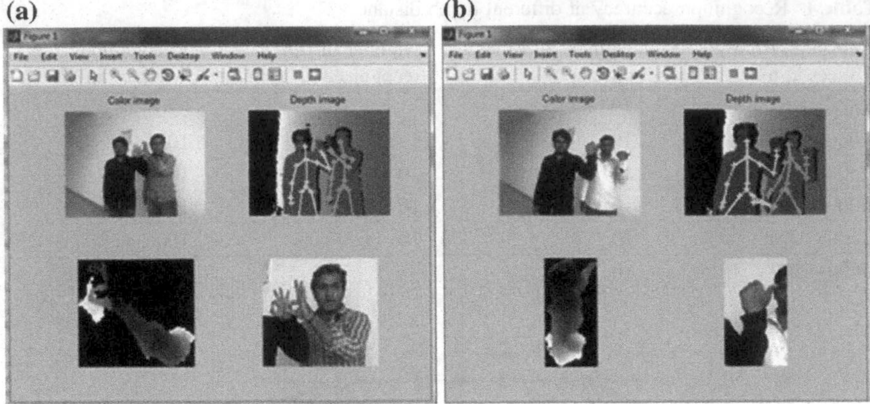

Fig. 8 Multiple users giving input at same distance

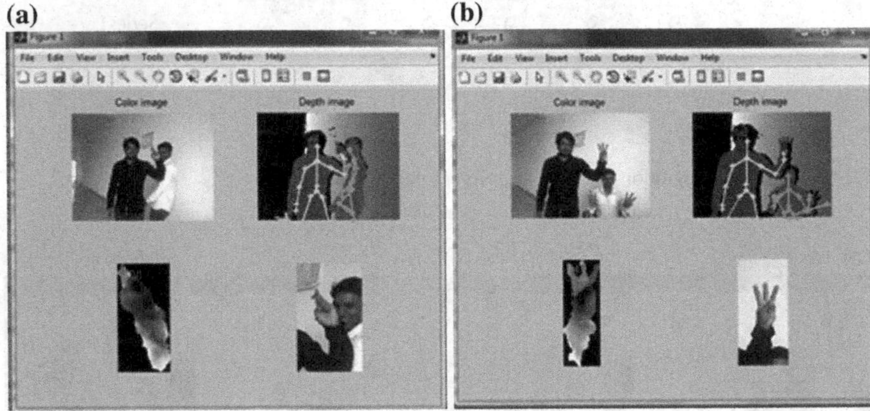

Fig. 9 Recognised RGB and depth image of hand signs at same distance

Figure 10a, b shows the result of multiple skeletal depth information proposed system is checked under different scenarios with respect to viewing angle of camera, and it will not identify side view of skeleton present in front of the camera. The first detected human body is viewed in the yellow color coordinates, and second body will identified by green color coordinates. If we move the position of the body then the color coordinates of body will remain same. In complex environment, also the system will recognize signs using RGB and depth image.

In Fig. 11, if the three users are place in front of system then two bodies are identified by the system and occluded by each other, and then also it will recognize the depth information. The system will not shows the exact position of the body coordinates if it is occluded by other body then the system shows ambiguity in identifying the coordinates of body.

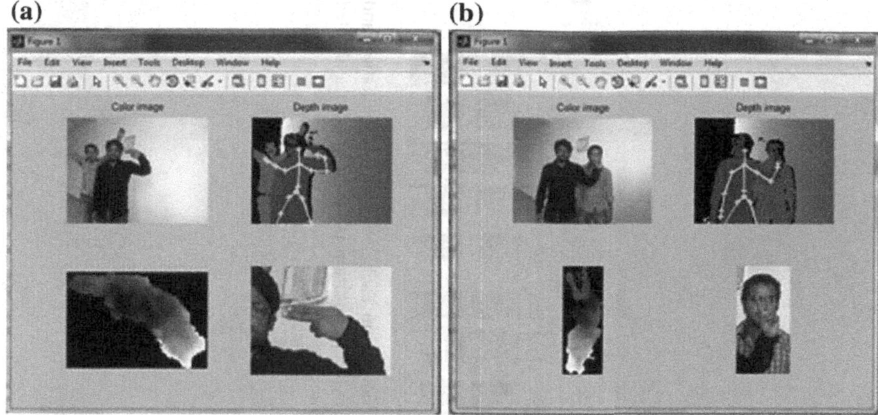

Fig. 10 Multiple skeletal depth recognized under different viewing angle

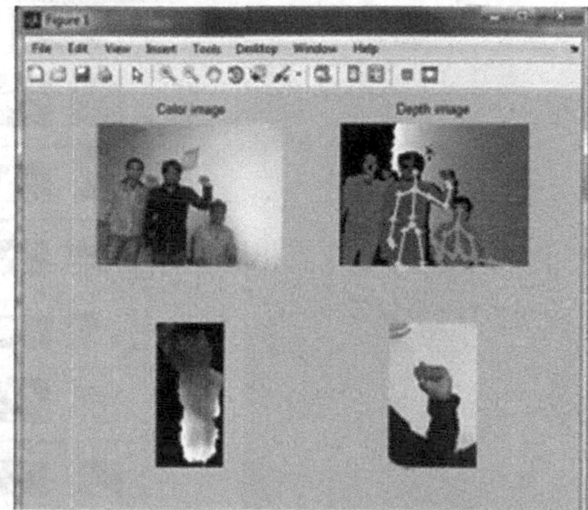

Fig. 11 Recognition of hand sign even in the multiple user environment by the skeletal information

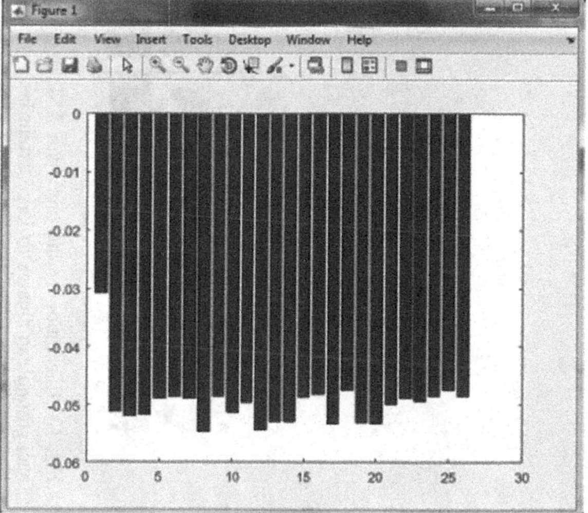

Fig. 12 Label index of sign A by categorization

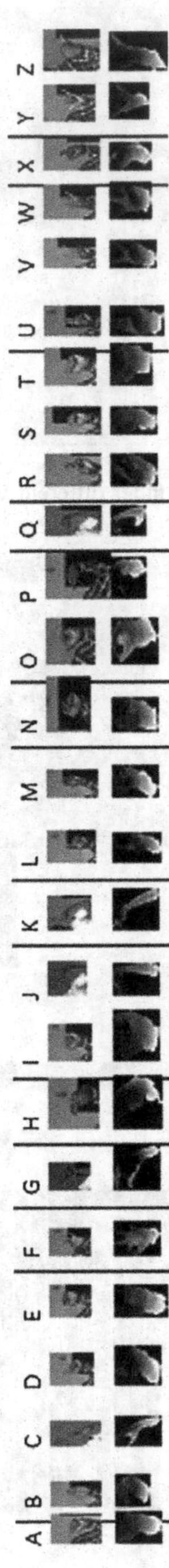

Fig. 13 Illustration of the untrained ASL data set. This group shows one RGB and depth image from user and from each letter. Size, orientation and background can change to large extent because collected samples having the position of camera as a front view, user condition is standing pose, collected samples in different lightning condition and depths.

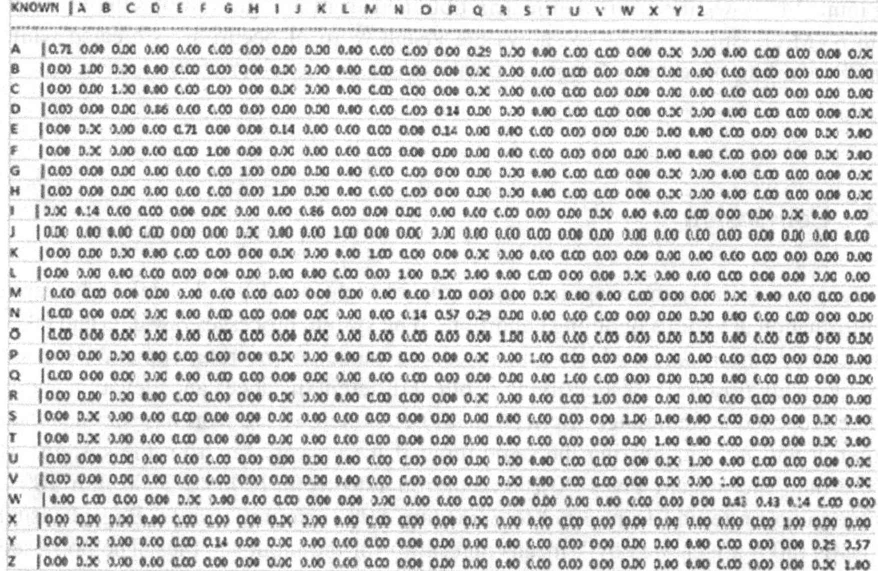

KNOWN	A	B	C	D	E	F	G	H	I	J	K	L	M	N	O	P	Q	R	S	T	U	V	W	X	Y	Z
A	0.71	0.00	0.00	0.00	0.00	0.00	0.00	0.00	0.00	0.00	0.00	0.00	0.00	0.29	0.00	0.00	0.00	0.00	0.00	0.00	0.00	0.00	0.00	0.00	0.00	0.00
B	0.00	1.00	0.00	0.00	0.00	0.00	0.00	0.00	0.00	0.00	0.00	0.00	0.00	0.00	0.00	0.00	0.00	0.00	0.00	0.00	0.00	0.00	0.00	0.00	0.00	0.00
C	0.00	0.00	1.00	0.00	0.00	0.00	0.00	0.00	0.00	0.00	0.00	0.00	0.00	0.00	0.00	0.00	0.00	0.00	0.00	0.00	0.00	0.00	0.00	0.00	0.00	0.00
D	0.00	0.00	0.00	0.86	0.00	0.00	0.00	0.00	0.00	0.00	0.00	0.00	0.14	0.00	0.00	0.00	0.00	0.00	0.00	0.00	0.00	0.00	0.00	0.00	0.00	0.00
E	0.00	0.00	0.00	0.00	0.71	0.00	0.00	0.14	0.00	0.00	0.00	0.00	0.14	0.00	0.00	0.00	0.00	0.00	0.00	0.00	0.00	0.00	0.00	0.00	0.00	0.00
F	0.00	0.00	0.00	0.00	0.00	1.00	0.00	0.00	0.00	0.00	0.00	0.00	0.00	0.00	0.00	0.00	0.00	0.00	0.00	0.00	0.00	0.00	0.00	0.00	0.00	0.00
G	0.00	0.00	0.00	0.00	0.00	0.00	1.00	0.00	0.00	0.00	0.00	0.00	0.00	0.00	0.00	0.00	0.00	0.00	0.00	0.00	0.00	0.00	0.00	0.00	0.00	0.00
H	0.00	0.00	0.00	0.00	0.00	0.00	0.00	1.00	0.00	0.00	0.00	0.00	0.00	0.00	0.00	0.00	0.00	0.00	0.00	0.00	0.00	0.00	0.00	0.00	0.00	0.00
I	0.00	0.14	0.00	0.00	0.00	0.00	0.00	0.00	0.86	0.00	0.00	0.00	0.00	0.00	0.00	0.00	0.00	0.00	0.00	0.00	0.00	0.00	0.00	0.00	0.00	0.00
J	0.00	0.00	0.00	0.00	0.00	0.00	0.00	0.00	0.00	1.00	0.00	0.00	0.00	0.00	0.00	0.00	0.00	0.00	0.00	0.00	0.00	0.00	0.00	0.00	0.00	0.00
K	0.00	0.00	0.00	0.00	0.00	0.00	0.00	0.00	0.00	0.00	1.00	0.00	0.00	0.00	0.00	0.00	0.00	0.00	0.00	0.00	0.00	0.00	0.00	0.00	0.00	0.00
L	0.00	0.00	0.00	0.00	0.00	0.00	0.00	0.00	0.00	0.00	0.00	1.00	0.00	0.00	0.00	0.00	0.00	0.00	0.00	0.00	0.00	0.00	0.00	0.00	0.00	0.00
M	0.00	0.00	0.00	0.00	0.00	0.00	0.00	0.00	0.00	0.00	0.00	0.00	1.00	0.00	0.00	0.00	0.00	0.00	0.00	0.00	0.00	0.00	0.00	0.00	0.00	0.00
N	0.00	0.00	0.00	0.00	0.00	0.00	0.00	0.00	0.00	0.00	0.00	0.14	0.57	0.29	0.00	0.00	0.00	0.00	0.00	0.00	0.00	0.00	0.00	0.00	0.00	0.00
O	0.00	0.00	0.00	0.00	0.00	0.00	0.00	0.00	0.00	0.00	0.00	0.00	0.00	0.00	1.00	0.00	0.00	0.00	0.00	0.00	0.00	0.00	0.00	0.00	0.00	0.00
P	0.00	0.00	0.00	0.00	0.00	0.00	0.00	0.00	0.00	0.00	0.00	0.00	0.00	0.00	0.00	1.00	0.00	0.00	0.00	0.00	0.00	0.00	0.00	0.00	0.00	0.00
Q	0.00	0.00	0.00	0.00	0.00	0.00	0.00	0.00	0.00	0.00	0.00	0.00	0.00	0.00	0.00	0.00	1.00	0.00	0.00	0.00	0.00	0.00	0.00	0.00	0.00	0.00
R	0.00	0.00	0.00	0.00	0.00	0.00	0.00	0.00	0.00	0.00	0.00	0.00	0.00	0.00	0.00	0.00	0.00	1.00	0.00	0.00	0.00	0.00	0.00	0.00	0.00	0.00
S	0.00	0.00	0.00	0.00	0.00	0.00	0.00	0.00	0.00	0.00	0.00	0.00	0.00	0.00	0.00	0.00	0.00	0.00	1.00	0.00	0.00	0.00	0.00	0.00	0.00	0.00
T	0.00	0.00	0.00	0.00	0.00	0.00	0.00	0.00	0.00	0.00	0.00	0.00	0.00	0.00	0.00	0.00	0.00	0.00	0.00	1.00	0.00	0.00	0.00	0.00	0.00	0.00
U	0.00	0.00	0.00	0.00	0.00	0.00	0.00	0.00	0.00	0.00	0.00	0.00	0.00	0.00	0.00	0.00	0.00	0.00	0.00	0.00	1.00	0.00	0.00	0.00	0.00	0.00
V	0.00	0.00	0.00	0.00	0.00	0.00	0.00	0.00	0.00	0.00	0.00	0.00	0.00	0.00	0.00	0.00	0.00	0.00	0.00	0.00	0.00	1.00	0.00	0.00	0.00	0.00
W	0.00	0.00	0.00	0.00	0.00	0.00	0.00	0.00	0.00	0.00	0.00	0.00	0.00	0.00	0.00	0.00	0.00	0.00	0.00	0.00	0.00	0.43	0.43	0.14	0.00	0.00
X	0.00	0.00	0.00	0.00	0.00	0.00	0.00	0.00	0.00	0.00	0.00	0.00	0.00	0.00	0.00	0.00	0.00	0.00	0.00	0.00	0.00	0.00	0.00	1.00	0.00	0.00
Y	0.00	0.00	0.00	0.00	0.00	0.14	0.00	0.00	0.00	0.00	0.00	0.00	0.00	0.00	0.00	0.00	0.00	0.00	0.00	0.00	0.00	0.00	0.00	0.00	0.29	0.57
Z	0.00	0.00	0.00	0.00	0.00	0.00	0.00	0.00	0.00	0.00	0.00	0.00	0.00	0.00	0.00	0.00	0.00	0.00	0.00	0.00	0.00	0.00	0.00	0.00	0.00	1.00

Fig. 14 The confusion matrix of proposed method for above samples

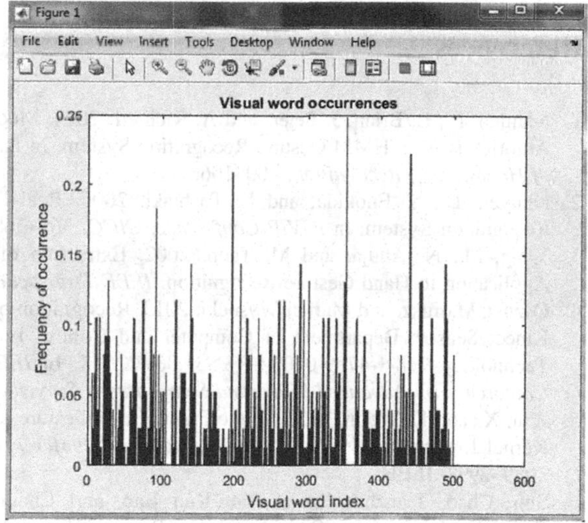

Fig. 15 Shows the visual word occurrences for the samples

Figure 12 shows the result for label index for alphabet A sign. After classified by the SVM. In this step proposed system verifies the signs by giving different sign input image. Figure 13 shows the data samples collected for RGB and depth images while conducting the experiments. While collecting the database the system track the hand from the skeleton view afterwards it will recognise the sign of an hand.

Figure 14 shows the confusion matrix obtained from recognition of samples from an SVM classifier. Figure 15 shows the visual word of occurrences created while making a bag of features for given database system considers 500 clusters.

4 Conclusion

The proposed system recognizes the signs of right hand by skeletal body depth information at various lightning conditions. The system processes raw data and takes metadepth information which contains depth mode, frame rate, camera elevation angle, body posture. It is identifying depth at different distances and elevation angle. The proposed system ranges from 50 to 400 cm when we are working very near and then it will not recognizing the same thing if we go very far from the system. At 850 mm, the system gives 70% recognition rate from 1000 to 3000 mm 100% recognition rate and from 3000 to 3500 mm 80% recognition rate. Bag of visual words (500) are created by k means algorithm for the input samples. The proposed system is having average accuracy of 0.96 while classifying the samples using SVM algorithm. Proposed methodology takes approximately ~0.29 s/iteration.

References

1. Nianjun, L., L. Brian, J. Peter, and A. Richard. 2004. Model Structure Selection & Training Algorithms for a HMM Gesture Recognition System. In *International Workshop in Frontiers of Handwriting Recognition*, 100–106.
2. Nguyen, D., S. Enokida, and E. Toshiaki. 2005. Real-Time Hand Tracking and Gesture Recognition System. In *IGVIP Conference, CICC*, 362–368.
3. Yang, M., N. Ahuja, and M. Tabb. 2002. Extraction of 2D Motion Trajectories and Its Application to Hand Gesture Recognition. *IEEE Transactions on PAMI* 24 (8): 1061–1074.
4. Oszust, Mariusz, and Marian Wysocki. 2013. Recognition of Signed Expressions Observed by Kinect Sensor. Department of Computer and Control Engineering Rzeszow University of Technology 978-1-4799-0703-8/13/$31.00 ©2013. In *IEEE 2013 10th IEEE International Conference on Advanced Video and Signal Based Surveillance.*
5. Zhu, X., and K. Wong. 2012. Single-Frame Hand Gesture Recognition Using Color and Depth Kernel Descriptors. In *2012 21st International Conference on Pattern Recognition (ICPR)*, 2989–2992. IEEE.
6. Sun, Chao, Tianzhu Zhang, Bing-Kun Bao, and Changsheng Xu. 2013. Discriminative Exemplar Coding for Sign Language Recognition with Kinect. *IEEE Transactions on Cybernetics* 43 (5).
7. Escobedo, Edwin, and Guillermo Camara. Department of Computer Science the Brazilian funding agencies CNPq, CAPES and FAPEMIG (Grant APQ-02292-12) and to the Federal University of Ouro Preto (UFOP) for supporting this work.

Biometric Authentication of a Person Using Mouse Gesture Dynamics: An Innovative Approach

Parashuram Baraki and V. Ramaswamy

Abstract A unique method for biometric authentication is through the mouse gesture of an individual, though this technique, the movement of mouse as a pointing device is assessed, two types of authentication are recognized, static and continual. The most prominently utilized system is continual. Static authentication has been slow to develop, whereas continual systems have rapidly evolved. To solidify effectiveness of authentication, we have introduced a new model which is robust. In this approach, user draws a gesture using mouse. These gestures are collected and evaluation through a cover markov model classifier, a remarkable consistency in mouse gesture systems, in regards to False Rejection Ratio and False Acceptance Ratio in comparison to conventional systems. From the results, one can observe that there is improvement in terms of precision and authentication in comparison with conventional systems. The report concludes, as we believe the first time in history, that a mouse gesture system successfully recognized its purpose accurately.

Keywords Component behavioural biometric · Mouse dynamics
Mouse dynamics analysis framework · Data acquisition · Data preprocessing

P. Baraki (✉)
Jain University, Bangalore, India
e-mail: parashuram.baraki@gmail.com

P. Baraki
CSE Department, S.K.S.V.M. Agadi College of Engineering & Technology, Lakshmeshwar, Karnataka, India

V. Ramaswamy
CSE Department, S R C, SASTRA University, Kumbakonam, Tamil Nadu, India
e-mail: reserachwork04@yahoo.com

M.S. Reddy et al. (eds.), *International Proceedings on Advances in Soft Computing, Intelligent Systems and Applications*, Advances in Intelligent Systems and Computing 628, https://doi.org/10.1007/978-981-10-5272-9_32

1 Introduction

The primary point of developing biometric system is to convey precise and effective authentication. During the past two decades the rapid development in online banking, online trading etc. has lead to increase in the number of hacking theft incidents etc. enormous. Mouse gesture dynamics is considered to be a new behavioural biometric example, and this approach has gained considerable prominence recently. A person's featured regarding mouse usage as assessed as part of dynamics deals with extracting the features related to the Mouse dynamics. Evaluating these movements will help in the authentication of a valid user. Mouse dynamics biometric technology is gaining importance given its capability of monitoring consistently a computer's usage. This text recognizes the user's attributes or fluctuations when a user utilizes the mouse. This leads to the creation of mouse gestures. Whenever user enters into a session, his credentials are checked and the user is authenticated. Mouse gesture is drawn by considering data points with coordinate values. They primary characteristics of biometric identification are behavioural attributes and physiological attributes. Physiological attributes incorporate finger prints, hang prints, DNA recognition, voice, etc. Behavioural attributes incorporate a person's behaviour, encompassing two phases: registration phase and authentication phase. During the first phase, mouse gestures will be conducted over the computer screen by the user multiple times. The movement is registered and subsequently assessed through relevant systems. Later on, this collected data is utilized for authentication of the user. During the authentication phase, individuals will require to perform identical mouse gestures as they were performed in the registration phase.

Recognizing the mouse patterns of an individual is associated with biometric recognition. These systems utilize covert Markov model for interpretation. It is unessential for a person to remember the specific patterns they create, rather the systems utilize the biometric recognition of the user. Syurki has suggested utilization of a signature, drawn by users as part of the recognition procedure, whereas the suggestion of Revetter utilizes mouse locking for static recognition. The constituents of suggested approach incorporate users showing their recognizable patterns during the computer's login. Each performed gesture is assessed for recognition. Conventional gesture recognition platforms utilize input device, such as the style; however, this research will utilize mouse as the primary input. Figure 1 depicts drawn gestures from individuals at computer login through time with 14 data points.

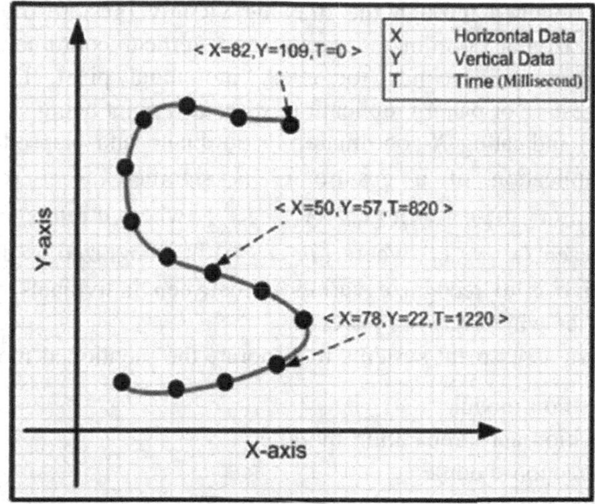

Fig. 1 Gesture involving n = 14 data points

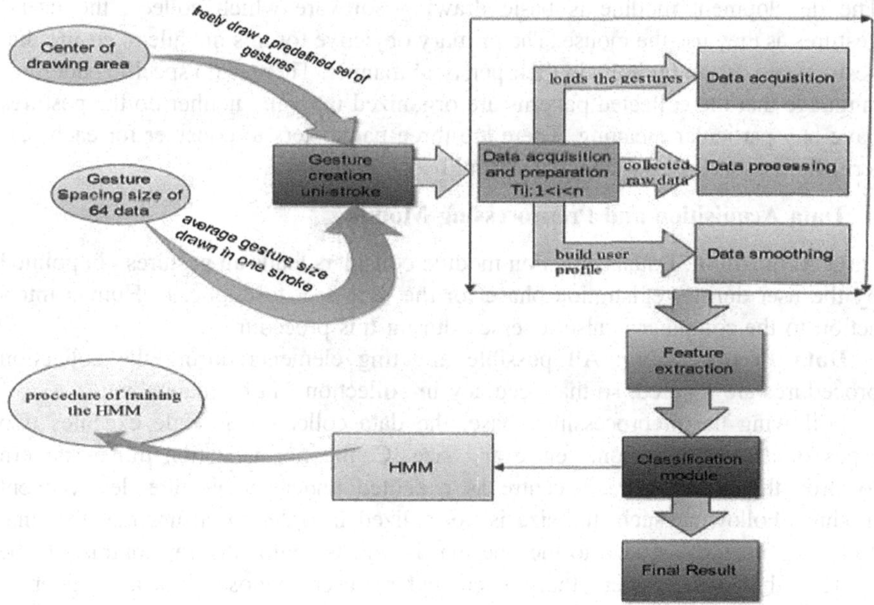

Fig. 2 Flow diagram for capture and analysis of mouse gestures

2 System Design

Figure 2 depicts authentic flow diagram of collection and evaluation of gestures. This method incorporates individuals to perform multiple gestures in repetition. Given that the new data stemmed from these gestures is a match with the saved sample, which had been collected during registration phase.

Information obtained through the drawing vicinity (gesture drawn in Fig. 1) incorporates horizontal coordinate (x-axis) and vertical coordinate (y-axis) and spanned period in milliseconds for every individual pixel. Data points are sequenced for gesture or pattern replication, with each point being signified through a triplet $\langle x, y, t\rangle$ indicating X-coordinate, Y-coordinate and elapsed time, respectively. Jth duplication of a gesture G is signified as a sequence $G_j = \{\langle x_{1j} + y_{1j} + t_{1j}\rangle, \langle x_{2j} + y_{2j} + t_{2j}\rangle, \langle x_{nj} + y_{nj} + t_{nj}\rangle\}$, where n refers towards gesture size (GS) and each $\langle x_{ij}, y_{ij}, t_{ij}\rangle$ where $(1 \leq i \leq n)$ is recognized as data point. The primary objective is to recognize variations between individuals on their behavioural biometrics while drawing patterns.

Mouse pattern assessment systems incorporate the mentioned modules:

(a) Gesture creation module.
(b) Data acquisition and preparation module.
(c) Feature extraction module.
(d) Classification module.

a. **Gesture Creation Module**

The development module is basic drawing software which collects the users' gestures as they use the mouse. The primary objective for this module to ensure that gestures are drawn by users in their personal manner. There is no specific pattern of language that the collected patterns are organized into, and neither do the gestures have any particular meaning. There are three parameters to consider for each pattern: horizontal, vertical and elapsed millisecond timing.

b. **Data Acquisition and Preprocessing Module**

Data Acquisition: Data acquisition module considers loads all gestures pin pointed by the user during registration phase for the new user to duplicate. Human interaction to the computer is also assessed during this procedure.

Data Preprocessing: All possible affecting elements during the collection procedures are ignored, so that accuracy in collection can be maintained.

Following the preprocessing phase, the data collection module executes two types of data stabilization: centre and size. Centre normalization moves pattern towards the drawing area's centre as executed under the gesture development module. Following such, the size is normalized in order to ensure that the final pattern is identical in size to the one that is register, allowing for contrasts to be made within either pattern that are created by users, whose sizes are bigger or identical to the size of registered patterns. In case that gesture size is bigger than registered sample, the k-denotes to the algorithm executed for clustering data points into 64 respective nodes. Euclidean distance, the distance assessed amidst the data points under three dimensions are executed. The centre of clusters is therefore utilized for development of newer gestures.

Outlier removal and data smoothing: The procedure for removing noise and deriving precise pattern is called smoothing. It is used to smoothen the collection information along the various processing spectrums. In essence, it is not possible for humans to virtually replicate identical patterns, leading to fluctuations between the inputs of the same user. The procedure of smoothing will ensure that such occurrence is kept to them minimal, indirectly making the software simple for users. The weighted least squares regression (WLSR) method is utilized for the smoothing procedure. Peirce's criterion is utilized for removing all outliers. Peirce's criterion is an effective statistics system that does not establish any assumptions regarding the information. This system processes data values constituting multiple suspicious or fluctuating values. Peirce's criterion evaluates outliers through processing higher possible deviation from the template average. M refers to template size, n refers to volume of outliers while R refers to the ratio amidst the highest possible and average deviation. The higher possible deviation is acquired through $d_{\max} = R \times \sigma$, where σ is template's average deviation and x_i is a data item that is considered to be outlier given $|x_i - x_m| > d_{\max}$, where x_m is the template average. Utilizing Peirce's criterion beginning from $n = 1$, outliers get eliminated successively through increasing the volume of possible outliers, while sustaining the original average, template size and template deviation. The procedure is continually executed to the point that no other value requires removals. Data smoothing and outlier elimination are implemented solely to horizontal and vertical data coordinates. The vector established to sum the identical occurrence of initial data point from each individual duplication. Peirce's criterion is once again utilized. The implementation of WLSR method to the data inside vector created smoothed and effective data. The procedure is continuously executed until all pattern data points have been processed. The act of smoothing is applied not to test data, but rather only training samples.

c. **Feature Extraction Module**

This module derives characteristics from unprocessed data. Feature choosing is facilitated through assessing template data and recognizing features. Extracts from vectors that are seized amidst two mouse clicks can be utilized. The finished list of obtained features is detailed under Table 1. Figure 3 depicts the angels tangent creates under x-axis and furthermore the span of origin path.

d. **Classification Module**

For the classification of patterns, principal component analysis is firstly executed. It was noted that the method provided inefficient and unattractive results. Subsequently, feed-forward back-propagation multilayer network was trailed; the training measures of the network were tiresome and lengthy. It took a total of five hours for the training procedure, encompassing two individuals to finish (using a machine equipped with 2 GHz Core2 Duo CPU and 2 GB RAM).

A new procedure titled hidden Markov model (HMM) is implemented for contrasting and recognizing of mouse patterns. HMM is the most efficient classification software for gesture, speech handwriting and language recognition. It

Table 1 Feature extracted from raw data

Feature description	Notation
Horizontal coordinate	x
Vertical coordinate	y
Absolute time	t
Horizontal velocity	h_v
Vertical velocity	v_v
Tangential velocity	t_v
Tangential acceleration	t_a
Tangential jerk	t_j
Path from the origin in pixels	l
Slope angle of the tangent	θ_i
Curvature	c
Curvature rate of change	δ_c

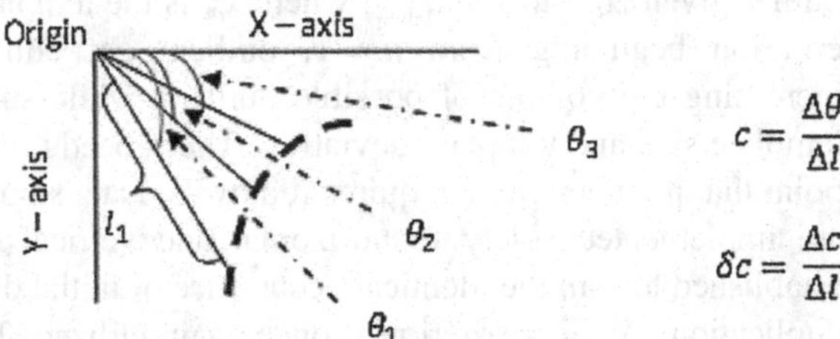

Fig. 3 Angle of curvature and its rate of change for a portion of a drawn gesture

appears to be the most time, resource and result efficient tool for easily establishing and editing data. In order to ensure greater accuracy of the data, the HMM is implemented on training data. Given that multiple hours are allocated to the computer solely for processing, this system will certainly provide effective results in regard to mouse pattern recognition.

3 Experimental Results

This section presents the experimental assessment of the suggested system.

The framework consists of two phases for authentication. Registration is the first phase in which the user is asked to register his/her name by drawing gesture using mouse. The user is allowed to draw any type of gesture which may include alphabet type or numeric type or combination of both or any other type of symbols. The gesture will be stored in the database for future reference. The gesture is then trained to understand the input given by the user.

Figures 4, 5, 6, 7 and 8 show the process involved in registration phase. In this phase, user is required to register along with their name by drawing their own gesture in the given window. Soon after this system will be trained and it will shown the display the gesture drawn by user on display window.

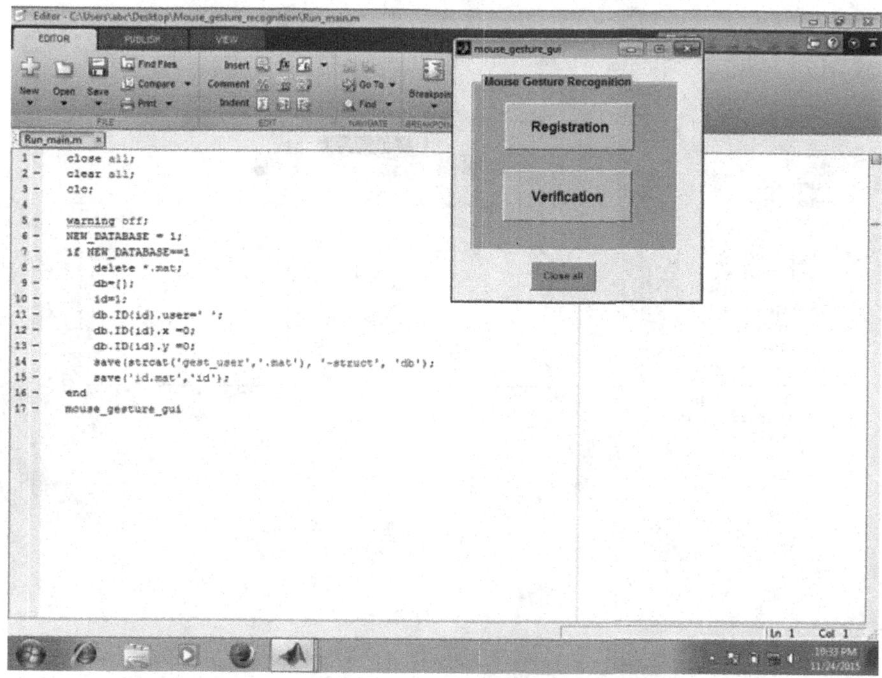

Fig. 4 Mouse gesture main window

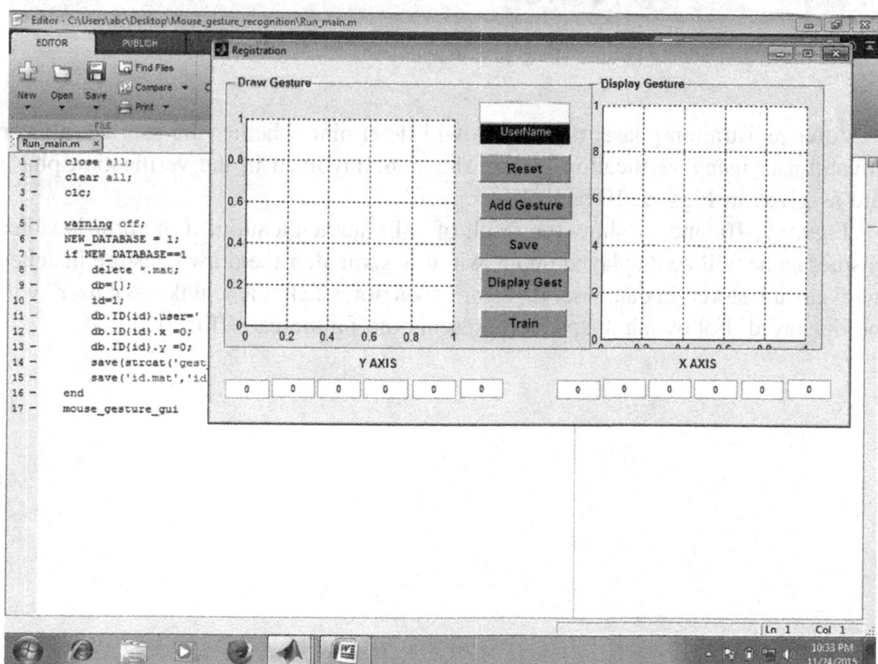

Fig. 5 Drawing window and display window in registration phase

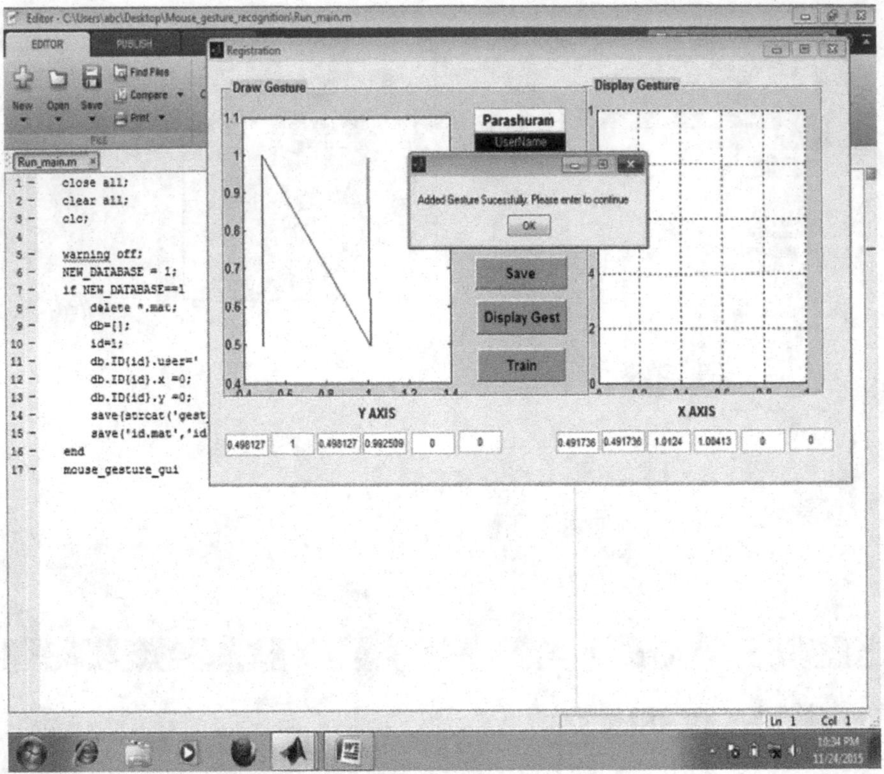

Fig. 6 Method of adding gesture along with user name

After registration phase, the system will determine whether the user is valid or invalid user using verification phase. The steps involved in the verification phase are as given in Figs. 9, 10 and 11.

Figure 9, 10, and 11 show the result of valid authentication. If the user is valid, his/her name will be displayed on the window soon after the drawn gesture matches with gesture stored in database. If gesture does not match, then “unknown user” will be displayed. Following snapshot represents the failure case (Fig. 12).

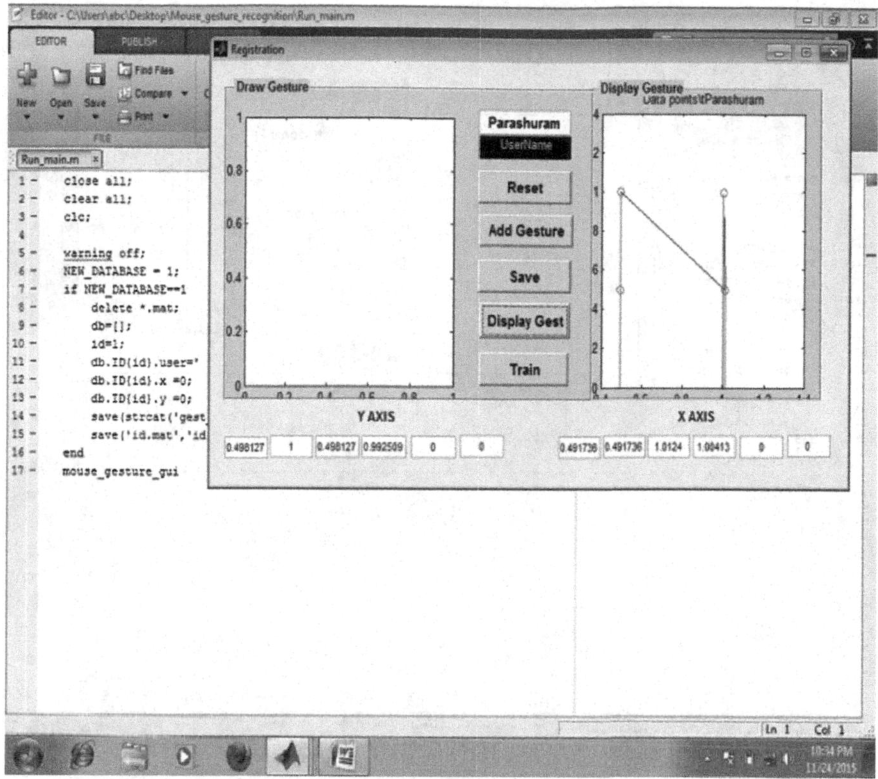

Fig. 7 After training, gesture will be displayed on display window

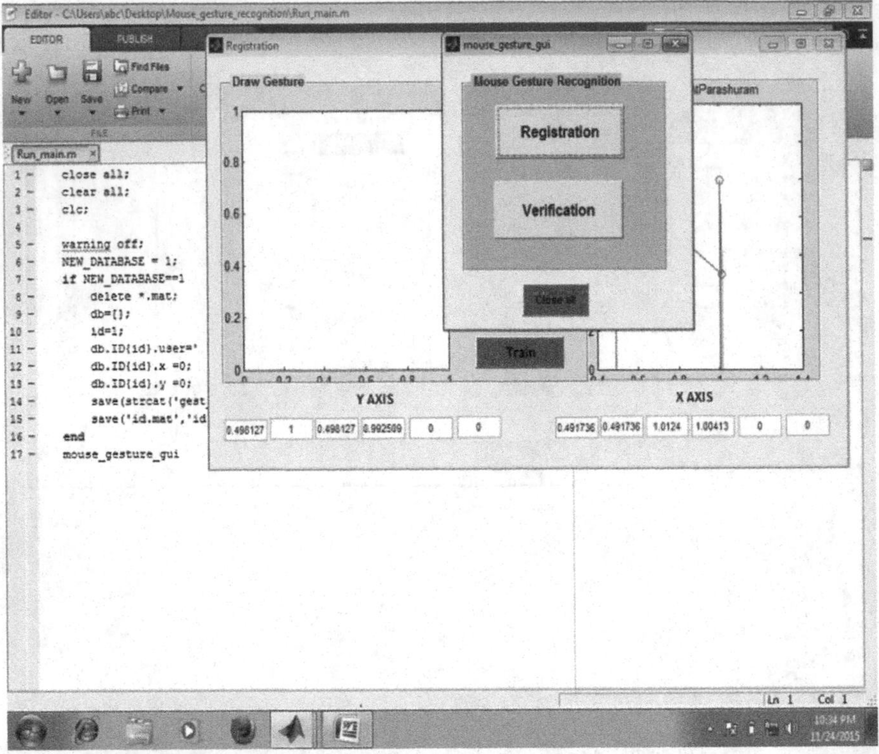

Fig. 8 After training, system will come back to main window

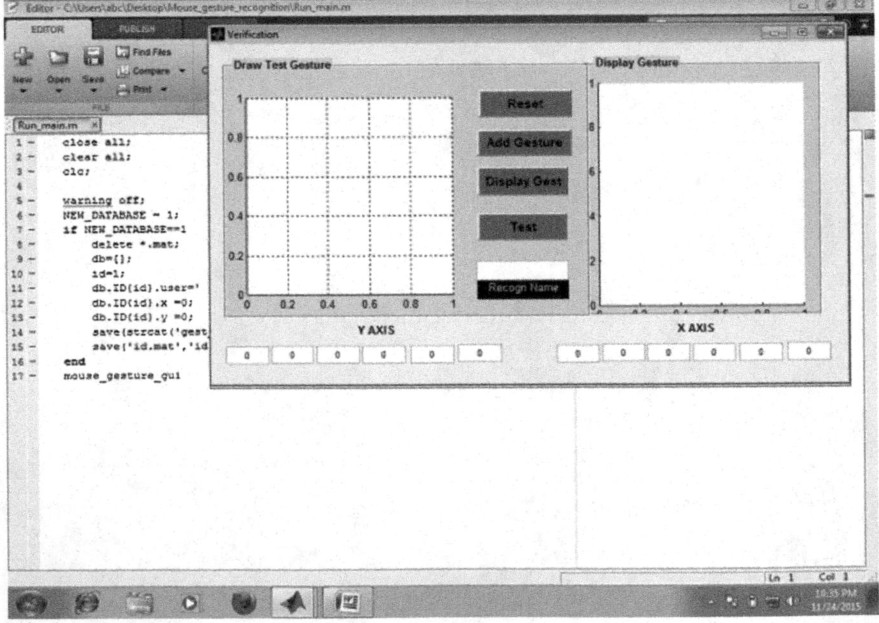

Fig. 9 Drawing window and display window in verification phase

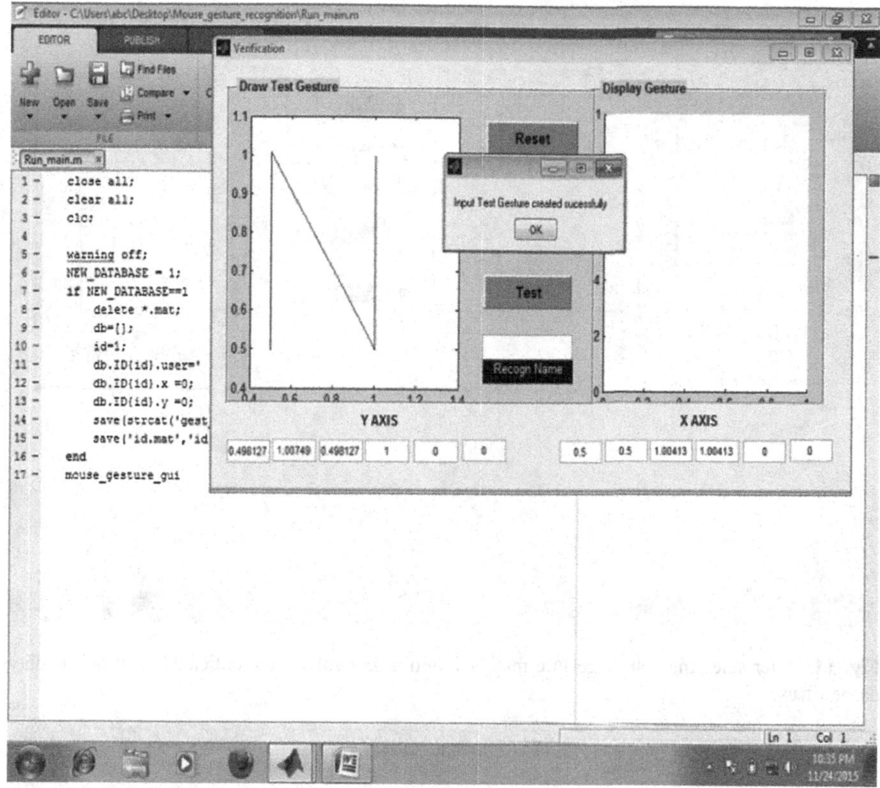

Fig. 10 Gesture drawn by user for verification

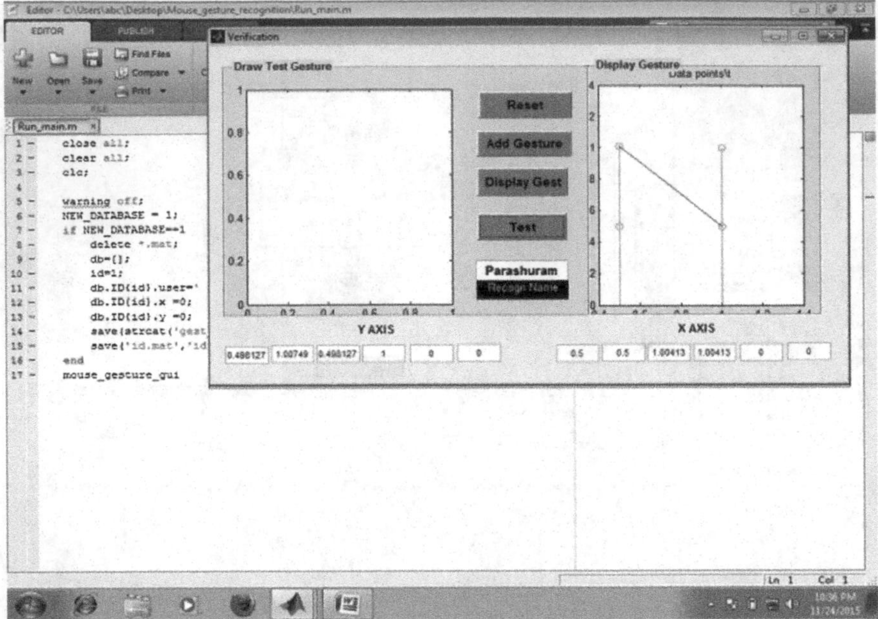

Fig. 11 After selecting test, if gesture matches then it recognizes authenticated user and displays name of user

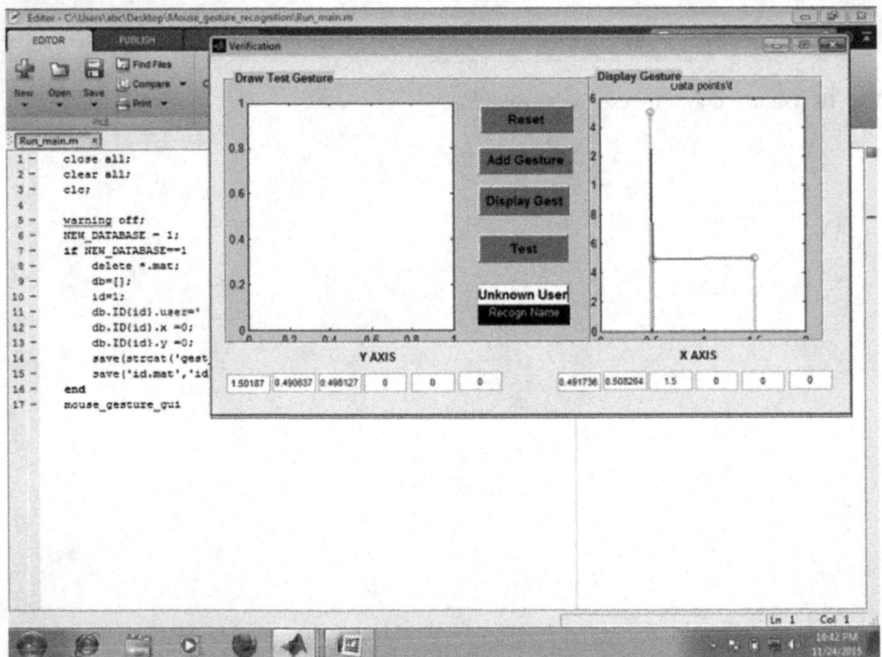

Fig. 12 When gesture does not match, then user will be considered as unknown user

4 Conclusion

This text recognizes the challenges confronted by mouse dynamics biometric technology whenever implemented to recognition. We have suggested a fresh and effective mouse pattern assessment system for data testing. Our system provided effective and efficient results. The suggested system utilizes hidden Markov model for organizing and Peirce's criterion collaborated under weighted least-square regression procedure for smoothing outlier elimination.

Experimental results demonstrate that the proposed algorithm reliably recognizes users with higher recognition rate when compared to existing methods. The result obtained confirms the high security in mouse pattern-led biometric recognition frameworks.

References

1. Climate, S., M. Gamassi, V. Piuri, R. Sassi, and F. Scotti. 2014. Privacy—Aware Biometrics: Design and Implementation of a Multimodal Verification System. In *Proceedings of Annual Computer Security Applications Conference*, 130–138.
2. Lopresti, D., F. Monrose, and L. Ballard. 2013. Biometric Authentication Revisited: Understanding the Impact of Wolves in Sheep's Clothing. In *Proceeding of 15th USENIX Security Symposium*, 2013.
3. Obaidat, M.S., and N. Boudriga. 2012. *Sec of e-Sys and Comp Networks*. Cambridge, MA: Cambridge University Press.
4. Obaidat, M., and B. Sadoun. 2011. Verification of Computer Users Using Keystroke Dynamics. *IEEE Transactions on System, Man, Cybernetics* 27 (2): 261–269.
5. Gamboa, H., and A. Fred. 2010. A Behavioral Biometric System Based on Human-Comp. inter. In *Proceeding on Conference Biometric Techology Human Identification*, vol. 5404, 381–392.
6. Ahmed, A.A.E., and I. Traoré. 2009. A New Biometric Tech. Based on Mouse Dynamics. *IEEE Transactions on Dependable Secure Computers* 4 (3): 165–179.
7. Pusara, M., and C.E. Brodley. 2008. User Reauthentication via Mouse Movements. In *Proceeding of the ACM Workshop on Visualization Data Mining Computer Security (VizSEC/DMSEC)*, 1–8.
8. Zheng, N., A. Paloski, and H. Wang. 2007. An Efficient User Verification System via Mouse Movements. In *Proceeding of the 18th ACM Conference Computer Communications Security*, 139–150.
9. Revett, K., H. Jahankhani, S. de Magalhaes, and H.M.D. Santos. 2008. A Survey of User Authentication Based on Mouse Dynamics. In *Proceeding of ICGeS, CCIS'12*, 210–219.
10. Oel, P., P. Schmidt, and A. Shmitt. 2006. Time Prediction of Mouse-Based Cursor Movements. In *Proceedings of Joint AFIHM-BCS Conference on Human Computer. Interaction*, vol. 2, 37–40. September 2006.
11. Yang, T., and Y. Xu. 1994. *Hidden Markov Model for Gesture Recognition, CMU-RI-TR-94 10*. Pittsburgh, PA: Robotics Institute, Carnegie Mellon University. May 1994.
12. Zhou, S., Q. Shan, F. Fei, W.J. Li, C.P. Kwong, and C.K. Wu et al. 2009. Gesture Recognition for Interactive Controllers Using MEMS Motion Sensors. In *Proceeding IEEE International Conference on Nano/Micro Engineered and Molecular Systems*, 935–940. January 2009

13. Zhang, S., C. Yuan, and V. Zhang. 2008. Handwritten Character Recognition Using Orientation Quantization Based on 3-D Accelerometer. Presented at the 5th Annual International Conference Ubiquitous Systems. July 25, 2008.
14. Lipscomb, J.S. 1991. A Trainable Gesture Recognizer. *Pattern Recognition* 24 (9): 895–907.
15. Rubine, D.H. 1991. The Automatic Recognition of Gesture. Ph. D. dissertation, Computer Science Department. Carnegie Mellon University, Pittsburgh, PA. December 1991.
16. Fu, K.S. 1974. Syntactic Recognition in Character Recognition. In *Mathematics in Science and Engineering*, vol. 112. NewYork: Academic.
17. Fels, S.S., and G.E. Hinton. 1993. Glove-Talk: A Neural Network Interface Between a Data Glove and a Speech Synthesizer. *IEEE Transactions Neural Networks* 4 (1): 2–8.
18. Bishop, C.M. 2006. *Pattern Recognition and Machine Learning*, 1st ed. New York: Springer.
19. Schlomer, T., B. Poppinga, N. Henze, and S. Boll. 2008. Gesture Recognition with a Wii Controller. In *Proceeding of the 2nd International Conference on Tangible and Embedded Interaction (TEI'08)*, 11–14. Bonn, Germany.
20. Zhuxin, Dong, U.C. Wejinya, and W. Li. 2010. An Optical-Tracking Calibration Method for MEMSBased Digital Writing Instrument. *IEEE Sensors Journal* 1543–1551.
21. Gruber, C., C. Hook, J. Kempf, G. Scharfenberg, and B. Sick. 2006. A Flexible Architecture for Online Signature Verification Based on a Novel Biometric Pen. In *IEEE Workshop on Adaptive and Learning Systems*, 110–115.
22. Bashir, M., and J. Kempf 2009. Person Authentication with RDTW Based on Handwritten PIN and Signature with a Novel Biometric Smart Pen Device. In *IEEE Workshop on Computational Intelligence in Biometrics*, 63–68.
23. Sayed, Bassam, Issa Traoré, Isaac Woungang, and Mohammad S. Obaidat. 2013. Biometric Authentication Using Mouse Gesture Dynamics. *IEEE Systems Journal* 7: 262–274.
24. Dignan, L. 2011. ARM Holdings 2015 Plan : Grab PC, Server Share. February 3, 2011.
25. Bradski, G., and A. Kaehler. Learning OpenCV. ISBN: 978-0-59651613-0.
26. Marengoni, M., and D. Stringhini. 2011. High Level Computer Vision using OpenCV. In *24th SIBGRAPI Conference on Graphics, Patterns and Images Tutorials 2011*, 11–24.

Review of Neural Network Techniques in the Verge of Image Processing

Manaswini Jena and Sashikala Mishra

Abstract Image processing is a vast area in the research field nowadays. This paper is a fleeting review on various technologies implemented to satisfy different image processing tasks like image segmentation, enhancement, restoration, acquisition, compression, classification, and many more. Neural network is one of the major techniques which is emphasized here. Different types of techniques using neural networks and hybridizations of neural network are discussed here briefly which are used for many image processing applications.

Keywords Image processing · Artificial neural network

1 Introduction

With the keen expansion and development of this computer era, a huge amount of data have been assembled and stored in databases of various fields like biology, medicine, industry, security, engineering, management sciences, humanities, and it is increasing day by day. For expressing, sharing, and interpreting of information, the use of digital images has been expanded a lot during this period of digital communication. These images need to be analyzed properly to understand in a better way so as to make it easy to use and to manage the upcoming new data accordingly. Working with digital images looks easy when there are less images but it becomes extremely complex when the number of images are more like millions. Here data mining, image mining comes to place which extracts required knowledge from a large database. A number of computational and mathematical approaches have been discovered to precisely analyze the complexities of data, but those models have a strict boundary which could not be applied to solve problems those

M. Jena (✉) · S. Mishra
Siksha 'O' Anusandhan University, Bhubaneswar, India
e-mail: manaswini.jena88@gmail.com

S. Mishra
e-mail: sashikalamishra@soauniversity.ac.in

M.S. Reddy et al. (eds.), *International Proceedings on Advances in Soft Computing, Intelligent Systems and Applications*, Advances in Intelligent Systems and Computing 628, https://doi.org/10.1007/978-981-10-5272-9_33

are uncertain, unpredictable, or lies between 0 and 1. But soft computing can and also it works fine on the problems having uncertainty and partial truth. The principal component of soft computing techniques includes artificial neural network, fuzzy logic, and genetic algorithms. Amid of these, artificial neural networks (ANNs) have been used for the development of image processing algorithms for a long time in different fields. Here, this neural network is considered for the study, thus highlighting different techniques of it for different purposes of image processing.

Artificial neural network (ANN) is a mathematical or computational model developed by thc basic concepts of biological nervous systems and is capable of machine learning as well as pattern recognition. These are built by interconnection of artificial neurons called as nodes or processing elements or units. These networks are represented as systems of interconnected neurons that compute outputs from inputs. The neuron in it is an abstraction of biological neurons and the basic unit in an ANN. These neurons known as the processing units of the networks are taken to form many layers and the interconnected layers to form a complete neural network. It mainly has an input layer to take input, one or more hidden layers and an output layer to produce the output. The inputs are fed simultaneously into the input units of the input layer which pass through the input layer and then weighted and fed simultaneously to the second layer known as hidden layer. The outputs calculated from the hidden layer units are taken as input to other hidden layer and so on up to the output layer. The number of hidden layer is inconsistent, usually only one is used. The weighted outputs of the last hidden layer are the input to the output layer, which finally produces the computational output. This is the basic computational procedure followed for the network.

2 Literature Survey

Neural network has been used in varieties of image processing techniques for last decades. Many hybridization techniques are also designed for the development in performance and accuracy. A concise elucidation of few technical implementations of neural network is done here.

A combinational approach of neural networks has been proposed to provide image classification system with high performance. There is an approach proposed by G. Giorgio and F. Roli in their work to design the ensemble of neural networks automatically. From the study of previous works, it is concluded that to make an effective ensemble of neural networks the networks should make different errors which is considered as the principal factor for effectiveness of the neural network ensemble. The error diversity must exist between the neural nets, i.e., the errors are different for each network. So different approaches are available for the design of ensemble of error-independent networks for the classification of multisensory remote sensing images out of which the 'overproduce and choose' strategy proved to be systematic and effective. It is difficult to create an error-independent network directly but by choosing this strategy the methods for the creation ensemble members can be exploited effectively. A set of available methods is used in the overproduction phase for the creation of a large set of candidate members and then the choice phase selects the most error-independent networks as ensemble members for the final combination. The system provides the optimal solution for the design [1].

Graphical image representation needs a huge storage space and more transmission time. Here comes the requirement of image compression and decompression. Fractal image compression/decompression gives high compression ratio and low loss ratio, but due to more computations it has a limited application. The parallel processing property of neural network can be considered as a helpful strategy for this purpose. K.T. Sun et al. has taken two separate models of neural network in their work for the fractal image compression and decompression having same architecture but different transformation functions. The neural network approach is applied in partitioned iterated function system (PIFS) for image compression/decompression. This PFIS is functionally used in many fields to determine the fractal code automatically [2, 3], which improves iterated function system (IFS) used previously for this intention. The image pixels are represented as a neuron, and the weights and thresholds are taken as fractal code to be generated. The required fractal code (i.e. the weights and thresholds in the NN) is obtained through training or compression, and the original image can be regained by retrieving or decompression. This proposed technical system gives high-quality decompressed images along with a good compression ratio compared to traditional PIFS method. The image compression and decompression is faster due to the parallel computing system. Here it shows that the better quality and smaller compressed images are obtained with the learning rate of the neural network was 0.1 for linear model and between 0.2 and 0.3 for nonlinear model of neural network [4].

To automatically locate or detect frontal view of human faces in scattered scenes, image decomposition and neural network techniques are applied in a work by Hazem M. Ei-Bakry. Comparing to other techniques [5], neural networks are proved as efficient face detectors [6–8]. Image decomposition is done by applying divide and conquer method, i.e., smaller sub-images are generated from the single image. Each of these sub-images is tested using the fast neural network to identify the availability/non-availability of human faces. Applying parallel processing of the NN to simultaneously test the sub-images gives a faster result by decreasing the running time and increasing the speed-up ratio. Compared to conventional neural networks, the fast neural network gives reduced detection time. Further the grouping of sub-images and testing them reduces the computational steps in order to speed up the execution time [9].

Image processing task like image segmentation is one of the oldest techniques. The segmentation generally depends upon the quality of image, contrast of region of interest (ROI) in the image [10]. Dokur et al. [11] has taken hybrid network structure called intersecting spheres (InS) neural network for the segmentation of ultrasound images in their work. This is a dynamic structured network having incremental learning whose synaptic weights and network arrangement are automatically determined during the training procedure. The feature vectors are produced by using the discrete cosine transform of image pixel intensities in the ROI. The dimension and the elements of the feature vectors are determined in view of two parameters, i.e., the amount of ignored coefficients and the dimension of the ROI. The proposed network is having three hidden layers, and the first hidden layer is formed by using genetic algorithms (GA) and back propagation is used to train all the layers. The nodes of the first layer network represent hyperspheres (HSs) in the feature space whose location and radius are found by GA. The difference in the proposed neural network and the previous studies is different in the sense of partitioning the fcature space. Here the feature space is partitioned by intersecting HSs to represent the distribution of classes. This network is compared to three other networks out of which the Modified Restricted Coulomb energy (MoRCE) network is one where intersection of the HS is not allowed. Restricted Coulomb energy (RCE) network, modified RCE network, multilayer perceptron (MLP), and the proposed hybrid neural network are comparatively experimented for the segmentation of ultrasound images. The classification performances of the four neural networks for the segmented images are analyzed, and the proposed InS neural network is outstanding among them.

In another exertion by Coppini et al. [12] an architecture is proposed and tested for lung nodule detection that narrates a neural network-based system which detects lung nodules in chest radiograms with the help of computer. It is based on multiscale processing and artificial neural networks (ANNs). The computational scheme is related to multiscale processing which is principally implemented by a feed-forward artificial neural network (FFANN). For the development and testing phases, various aspects are analyzed on the public JSRT database along with additional cross validation with the images from UNIFI database. Different points of view on image processing techniques like conventional pattern recognition methods and artificial neural networks (ANNs) are archived. Here the first one locates possible nodular patterns while the second, implemented by a convolutional neural network, differentiates nodules from non-nodules [12]. The feed-forward neural networks are also used in the system proposed by Penedo et al. There, dubious regions are first detected in a low-resolution image, and at the same time local image curvature is analyzed to locate nodules. Here the problem of nodule detection is solved by using two-layer architecture. At first to locate possible nodular regions securing high sensitivity, an attention-focusing subsystem that processes whole radiography is applied and then at second a validation is done, i.e., a sub-system is followed which reduces false alarms and increase detection specificity by evaluating the prospect of the presence of a nodule by processing regions of interest. For the enhancement of image features, biologically inspired filters are used. The FFANN allows an efficient use of the previous knowledge about the shape of nodules and the background structure. Experimental results are narrated by ROC/FROC analysis. The noted system performances support the undertaking of system validation in clinical settings. The intense changeability of chest radio graphs and the variation in nodular patterns are a prime concern in designing CAD (computer-aided diagnostic scheme) systems. Though the extended representation of data sets is still a vital issue, the winning approach is to use machine learning methods [13].

One more image processing technique using artificial neural network has been developed by Kenji Suzuki et al. for the detection of lung nodules. Here by repressing the contrast of ribs and clavicles in chest radio graphs, a technique is developed for the lung nodule detection. The lung nodules and lung vessels are made prominent or clear by suppressing the rib components to a large extent by virtue of multiresolution massive training artificial neural network (MTANN). Three MTANNs for three different resolution images are developed, each having decomposition/composition techniques for the effective suppression of ribs having different spatial frequencies. MTANN is a highly nonlinear filter and is trained by using input chest radio graphs and the corresponding "teaching" images. The "bone" images obtained by use of a dual-energy subtraction technique are employed as the teaching images. After training with input chest radio graphs and its corresponding dual-energy bone images, the multiresolution MTANN provides "bone-image-like" images which were similar to the teaching bone images. By subtracting the bone-image-like images from the corresponding chest radio graphs, "soft-tissue-image-like" images are produced where ribs and clavicles were substantially suppressed. Thus this image processing technique for rib suppression with the help of a multiresolution MTANN can be potentially useful for radiologists and CAD schemes to detect the lung nodules on chest radio graphs [14].

Soo Beom Park et al. initiated a content-based image classification method using artificial neural network in 2004. At first, a region segmentation technique is used. Feature information is extracted from images by using the wavelet transform and the sliding window-based feature extraction. A neural network classifier using back-propagation learning algorithm is created using the texture features to reflect shape of an object. The feature vectors are used as input values in the training process to construct a classifier of the object unit. A higher classification rate is achieved by this. It is applicable for raising the performance of content-based image indexing or retrieval systems, and it can be used to classify various object images, also it can automatically perform all processing for object image classification. It shows the capacity to retrieve images more efficiently by an automatic classification system and is suitable for many practical applications. A test with 300 training data and 300 test data formulated of 10 images from each of 30 classes exhibits classification rates of 81.7 and 76.7% correct, respectively [15].

In image processing techniques like handwriting recognition, text recognition, fingerprint classification, data compression, etc. binary image thinning is a primary part. It represents the structural shape of original images by using less data. A good thinning method produces skeletons including the shape information of original object so that it can be suitably used. Gu et al. proposed a new approach for binary image thinning by using the pulse-coupled neural network (PCNN). Pulse-coupled neural network (PCNN) models are bio-inspired networks based on the experimental analysis of synchronous pulse bursts in the cat visual cortex which is not similar as traditional artificial neural networks. PCNN can be applied in many fields, such as image processing, image recognition, and optimization. Here the pulse parallel transmission characteristic of PCNN is used for binary image thinning. By using this algorithm, including PCNN noise-reducing process, thin binary image can be generated and skeletons are produced with accuracy. The proposed procedure is said to be faster compared to other methods [16]. Another approach using PCNN is proposed by Xiaodong Gu et al. for the purpose of image shadow removal. Image shadow is taken as the reduction of image intensity. Two steps are followed to remove the image shadow. At first the shadowed image is segmented using PCNN that results in a multivalued segmented image. In the second step, the original shadowed image is divided by the multivalued segmented image where the quotient image comes as a result. In the quotient image, the object is more obvious and easy to find as it keeps the information of the original image. The results shows that shadows are removed completely, and the shadow-removed images are almost as the same as the original non-shadowed images. So this approach can be efficiently used for removing shadows which does not have random noise [17].

Among different image compression methods applied for medical image compression, Meyer-Bäse [18] has suggested a new method using topology-preserving neural networks. It is quite advantageous in a way that it can be applied to larger image blocks in order to obtain a low bit rate digital representation of the image with reduced image data without any impelling loss of the image quality. A combination of transformation-based linear neural network for PCA analysis and a vector-quantization neural network or neural-gas network as data compressor is applied. The linear neural network performs three types of PCA analysis by three algorithms such as Generalized Hebbian Algorithm (GHA), Oja's symmetric algorithm (OJA), and nonlinear PCA (NLPCA). The image blocks are taken as data vectors on which the PCA transformations are applied and projected on a low-dim space. The compression ratio comparisons shows an improvement in the image quality, and OJA algorithms turns out the best for it while nonlinear PCA turns out as fastest among all. PCA combined with neural vector quantizer can be taken as primary technique for image compression after studying the efficiency by blending mathematical phantom features into clinically proved cancer free mammograms [18]. Another hybrid neural model called direct classification model is proposed for image compression by Soliman and Omari [19]. Here the model is developed as a hybridization of two technique, i.e., Self-Organizing Kohonen (SOK) model and Adaptive Resonance Theory (ART) model. The features like accuracy and fastness of SOK model and ART model, respectively, generates an effective and efficient hybrid neural model for the purpose of image compression so as to achieve high image quality at high compression ratio. It provides better result than the traditional peer techniques like JPEG2000 and DjVu wavelet technology specifically for the colored images and immobile satellite images [19].

Artificial neural network is extensively applied in pattern recognition. Fadzilah Siraj et al. have taken this technique for the emotion classification. For the purpose of communication, emotion has become a prime interface between human and machine to play a primary role in rational decision-making, perception, learning along with various cognitive tasks. For emotion detection, the application of pattern recognition technology has raised a lot. Based on the physiological mensuration, facial expression, and vocal recognition, emotions are detected in human being. A human being shows the similar facial muscles while expressing a particular emotion, from which the emotion can be quantified. There are six primary emotions like anger, happiness, surprise, disgust, fear, and sadness, which were classified using neural network. For neural network training and testing real dataset of facial expression, images were captured and processed. By using multilayer perceptron network with back-propagation learning algorithm and regression analysis, the data are tested. The experimental result shows that neural network has a misclassification rate of 2.5%, while regression analysis gains a misclassification rate of 33.33%. The emotion classification model developed in by Fadzilah Siraj et al. can support the development of intelligent tutoring system in particular, and E-learning system in general [20].

To estimate the total suspended matter (TSM) concentration from remotely sensed multispectral data in a particular area of the Portuguese coast, different methodologies are applied by Ana C. Teodoro et al. in their work. These techniques based on single-band models, multiple regressions, and artificial neural networks (ANN) were evaluated by error estimation to find out the more accurate methodology. The root-mean-square errors by both the linear and nonlinear models are analyzed and found out that they support the hypothesis that the relationship between the seawater reflectance and TSM concentration is clearly nonlinear. For estimating the TSM concentration estimation, the ANNs are found to be more useful from reflectance of visible and bands of different sensors used. The ANN which is implemented here is with ten units in the hidden layer and is able to model the transfer function better than multiple regressions [21].

A large review says that artificial neural networks hold a huge role to amplify the performance of classification or segmentation. Mehmet Nadir Kurnaz et al. has proposed an unsupervised incremental neural network for segmentation of tissues in ultrasound images and compared its performances with another unsupervised neural network known as Kohonen neural network. Kohonen neural network is a non-incremental unsupervised neural network for which the topology is defined before while for incremental neural network is not essentially predefined. Trial and error method is used to determine the number of output nodes in the KNN while in incremental NN they are selected by analyzing histograms and the image to be processed. To extract the feature from the images, 2D-DFT (Discrete Fourier Transform) and 2D-DCT (Discrete Cosine Transform) are used. The incremental neural network gives better result compared to Kohonen network in terms of both result and time. Some works also describes that incremental neural network gives enhanced performance in terms of quality of the reconstructed images along with the compression rates. The proposed network takes less number of nodes so that less time to perform and giver better performance. The neural network is a two-layer incremental neural network which can be used further without any modifications easily only by forming the training sets to resolve the topology of the network accordingly [22, 23].

In the medical science and orthopedic community, the identification of the spinal deformity classification is an important topic. The artificial neural network

(ANN) can also be used to identify the classification patterns of the scoliosis spinal deformity. Lin Hong in his paper has used a multilayer feed-forward neural network with back propagation (MLFF/BP) for the classification. At first based on the coronal and sagittal X-ray images, the simplified 3-D spine model was constructed and the features of the central axis curve of the spinal deformity patterns were extracted by the total curvature analysis. The discrete form of the total curvature, including the curvature and the torsion of the central axis of the simplified 3-D spine model, was derived from the difference quotients. The total curvature values are taken as input to the MLFF/BP ANN and five neurons are at output layer representing five King classification types. About 67% of the data is taken for training and the rest for testing. Two types of network architecture are taken into consideration: one with only one hidden layer and the other network having two hidden layers. The result was found that the two-layer hidden neural network performs better, which can be further improved by increasing the number of training datasets or by participation of more experienced observers [24].

A fusion of neural network is taken as a classifier for image classification problems here by Sanggil Kang et al. The input features such as color layout (CL), edge histogram (EH), and region-based shape (RS) are extracted from different MPEG-7 descriptors. The fusion of input features which are extracted from multiple descriptors gives better performance than the features extracted from single descriptor. The networking system has two parts: a feature extraction module and a classification module. The conclusive result of the experiment says that this method provides robust training performance compared to conventional neural network classifier. It is useful in cases where fusion of different dimension features is used for neural network classifier. The disadvantage of using this method is complexity, i.e., the proposed classifier is more complicated than the conventional neural network for its functionality [25].

In the field of remote sensing for complex retrieval tasks, artificial neural networks (ANNs) are proved to be very effective technique. For solving the unmixing problem in hyperspectral data, Licciardi and Frate have used artificial neural network. The neural network have two stages of processing: the first stage is used to reduce the dimension of the input vector, and the second stage maps the reduced input vector to the abundance percentages. For the dimensionality reduction, auto-associative N's is used. Both the dimensionality reduction procedure and the final unfixing are performed by the developed neural network model. The final scheme of the model is a single architecture of NN sequencing the two operations in an automatic mode. Different sets of experimental data are taken for the performance estimation. The unmixing results show that the reduced vector helps to yield accurate pixel abundance estimation. The result shows that it is effective in terms of dimensionality reduction as well as accuracy in the final estimation. The impact of the applied technique is quite advantageous and could be more significant for future use as the satellite multiconfiguration data is continuously increasing day by day [26].

In recognition of handwritten digits or traffic signs, machine learning methods do not perform well all times while the wide deep artificial neural network

(DNN) gives satisfactory results many times. A large depth network is obtained by minimal receptive fields of convolutional winner-take-all neurons, which results in many sparsely connected neural layers and only winner neurons are need to be trained. Dan Ciresan et al. have taken a DNN having two-dimensional layers of winner-take-all neurons with overlapping receptive fields whose weights are shared. DNN columns are combined to make multicolumn DNN (MCDNN) inspired by microcolumns of neurons in the cerebral cortex and give a far better performance compared to single DNN. The method is fully supervised and does not use any additional unlabeled data source. This proposed method improves the state of the art by 30–80% over many image classification datasets. Drastically improvement is recognized on MNIST, NIST SD 19, Chinese characters, traffic signs, CIFAR10, and NORB datasets [27].

Along with different usage, neural networks also show their applicability in the agricultural field. A classification technique has been developed in Malaysia by Z. Husin et al. to recognize and to classify the herbal plants. They developed a device capable of recognizing herbs species by classification technique based on structural characteristics of the leaves. Generally, these are done by human directly which is time-consuming so turned to be ineffective and inefficient. So to identify the herbs and agricultural plants, different image processing methods are used successfully. The picture samples of leaves were collected from the Agricultural Department of Malaysia Perlis University on which the image processing techniques are applied and tested. The RGB image is converted to a gray scale image which is again converted to a binary image from which features are extracted. These features are gone through processing using morphological technique and SVD function to get the input for the neural network. The neural network used here is back-propagation neural network (BPNN). The inputs for the neural network are the individual pixels of a leaf image developed through appropriate image processing steps. A two-layer BPNN is used which has 20 hidden neurons, 4800 input neurons and 20 output neurons and threshold value is set to be 0.5. Here each output neuron is a type of plant species to be identified. Most outstandingly, the system is capable of identifying the herbs leaves species even though they are dried, wet, torn, or deformed. The average correct recognition rate is found to be 98.9% which is quite appreciable [28].

Dan Ciresan et al. from Switzerland have developed a system using multicolumn deep neural network (MCDNN) for the traffic signal recognition. This is essential for the automotive industry and for many traffic associated applications. The MCDNN is developed by combining several deep neural networks (DNN), which are training by several pre-processed data. The DNN contains a sequence of convolutional and max-pooling layers and is of feed forward type. This creates feature vectors from the pixel intensities and the adaptable parameters are optimized through minimization of the misclassification error over the training set. Eventually, the MCDNN is formed by averaging the output activations of several DNN columns. Compared to single DNN, the combination of DNNs gives a better improved result robust to noise. This particular proposition has won the final phase of the German traffic sign recognition benchmark and achieved a recognition rate of 99.46% which is far better than human recognition rate [29] (Table 1).

Table 1 Comparison of neural network-based image processing techniques

Year/author/publisher	Type of image data and data source	Applied method/technique	Findings
2001, G. Giacinto, F. Roli, Image and vision computing (Elsevier)	Multisensor remote sensing image (related to agricultural area)	An ensemble of neural network model is applied. For choosing the networks, 'overproduce and choose' strategy is used	Provides better results as compared to other models and methods
2001, K.T. Sun, S. J. Lee, and P.Y. Wu, Neurocomputing (Elsevier)	Lena or Lenna images Source: http://sipi.usc.edu	Neural network in partitioned iterated function system (PIFS) is used	Gives good compression ratio compared to traditional PIFS method
2002, Hazem M. El-Bakry, Neurocomputing (Elsevier)	–	Image decomposition and fast neural networks	Better result is obtained with decreasing running time, thus increasing speed-up ratio compared to conventional neural network
2002, Zumray Dokur, Tamer Olmez, Pattern recognition Letter (Elsevier)	Ultrasound images of bladder and kidney cyst are taken	Intersecting spheres (InS) neural network	Classification performance is better than RCE, MoRCE, and MLP
2003, G. Coppini et al., IEEE Transactions on Information Technology in Biomedical	Chest radiograph images (medical image data) Source: Japanese Society of Radiological Technology (JSRT) and Department of Physiopathology of University of Florence	Feed-forward ANN system is used based on multiscale processing, and a convolutional NN is used for discrimination	Performs better than other models
2004, Soo Beom Park et al., Pattern recognition Letter (Elsevier)	Different object images are taken from Internet	Back-propagation neural network	A higher classification rate is achieved

(continued)

Table 1 (continued)

Year/author/publisher	Type of image data and data source	Applied method/technique	Findings
2004, X.D. Gu, D.H. Yu, L.M. Zang, Pattern recognition Letter (Elsevier)	Binary images including English words, Chinese words, and other images	Pulse-coupled neural network (PCNN)	Performs better compared to traditional parallel algorithm
2005, X.D. Gu, D.H. Yu, L.M.Zang, IEEE Transactions on neural network	Gray images and a color image are taken with shadow	Pulse-coupled neural network (PCNN)	PCNN can be used for the shadow removal technique
2005, A.M. Base et al., Engineering applications on artificial intelligence (Elsevier)	Mammogram images Source: MIAS database at http://skye.icrac.uk/misdb/miasdb.html	Topology-preserving neural network	This neural network approach provides better probability distribution estimation method
2006, H.S. Soliman, M. Omari, Applied soft computing (Elsevier)	Several satellite images and Lena images are used Source: http://sipi.usc.edu	A hybrid neural network called direct classification neural network	DC model performed well for image compression compared to the peer state-of-the-art models
2006, S. Kenji, H. Abe, H. Mecmohan, IEEE transactions on medical imaging	Chest Radio graph Images Source: Digital Image database developed by Japanese Society of Radiological Technology (JSRT) and FCR 9501ES; Fujifilm medical Systems, Stanford, CT	Nonlinear filter MTANN is used for suppressing the contrast of ribs and clavicles in chest radio graphs	Proved to be a useful technique for radiologists and CAD schemes in detection of lung nodules
2006, F. Siraj, N. Yusoff, L.C. Kee, Computing and informatics (IEEE conference)	Image data Captured by digital camera	Multilayer perceptron with back-propagation learning algorithm and regression analysis is used	Neural network gives less error, i.e., NN performs better than regression analysis
2007, Ana. C. Teodoro, F.V. Gomes, H. Goncalves, IEEE Transactions on Geo-science and Remote sensing	Remote-sensed multispectral data	Multiple regression and artificial neural network are used	ANN performs better than multiple regression

(continued)

Table 1 (continued)

Year/author/publisher	Type of image data and data source	Applied method/technique	Findings
2007, Mehmet Nadir Kurnaz, Zu mray Dokur, Tamer Olmez, Computer methods and programs in biomedicine (Elsevier)	Phantom ultrasound bladder image and an original bladder image are taken Source: www.fantom.suite.dk http://drgdiaz.com	An incremental neural network is used	Incremental neural network performs better than Kohonen neural network
2008, Hong Lin, IEEE Transactions on biomedical engineering	3-D spine models are constructed based on coronal and sagittal spinal images	Total curvature analysis is used for feature extraction, and multilayer feed-forward neural network with single and double hidden layers are used	Neural network with two hidden layers performs better
2009, S. Kang, S. Park, Pattern recognition Letter (Elsevier)	Sports image data collected from Internet	Feature extraction using two MPEG-7 descriptors EH and RS. Neural network and a fusion of neural network is used	Proposed fusion neural network is better than conventional neural network
2011, G.A. Licciardi, F.D. Frate, IEEE Transactions on Geo-science and remote sensing	Airborne and space-borne hyperspectral scanning images Source: INTA-AHS instrument dataset from European Space aging (ESA), CHRIS-PROBA images and AVIRIS images from http://avairis.jpe.nasa.gov	Auto-associative neural network is used	Though auto-associative NN is used for airborne images before, it can also be used for space-borne images
2012, C. Dan et al. Computer vision and pattern recognition (IEEE conferences)	MNIST, NIST SD 19, Chinese characters, Traffic Signs, CIFARIO NORB databases are used	Multicolumn deep neural network (MCDNN)	Improved classification is observed
2012, Z. Husin et al., Computers and electronics in agriculture (Elsevier)	Leaf of different species are taken Source: Agricultural Department of Malaysia Perlis University	Back-propagation neural network	Recognition rate of 98.9% is achieved
2012, Dan Cireşan et al., Neural networks (Elsevier)	Traffic sign imagers are taken	Multicolumn deep neural network (MCDNN)	Recognition rate of 99.46% is achieved

3 Different Type of Neural Networks

Neural networks have been applied for solving a large variety of tasks that are generally not easy to resolve using ordinary rule-based or traditional programming. Many types of neural network models are developed which has been used in many fields for solving different problems, and some of the types are briefly discussed below.

3.1 Feed-Forward Artificial Neural Network (FFANN)

The feed forward neural network is the simple artificial neural network. It is said to be feed-forward type or acyclic in nature as they do not have any feedback loop or self-feedback links between the layers, i.e., here the processing units, the neurons are only connected forward. In this network, the information moves in a single (forward) direction, i.e., from input units, through the hidden units to the output units. It can be of single layered or multilayered called as multilayer feed-forward neural network. Such a network is fully connected if each node in layer I is connected to all nodes in layer $i + 1$ for all I.

When a training tuple is fed to the input layer of the network, the inputs pass through the input units, unchanged in the input layer first. That is, for an input unit, j, its output, O_j is equal to its input value, I_j. Then the net input and output of each unit in the hidden and output layers are computed. The net input to a unit in the hidden or output layers is computed as a linear combination of its inputs. Each connection has a weight. To compute the net input to the unit, each input connected to the unit is multiplied by its corresponding weight, and the summation is calculated.

Given a unit j in a hidden or output layer, the net input, I_j, to unit j is

$$I_j = \sum w_{ij} O_j + \theta_j \tag{1}$$

where w_{ij} is the weight of the connection from unit i in the previous layer to unit j, O_j is the output of unit i from the previous layer, and θ_j is the bias of the unit.

Each unit in the hidden and output layers takes its net input and then applies an activation function to it. The function symbolizes the activation of the neuron represented by the unit. Given the net input I_j to unit j, then O_j, the output of unit j, is computed as

$$O_j = 1/1 + e^{-I_j} \tag{2}$$

Thus, the output of the network is calculated and compared with the target output for the training purpose.

3.2 Multilayer Perceptron (MLP)

A multilayer perceptron is a feed-forward artificial neural network which maps the sets of input data to a set of appropriate outputs. It has multiple layers of nodes in a directed graph, where every layer is fully connected to the next layer. The calculations are similar to feed-forward neural networks as described above.

3.3 Massive Training Artificial Neural Network (MTANN)

The massive training artificial neural network (MTANN) is a modified multilayer ANN, which can directly handle input gray levels and output gray levels. Here for each layer, a different activation function is selected. The activation functions of the input, hidden, and output layers are a linear, a sigmoid, and a linear function, respectively. In this network, image processing or pattern recognition is performed by scanning an image with the modified ANN. The MTANN consists of a linear-output multilayer ANN model for which it is capable of operating on image data directly. The MTANN uses a linear function as the activation function in the output layer that significantly improves the characteristics of an ANN.

3.4 Deep Artificial Neural Network and Multicolumn Deep Artificial Neural Network

A deep neural network (DNN) is an artificial neural network that contains many hidden layers of units between the input and output layers. Generally, DNNs are designed as feed-forward networks, but can be designed as recurrent neural network recently for the applications such as language modeling. A DNN can be trained with the standard back-propagation algorithm. Here the weight updating can be done using the following equation:

$$w_{ij}(t+1) = w_{ij}(t) + \eta \frac{\partial C}{\partial w_{ij}} \tag{3}$$

Here η is the learning rate and C is the cost function. The choice function depends on the type of learning and activation function chosen.

3.5 *Pulse-Coupled Neural Network (PCNN)*

PCNN is a single-layered, two-dimensional artificial neural network developed by Johnson et al. In this network, each neuron corresponds to one pixel of the input image. It has three parts mainly, i.e., input, linking, and the pulse generator. PCNN receives the input stimulus through both feeding and linking connections that are combined in an internal activation system and accumulates the stimuli until it exceeds a dynamic threshold, resulting in a pulse output. It does not need any pre-training. Through iterative computation, PCNN neurons produce temporal series of pulse outputs that contain information of input images to be employed for miscellaneous applications of image processing.

4 Conclusion and Future Direction

Here many applications of artificial neural network have been elaborately discussed. The discussion shows a wide usability of the network in various image processing techniques. Its role in image processing can be primary or secondary, or can be used as a part of different combination of techniques. Also it can be used in supervised/unsupervised or parametric/nonparametric or linear/nonlinear regression functions or feature extractions and many more. Every applications of neural network are unique and no technique is better than another, each has its own strength and weaknesses. Though it has been applied and proved to be very usable for a wide range of applications, it has also been seen that the combination of different models/techniques works more effectively. Therefore, it can lead to the development of many hybrid models in future. The neural network performance depends on different parameters of the models which need to be properly decided and optimized so as to design some new helpful models.

References

1. Giorgio, Giacinto, and Fabio Roli. 2001. Design of Effective Neural Network Ensembles for Image Classification Purposes. *Image and Vision Computing* 19 (9): 699–707.
2. Jacquin, A.E. 1992. Image Coding Based on a Fractal Theory of Iterated Constructive Image Transformations. *IEEE Transactions on Image Processing* 1 (1): 18–32.
3. Jacquin, A.E. 1993. Fractal Image Coding a Review. *Proceedings of the IEEE* 81 (10): 1451–1465.
4. Sun, K.T., S.J. Lee, and P.Y. Wu. 2001. Neural Network Approaches to Fractal Image Compression and Decompression. *Neurocomputing* 41 (1): 91–107.
5. Schneiderman, H., and T. Kanade. 1998. Probabilistic Modeling of Local Appearance and Spatial Relationships for Object Recognition. In *Proceedings of IEEE Conference on Computer Vision and Pattern Recognition (CVPR)*, Santa Barbara, CA, 45–51.

6. Feraud, R., O. Bernier, J.E. Viallet, and M. Collobert. 2000. A Fast and Accurate Face Detector for Indexation of Face Images. In *Proceedings of Fourth IEEE International Conference on Automatic Face and Gesture Recognition*, Grenoble, France, 77–82.
7. El-Bakry, H.M. 2001. Automatic Human Face Recognition Using Modular Neural Networks. *International Journal of Machines and Graphics* 10 (1): 47–73.
8. Rowley, H.A., S. Baluja, and T. Kanade. 1998. Neural Network-based Face Detection. *IEEE Transactions on Pattern Analysis and Machine Intelligence* 20 (1): 23–38.
9. El-Bakry, Hazem M. 2002. Face Detection Using Fast Neural Networks and Image Decomposition. *Neurocomputing* 48 (1): 1039–1046.
10. Alison, Noble J., and Djamal Boukerroui. 2006. Ultrasound Image Segmentation: A Survey. *Medical Imaging, IEEE Transactions.* 25 (8): 987–1010.
11. Zümray, Dokur, and Tamer Ölmez. 2002. Segmentation of Ultrasound Images by Using a Hybrid Neural Network. *Pattern Recognition Letters* 23 (14): 1825–1836.
12. Coppini, Giuseppe, et al. 2003. Neural Networks for Computer-aided Diagnosis: Detection of Lung Nodules in Chest Radiograms. *Information Technology in Biomedicine, IEEE Transactions* 7 (4): 344–357.
13. Penedo, M., M. Carreira, A. Mosquera, and D. Cabello. 1998. Computer-aided Diagnosis: A Neural-network Based Approach to Lung Nodule Detection. *IEEE Transactions on Medical Imaging* 17: 872–880.
14. Kenji, Suzuki, Hiroyuki Abe, and Heber MacMahon. 2006. Image-processing Technique For Suppressing Ribs in Chest Radiographs by Means of Massive Training Artificial Neural Network (MTANN). *Medical Imaging, IEEE Transaction* 25 (4): 406–416.
15. Park Soo Beom, Jae Won Lee, and Sang Kyoon Kim. Content-based image classification using a neural network. *Pattern Recognition Letters* 25 (3): 287–300.
16. Xiaodong, Gu, Yu, Daoheng, and Liming Zhang. 2004. Image Thinning Using Pulse Coupled Neural Network. *Pattern Recognition Letters* 25 (9): 1075–1084.
17. Xiaodong, Gu, Yu, Daoheng, and Liming Zhang. 2005. Image Shadow Removal Using Pulse Coupled Neural Network. *Neural Networks, IEEE Transactions* 16 (3): 692–698.
18. Meyer-Bäse, Anke, et al. 2005. Medical Image Compression Using Topology-Preserving Neural Networks. *Engineering Applications of Artificial Intelligence* 18 (4): 383–392.
19. Soliman, Hamdy S., and Mohammed Omari. 2006. A Neural Networks Approach to Image Data Compression. *Applied Soft Computing* 6 (3): 258–271.
20. Siraj Fadzilah, Nooraini Yusoff, and Lam Choong Kee. Emotion classification using neural network. In *Computing & Informatics, ICOCI'06, International Conference on IEEE*, 1–7.
21. Teodoro Ana C., Fernando Veloso-Gomes, and Hernâni Gonçalves. Retrieving TSM Concentration from Multispectral Satellite data by Multiple Regression and Artificial Neural Networks. *Geoscience and Remote Sensing, IEEE Transactions* 45 (5): 1342–1350.
22. Kurnaz, Mehmet Nadir, Zumray Dokur, and Tamer Olmez. An Incremental Neural Network for Tissue Segmentation in Ultrasound Images. *Computer Methods and Programs in Biomedicine* 85 (3): 187–195.
23. Dokur, Zumray. 2008. A Unified Framework for Image Compression and Segmentation by Using an Incremental Neural Network. *Expert Systems with Applications* 34 (1): 611–619.
24. Hong, Lin. 2008. Identification of Spinal Deformity Classification with Total Curvature Analysis and Artificial Neural Network. *Biomedical Engineering, IEEE Transaction* 55 (1): 376–382.
25. Sanggil, Kang, and Sungjoon Park. 2009. A Fusion Neural Network Classifier for Image Classification. *Pattern Recognition Letters* 30 (9): 789–793.
26. Giorgio, Licciardi, and Fabio Del Frate. 2011. Pixel Unmixing in Hyperspectral Data by Means of Neural Networks. *Geoscience and Remote Sensing, IEEE Transactions* 49 (11): 4163–4172.
27. Dan, Ciresan, Ueli Meier, and Jürgen Schmidhuber. 2012. Multi-column Deep Neural Networks for Image Classification. In *Computer Vision and Pattern Recognition (Cvpr), IEEE Conference*, 3642–3649.

28. Husin, Z., et al. 2012. Embedded Portable Device for Herb Leaves Recognition Using Image Processing Techniques and Neural Network Algorithm. *Computers and Electronics in Agriculture* 89: 18–29.
29. Dan, Cireşan, et al. 2012. Multi-column Deep Neural Network for Traffic Sign Classification. *Neural Networks* 32: 333–338.

Ensemble Methods for Improving Classifier Performance

Monalisa Panda, Debahuti Mishra and Sashikala Mishra

Abstract In this paper, ensemble methods for different base classifiers are proposed. An ensemble technique is a supervised learning algorithm that combines a group of classifiers in order to acquire an overall model with more exact decisions. The classifiers that are support vector machine (SVM), naive Bayes (NB), and back propagation neural network (BPNN) are trained and tested on different gene expression datasets using both random selection method and *k*-fold cross-validation method. Both binary-class and multi-class datasets are used for evaluation of effectiveness of the ensemble method. Various publicly available gene expression datasets have been used for experiments in order to find the accuracy and effectiveness of the ensemble technique. Performance of the different classification methods and ensemble methods has been compared by using the accuracy values. The results have shown that the accuracy for the gene expression datasets has been increased by using the ensemble methods.

Keywords Gene expression · Ensemble · SVM · NB · BPNN
K-fold cross-validation

1 Introduction

Micro-array data is now used in many fields of medical diagnosis that is used for the detection of breast cancer, lymphoma, leukemia, etc. In order to measure the changes in expression levels of huge number of genes, micro-array data is used.

M. Panda (✉)
Department of CSE, CAPGS, BPUT, Rourkela, India
e-mail: monalisapanda1989@gmail.com

D. Mishra · S. Mishra
Department of CSE, ITER, Siksha 'O' Anusandhan University, Bhubaneswar, India
e-mail: mishradebahuti@gmail.com

S. Mishra
e-mail: sashi.iter@gmail.com

M.S. Reddy et al. (eds.), *International Proceedings on Advances in Soft Computing, Intelligent Systems and Applications*, Advances in Intelligent Systems and Computing 628, https://doi.org/10.1007/978-981-10-5272-9_34

Classification is a supervised learning process used for predicting a class label to any unseen data on the basis of training set of data, whose class label is already known. Nowadays, many existing classifiers such as SVM, *k*-nearest neighbor, ANN, Bayesian classifier, decision tree, linear regression are present. Commonly, a single classification method is not sufficient enough to correctly identify the class level. An ensemble technique is a supervised learning algorithm technique which combines a group of models in order to obtain an overall model with more precise decisions [1]. The models prediction, classification performance is usually improved by using the ensemble techniques.

Hencc instcad of choosing just one model, if we combine the outputs of different models, then the risk of selection of a badly performing classifier can be reduced. Several ensemble methods are there like voting, bagging, boosting, Bayesian merging, stacking, distribution summation, Dempster–Shafer, density-based weighting [2, 3]. This work mainly contains various classification and ensemble strategies, the set of laws for selecting the reduced data from large data sets, the act of using different classification techniques, how the classification and ensemble technique can be applied over different gene expression data sets. Here, stacking is used as an ensemble technique; that is, it combines the decisions of the individual classifier by using majority voting fusion rule. Stacking is concerned with combining multiple classifiers obtained by using various learning algorithms on a particular data set [4].

Finally, a comparison is done among different base classifiers and ensemble methods, and it was found that the ensemble methods were demonstrated with much better performance.

The rest of the paper is organized as follows: the basic definition of classifier ensemble is described in Sect. 2. Section 3 depicts the model. Section 4 explains the general methods, concepts, and approaches that are used to find out the result. Section 5 describes the two different ensemble techniques that are used to improve the result. Through simulation on variety of datasets, the result of the proposed model is reported in Sect. 5.

2 Classifier Ensemble Analysis

Classification is prediction of a certain result based on a given input. A training set containing a set of attributes and the result, usually called goal or prediction is being processed in order to predict the result. Classification in other words is a data mining function that assigns items in a group to mark categories or classes. Generally Classification is a process of estimating to which of a set of examples a new example belongs to, on the basis of a training dataset, whose class label is already known [2]. The algorithm that implements this process is known as classifier, which is a mathematical function that maps a data to a category.

In general, the single classification technique is not sufficient enough to identify the class level properly. An ensemble is itself a supervised learning algorithm which combines a set of models in order to obtain a global model with more accurate and reliable decisions [1]. When more number of algorithms is used in a model it becomes expensive. Therefore, nowadays, the researchers are emphasizing on the ensemble techniques. These techniques use to reduce the error rate in classification tasks in comparison with single classifiers. Also, the amalgamation of various techniques to make a final conclusion makes the performance of the system more strong against the difficulties that each individual classifier may have on each data set. Ensemble is mainly done to improve the accuracy and efficiency of the classification system.

3 Proposed Model

As mentioned earlier, this work focuses on the second phase of the model, that is, classifier ensemble techniques. In phase one, random selection method is used for training and testing of data. Here, we have used three classifiers, namely naive Bayes, backpropagation neural network, and support vector machine. In the second phase, *k*-fold cross-validation technique is used to divide the data set into training and testing. The value of *k* depends on the data set. Then training and testing is done up to *k* times for all the classifiers iteratively and then classifier fusion technique that is Stacking and Majority Voting are used to combine the outputs of the individual classifiers (Fig. 1).

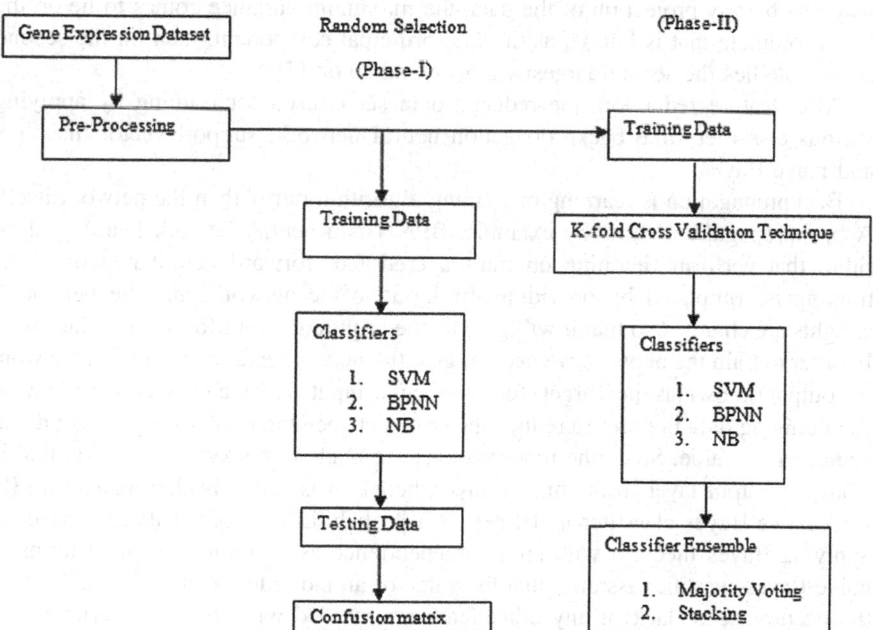

Fig. 1 Proposed model

4 Concepts, Methods, and Approaches

Initially, datasets need to be normalized. Data transformation such as normalization is a data preprocessing tool which is used in data mining system in order to remove the noisy data. An attribute of a dataset is getting normalized by scaling its values so that they fall within a specified range, such as 0.0–1.0 [5, 6]. Normalization is mostly useful for classification algorithms and clustering technique. Here, Min–Max normalization is used as a tool for preprocessing. Here, min A and max A are the minimum and maximum values of an attribute A. This technique can be calculated by using

$$V = (v - \min A)/(\max A - \min A). \tag{1}$$

After normalization, data reduction is done using PCA. The presence of large data sets can cause rigorous problems in an organization's decision support systems and database management systems. Micro-array data is high-dimensional data which can cause significant problems such as irrelevant genes, difficulty in constructing classifiers, and multiple missing gene expression values. In this paper, we have employed principal component analysis (PCA) as the feature reduction technique to extract the needful features, which can be used to train the classifiers. This feature reduced dataset is expected to provide a better classifier in terms of accuracy and efficiency. PCA is defined as a feature extraction method that transforms the data to a new coordinate system that is known as orthogonal linear transformation in such a way that by any projection of the data, the maximum variance comes to lie on the first coordinate that is known as the first principal component, then on the second coordinate lies the second largest variance and so on [7].

After feature reduction, the reduced data set is used for training by applying various classifiers like backpropagation neural network, support vector machine, and naive Bayes.

Backpropagation is learning or training algorithm rather than the network itself. A backpropagation learns by example. BPNN is a neural network learning algorithm that performs learning on multilayered feed-forward neural network. The training is completed by providing the input to the network, and the networks' weights are changed so that it will give us the required output for a particular input. In order to train the network we need to give the network examples of what we want the output (known as the Target) for a particular input. The weights are modified for each training data in order to reduce the error between the network's prediction and actual target value. Since the modifications are made in backward direction that is from the output layer to the hidden layer, hence, it is called backpropagation [8].

A naive Bayes classifier is defined as a probabilistic classifier that is based on applying Bayes theorem with some independence assumptions. In plain terms, a naive Bayes classifier assumes that the value of an individual feature is unrelated to the occurrence or lack of any other feature, provided with the class variable. An advantage of the naive Bayes classifier is that it only requires a small amount of training data to guess the parameters, that is, mean and variance of the variables that are necessary for the classification [9]. The Bayesian classification assumes a basic probabilistic model, and it allows to capture uncertainty about the model in a disciplined way by determining probabilities of the outcomes. It calculates explicit probabilities for hypothesis, and it is robust to noise in input data. Bayes theorem provides an approach to update the probability distribution of a variable based on information newly available by calculating the conditional distribution of the variable given the new information. The updated conditional probability distribution provides the new level of certainty about the variable. Posterior probability is calculated by updating the prior probability by using Bayes theorem. It uses the knowledge of prior events to predict the future events [10, 11]. Bayes theorem says:

$$P\left(\frac{\theta}{Y}\right) = P(\theta)^{*}P(Y/\theta)/P(Y), \tag{2}$$

where $P(\theta)$ and $P(Y)$ are the unconditional distributions of θ and Y. $P(\frac{\theta}{Y})$ is the posterior distribution of θ.

$P(Y/\theta)$ is the likelihood function, and it measures how closely Y is distributed around θ.

SVM is used as a mapping function that transforms data in input space to data in feature space in a linearly separable manner [12, 13]. In machine learning, support vector machines (SVMs) are supervised learning models with associated learning algorithms, which analyze data and recognize patterns used for classification [14]. A support vector machine represents points in space, where the examples can be separated into distinct categories by a clear wide gap. Based on their category, new groups are being classified into one of those groups. In order to transform the original training data into a higher dimension, a nonlinear mapping is used. Support vector machines find a hyperplane which would be able to separate both the plane by retrieving the support vectors. SVM separates the hyperplane of class levels +1, −1 that is situated in maximum distance from both the positive and the negative samples. From both the negative and the positive pair, feature vectors are being extracted which are assigned with the class label of +1 and −1 to know whether the pair is a interacting or a non-interacting pair.

5 Classifier Ensemble Methods

An ensemble is itself a supervised learning algorithm which combines a set of models in order to obtain a global model with more accurate and reliable decisions [2, 15]. Classifier combination is one of the most frequently explored methods in data mining in the recent years. These techniques use to reduce the error rate in classification tasks in comparison with single classifiers. Therefore, nowadays, the researchers are emphasizing on the ensemble techniques. In this paper, majority voting and stacking are used on various gene expression datasets.

In majority voting, an unlabeled example is classified in accordance with the class that obtains the highest number of votes. It can be represented as follows:

$$\text{Class}(X) = \arg\max_{c_i \in \text{dom}(y)} \sum \forall k c_i = \arg\max_{c_i \in \text{dom}(y) PM_k} \left(y = \frac{c_j}{x} \right), \quad (3)$$

where M_k denotes the classifier k and $_{PM_k}\left(y = \frac{c_j}{x}\right)$ denotes the probability of y obtaining the value of c at an instance x [16, 17].

Stacking is an ensemble method that is used for achieving the highest generalization accuracy. The reliability of the classifiers is judged on the basis of the meta-learner which learns from the outputs of the base learners. It uses the results of the base classifiers to produce a new record on which we need to apply a second learning algorithm [4]. This method allows us to maximize the utilization of the information contained in the training dataset. Normally to form a meta-learner training set, we divide the original training set into k disjoint subsets of equal size that is known as k-fold cross-validation technique [4, 18]. k will affect the overall accuracy boost and overall cost. The different base classifiers are trained and tested on different partitions of the training data. In the second level, again the classifiers are trained with the new class obtained from first level and the final accuracy is obtained. The results provided by this method were very good. The algorithm says as follows:

1. From the training set T, create k partitions from it and the cross-validation technique is used for all the base classifiers.
2. Machine learning is used to obtain second-level classifier.
3. A new class label is created and again uses the base classifiers to test the data and accuracy is found (Fig. 2).

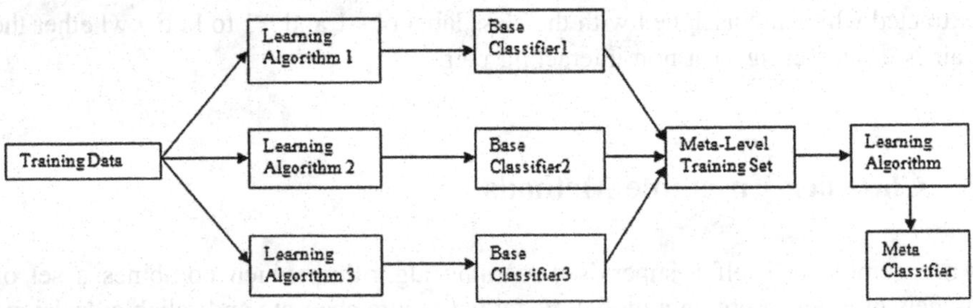

Fig. 2 Stacking technique

6 Results and Discussion

The set of experiments has been carried out using six datasets as shown in Table 1, such as breast cancer, lung cancer, iris, *E. coli*, yeast from UCI repository.

The proposed model has been tested with all the individual classifiers SVM, BPNN, NB, and the ensemble method that is stacking and majority voting for all five bench mark data sets as illustrated in Tables 2, 3 and 4. The threefold cross-validation test had been carried out, and the accuracy is measured. Entire algorithm is written and tested in MATLAB R2010a (Figs. 3, 4, 5, 6 and 7).

Table 1 Different datasets used for experimental evaluation

S. No.	Data set name	No. of instances	No. of attributes	No. of classes	References
1	Breast cancer	569	32	2	[19]
2	Lung cancer	32	56	4	[20]
3	Iris	150	4	3	[21]
4	*E. coli*	1484	9	10	[22]
5	Yeast	336	8	8	[23]

Table 2 Accuracy of different datasets using different classifiers using random selection method

S. No.	Data set name	Accuracy using BPNN	Accuracy using NB	Accuracy using SVM
1	Breast cancer	84.21	86.84	81.81
2	Lung cancer	84.28	85.71	86.84
3	Iris	83.33	85	84.84
4	*E. coli*	83.87	83.87	83.87
5	Yeast	86.80	87.81	86.80

Table 3 Accuracy of different datasets using stacking ensemble method

S. No.	Data set name	Accuracy using stacking BPNN	Accuracy using stacking NB	Accuracy using stacking SVM
1	Breast cancer	92.10	94.73	94.73
2	Lung cancer	97.14	100	94.28
3	Iris	93.33	95	95
4	*E. coli*	90.32	93.54	90.32
5	Yeast	95.17	93.44	95.86

Table 4 Accuracy of different datasets using majority voting ensemble method

S. No.	Data set name	Accuracy using majority voting	Execution time in s	Memory occupied in KB
1	Breast cancer	100	11.72	56.6
2	Lung cancer	98.57	12.70	28.6
3	Iris	100	11.86	11.7
4	*E. coli*	90.32	16.97	25.1
5	Yeast	98.62	93.43	92.6

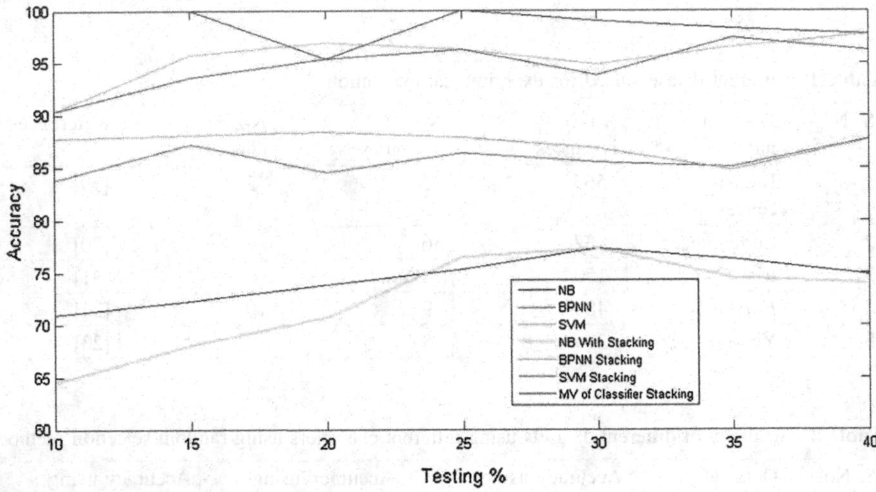

Fig. 3 Accuracy of classifiers and ensemble methods on breast cancer dataset

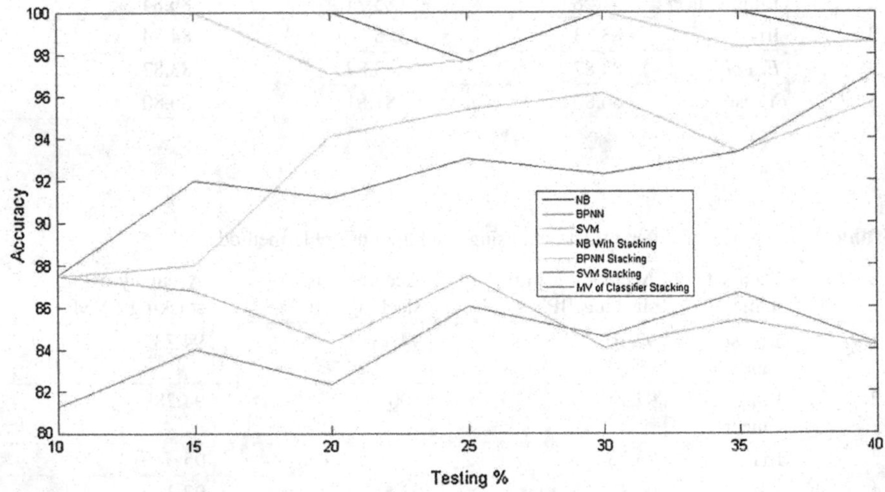

Fig. 4 Accuracy of classifiers and ensemble methods on lung cancer dataset

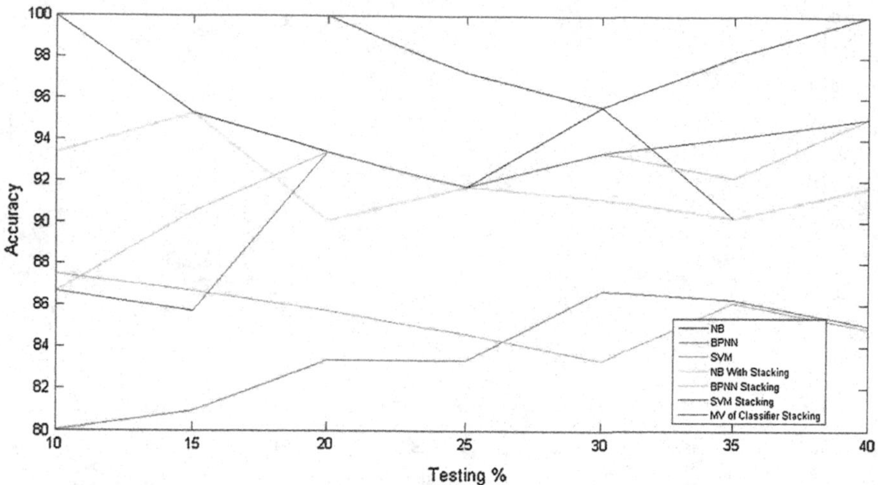

Fig. 5 Accuracy of classifiers and ensemble methods on iris dataset

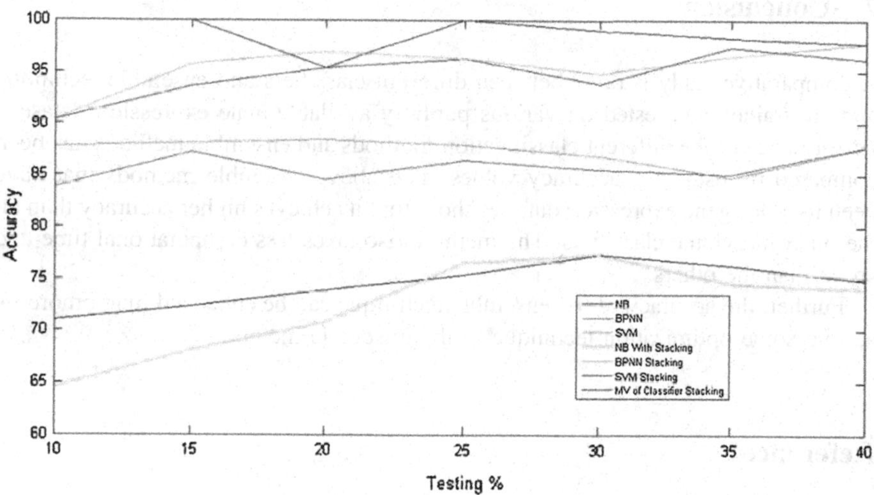

Fig. 6 Accuracy of classifiers and ensemble methods on *E. coli* dataset

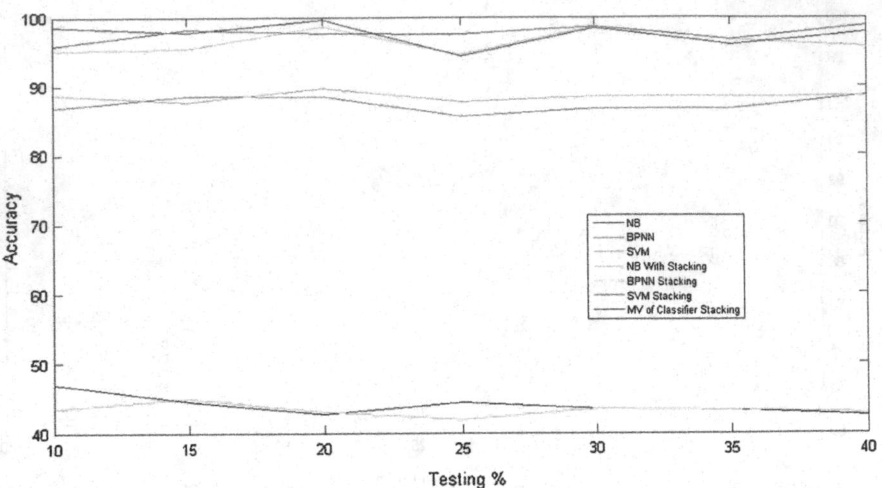

Fig. 7 Accuracy of classifiers and ensemble methods on yeast dataset

7 Conclusion

A comparative study is done between different classifiers and ensemble technique and are trained and tested on various publicly available gene expression datasets. Performance of the different classification methods and ensemble methods has been compared by using the accuracy values. The above ensemble methods that have been used for gene expression data set show that it achieves higher accuracy than all the other individual classifiers. The method also takes less computational time and space than the others.

Further, the accuracy of the ensemble technique can be enhanced much more by adding some optimization technique to the ensemble method.

References

1. Kun, M. 2013. A Vision-Based Hybrid Method for Eye Detection and Tracking. *International Journal of Security and Its Applications.*
2. Rokach, L. 2010. *Ensemble Methods in Supervised Learning*, vol. 33, 1–33. Springer.
3. Rokach, L. 2005. Ensemble Methods for Classifiers. *Data Mining and Knowledge Discovery Handbook*, Springer, US, 957–980.
4. Enriquez, F., F.L. Cruz, F. Javier Ortega, C.G. Vallego, and J.A. Troyano. 2013. A Comparative Study of Combination Applied to NLP Tasks. *Information Fusion* 14: 255–267.
5. Zhan, G.P. 2000. Neural Networks for Classification: A Survey. *IEEE Transactions on Systems, Man and Cybernetics-Part. C: Applications and Reviews* 30 (4): 451–446.

6. Ziadduin, S., and M.N. Dailey. 2008. Iris Recognition Performance Enhancement Using Weighted Majority Voting. *15th IEEE Inter-National Conference on Image Processing*, 227–280.
7. Isa, S.M., M. Ivan Fanany, W. Jatmiko, and A. Murni Arymurthy. 2011. Sleep Apnea Detection from ECG Signal: Analysis on Optimal Features. In *Principal Components and Nonlinearity, 5th International Conference on Bioinformatics and Biomedical Engineering.*
8. Kittler, J., M. Hatef, R.P.W. Duin, and J. Matas. 1998. On Combining Classifiers. *IEEE Transactions on Pattern Analysis and Machine Intelligence* 2 (3): 226–239.
9. Kim, Seoyoung, and Y. Kim. 2012. Application-Specific Cloud Provisioning Model Using Job Profiles Analysis. In *IEEE 14th Conference on High Performance Computing and Communication and IEEE 9th International Conference on Embedded Software and Systems.*
10. Luo, L., E.F. Wood, and M. Pan. 2007. Bayesian Merging of Multiple Climate Model Forecasts for Seasonal Hydrological Predictions. *Journal of Geophysical Research* 112: 1–13.
11. Ajitha, P., and G. Gunasekaran. 2014. Semantic Based Intuitive Topic Search Engine. *International Review on Computers and Software.*
12. Chen, Z., J. Li, L. Wei, W. Xu, and Y. Shi. 2011. Multiple-Kernel SVM Based Multiple-Task Oriented Data Mining System for Gene Expression Data Analysis. *Expert Systems with Applications* 38: 12151–12159.
13. Hansen, L., and P. Salamon. 1990. Neural Network Ensembles. *IEEE Transactions on Pattern Analysis and Machine Intelligence* 12: 993–1001.
14. Rokach. 2014. Decision Forests, Series in Machine Perception and Artificial Intelligence.
15. Helman, Paul, Robert Vero Susan, R. Atlas, and Cheryl Will-man. 2004. A Bayesian Network Classification Methodology for Gene Expression Data. *Journal of Computational Biology* 11 (4): 581–615.
16. Tsiliki, G., and S. Kossida. 2011. Fusion Methodologies for Biomedical Data. *Journal of Proteomics* 74: 2774–2785.
17. Kapp, M.N., R. Sabourin, and P. Maupin. 2012. A Dynamic Model Selection Strategy for Support Vector Machine Classifiers. *Applied Soft Computing* 12 (8): 2550–2565.
18. Dzeroski, S., and B. Zenko. 2004. Is Combining Classiers with Stacking Better Than Selecting the Best One? *Machine Learning* 54: 255–273.
19. Hong, Zi, and Jing-vu Yang. 1991. Optimal Discriminant Plane for a Small Number of Samples and Design Method of Classifier on the Plane. *Pattern Recognition* 24 (4): 317–324.
20. http://archieve.ics.uci.edu/ml/datasets/iris,2000-07-11.
21. http://archieve.ics.uci.edu/ml/datasets/yeast+dataset,1997-06-06.
22. http://archieve.ics.uci.edu/ml/datasets/ecoli,1997-06-06.
23. Seeja, K.R., and Shweta. 2011. Microarray Data Classification Using Support Vector Machine. *International Journal on Biometric and Bioinformatics* 5 (1): 10–15.
24. Shah, C., and A.G. Jivani. 2013. Comparison of Data Mining Classification Algorithms for Breast Cancer Prediction. In *Proceedings. 4th International Conference on Computing, Communication and Net-working Technologies*, 1–4.
25. ReboiroJato, M., F. Diaz, D. Glez-Pena, and F. Fdez-Riverola. 2014. A Novel Ensemble of Classifiers That Use Biological Relevant Gene Sets for Micro-array Classification. *Applied Soft Computing* 17: 117–126.
26. Opitz, D., and R. Maclin. 1999. Popular Ensemble Methods: An Empirical Study. *Journal of Artificial Intelligence Research* 11: 169–198.
27. Morrison, D., and L.C. De Silva. 2007. Voting Assembles of Spoken a ECT Classification. *Journal of Network and Computer Applications* 30: 1356–1365.
28. AliBagheri, M., Q. Gao, and S. Escalera. 2013. Logo Recognition Based on the Dempster-Shafer Fusion of Multiple Classifiers. *Advances in Artificial Intelligence Lecture Notes in Computer Science* 7884: 1–12.
29. Sohn, S.Y., and S. Ho Lee. 2003. Data Fusion, Ensemble and Clustering to Improve the Classification Accuracy for the Severity of Road Track Accidents in Korea. *Safety Science* 41: 1–14.

30. Hanczar, B., and A. BarHen. 2012. A New Measure of Classifier Performance for Gene Expression Data. *IEEE/ACM Transactions on Computational Biology and Bioinformatics* 9 (5): 1379–1386.
31. Tong, M., K. Hong Liu, C. Xu, and W. Ju. 2013. An Ensemble of SVM Classifiers Based on Gene Pairs. *Computers in Biology and Medicine* 43: 729–737.
32. Liu, H., L. Liu, and H. Zhang. 2010. Ensemble Gene Selection by Grouping for Microarray Data Classification. *Journal of Biomedical Informatics* 43: 81–87.
33. Reboiro-Jato, M., F. Diaz, D. Glez-Pena, and F. Fdez-Riverola. 2014. A Novel Ensemble of Classifiers That Use Biological Relevant Gene Sets for Microarray Classification. *Applied Soft Computing* 17: 117–126.
34. Nanni, L., and A. Lumini. 2007. Ensemblator: An Ensemble of Classifiers for Reliable Classification of Biological Data. *Pattern Recognition Letters* 28 (5): 622–630.
35. Lee, J., M. Park, and S. Song. 2005. An Extensive Comparison of Recent Classification Tools Applied to Microarray Data. *Computational Statistics and Data Analysis* 48 (4): 869–885.
36. Boulesteix, A., C. Strobl, T. Augustin, and M. Daumer. 2008. Evaluating Microarray Based Classifiers: An Overview. *Cancer Informatics* 6: 77–97.
37. Xu, L., A. Krzyzak and C.Y. Suen. 1992. Methods of Combining Multiple Classifiers and Their Applications to Handwriting Recognition. *IEEE Transactions on Systems, Man and Cybernetics* 22 (3): 418–435.
38. Chen, M.S., J. Han, and P.S. Yu. 1996. Data Mining: An Overview from a Database Perspective. *IEEE Transactions on Knowledge and Data Engineering* 8: 866–883.
39. Han, J., and M. Kamber. 2001. *Data Mining, Concepts and Techniques,* 67–120. Morgann Kaufmann Publishers.
40. Ester, M., H.P. Kriegel, J. Sander, and X. Xu. 1996. A Density-Based Algorithm for Discovering Clusters in Large Spatial Databases with Noise. In *Proceedings of 2nd International Conference on Knowledge Discovery and Data Mining*, vol. 96, 226–231.

Analysis on Big Data Techniques

Harleen and Naveen Garg

Abstract Nowadays, digitization data is used and there are more and more data that generates everyday about everything. Data sets whose size is complex or large that commonly used today. Management of this data which ensures that the data from varied sources is processed error free and is of good quality to perform analysis processing and sharing of such a large data by traditional methods is difficult by the use of traditional methods. So we need such systems that are more flexible, scalable, fault tolerance, compatible, and cheap to process large amount of data. Hadoop is designed to handle the extremely high volumes of data in any structure. There are various ways for Hadoop to run the job. The three programming approaches that are MapReduce, Hive, and Pig are used. In this paper, we present the comparison of MapReduce, Hive, and Pig. These three techniques are useful under different constraints.

Keywords Hadoop · MapReduce · Pig · Hive · Hadoop distributed file system ReStore

1 Introduction

Regularly, each individual made quintillion bytes of information and has 320* times information in their library. Data sets whose size is sufficiently huge that regularly utilized today, which can't be examined by traditional data processing application devices, for example, social databases, within the time important to make them useful. The difficulties incorporate the zones of capture, curation, storage, search, sharing, exchange, examination, and representation of this data. Large volume of structured and unstructured data is gathered by means of different

Harleen (✉) · N. Garg
Department of Information Technology, Amity University, Noida, India
e-mail: Harleenaujla18@gmail.com

N. Garg
e-mail: er.gargnaveen@gmail.com

M.S. Reddy et al. (eds.), *International Proceedings on Advances in Soft Computing, Intelligent Systems and Applications*, Advances in Intelligent Systems and Computing 628, https://doi.org/10.1007/978-981-10-5272-9_35

sources. Any association or organizations are routinely keep up each byte of their client's information which is 51% of information is organized and rest 49% is unstructured or semi structured. Different programs send differently formatted information.

1.1 Big Data

Data sets whose size is beyond the capacity of regularly utilized tools to process it with satisfactory time. Big data may be petabytes (1,024 TB) of information comprising of billions to trillions of records of a large number of individuals all from various sources. Importance of big data is adequately and productively caught, prepared, and analyzed, and organizations can pick up a more complete understanding of their business, clients, items, contenders, and so on which can prompt capability changes, expanded deals, lower costs, better customer organization, and/or upgraded things and administrations.

The big data application will handle countless and unstructured information. The information handling will include more than one information hub and finished in a shorter time frame.

Specialists utilize big data analytics to straightforwardly make an interpretation of data into knowledge, enhancing basic leadership and business execution.

1.2 Hadoop

Hadoop is not a type of database, but instead a software environment that takes into account enormously parallel computing. A staple of the Hadoop environment is MapReduce, a computational model that basically takes concentrated data strategies and spreads the estimation over a potentially unending number of servers (by and large alluded to as a Hadoop bunch). It has been a distinct point of preference in supporting the huge get ready needs of big data; a substantial information methodology which may take 20 h of preparing time on a concentrated social database framework may just take 3 min when distributed over an extensive Hadoop cluster of merchandise servers, all handling in parallel [1]. It has a several different applications, yet one of the top use cases is for large volumes of always showing signs of change information, for example, area-based information from climate or traffic sensors, online or online networking information, or machine-to-machine value-based information [2].

1.3 MapReduce

MapReduce is a programming model and a related execution for taking care of and making substantial data sets with a parallel, distributed calculation on a cluster [1].

A MapReduce system is made out of a two technique:

1. Map () technique that functions separating and sorting (for instance, sorting students starting with name into lines, sole line for every single name) [3], where an input data set is changed over onto an alternate arrangement of value/key pairs, or tuples [1].
2. Reduce () strategy that performs a rundown operation (e.g., counting the number of students in each queue, yielding name frequencies) [3], where a few of the "Map" are combined task to form a reduced set of tuples [1].

This is a programming paradigm that takes into consideration enormous occupations execution adaptability against thousand of servers [2] and accommodating excess and adaptation to noncritical failure. The key commitments of the MapReduce structure are not the genuine guide and reduce capacities, but rather the scalability and adaptation to noncritical failure accomplished for an assortment of uses by advancing the execution engine once [3].

1.4 Hive [1]

Hive is a "SQL-like" scaffold that permits ordinary BI applications to run queries against a Hadoop group. It was developed initially by Facebook, and, however, has been made open hot spot for quite a while, and it is a higher level abstraction of the Hadoop structure that permits anybody to make queries against data put away in a Hadoop group generally as though they were controlling an ordinary data store. It increases the range of Hadoop, making it better known for BI clients.

1.5 Pig

Pig is another extension that tries to convey Hadoop closer to the substances of designers and business clients, like Hive. Not at all like Hive, in any case, Pig comprises of a "Perl-like" language that takes into consideration query execution over information put away on a Hadoop cluster, rather than a "SQL-like" language. Pig was implemented by Yahoo!, and, much the same as Hive has additionally been made completely open source [2]. Pig comprises of a language and an execution situation. Pig's language, called as PigLatin, is an information stream dialect—this is the sort of language in which you program by associating things together. Pig can work on complex information structures, even those that can have levels of nesting. Not at all like SQL, Pig does not require that the information must have a pattern, so it is appropriate to prepare the unstructured information [1].

2 Literature Survey

Jasmin Azemovic and Denis Music represent with theory that present technique for putting away and recovering unstructured information is not proficient. Unstructured information is routinely secured outside the database, separate from its sorted out data. This partition can realize data organization complexities. On the other hand, if the data is associated with sorted out capacity, the record streaming limits and execution can be confined. Along these lines, they use the better approach for record system under database consistency which was great opportunity to test new advancement in various zones of putting away information. Advantage is getting most extreme execution in light of equipment and programming framework. Likewise frameworks where execution issue as of now exists, this model can distinguish bottlenecks and figure out how to enhance it [4] (Fig. 1).

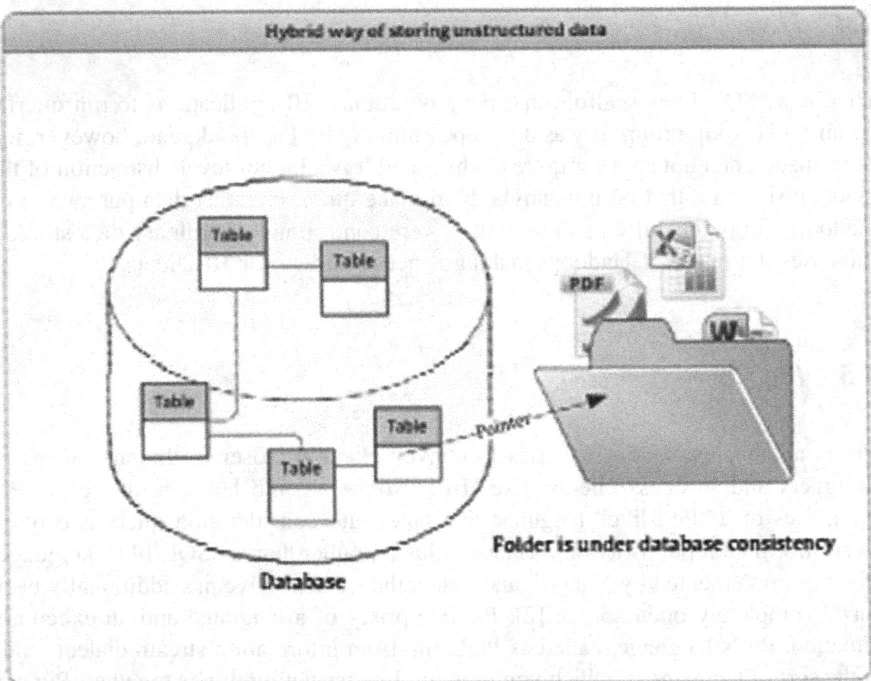

Fig. 1 Hybrid way of storing unstructured data

Sherif Sakr, Anna Liu, Nicta, and Ayman G. Fayoumi represented a study of the MapReduce group of methodologies for creating adaptable information handling frameworks and arrangements. When all is said in done they saw that in spite of the fact that the MapReduce structure, and its open source execution of Hadoop, they assume that it is fantastical that MapReduce will substitute database structures despite for data warehousing applications and expect that they will dependably exist together and supplement every others in various circumstances. They are, moreover, impacted that there is still space for further alter and movement in various courses on the extent of the MapReduce structure that is required to demonstrate the vision of giving immense scale information examination as a thing for learner end-clients [5].

Jyoti Nandimath, Ekata Banerjee, Ankur Patil, Pratima Kakade, Saumitra Vaidya, and Divyansh Chaturvedi, 2013, confronted issues of get-together substantial pieces of information, and they found that the information cannot be prepared utilizing any of the current concentrated design arrangements. Aside from time imperatives, the undertakings confronted issues of proficiency, execution, and elevated base expense with the information preparing in the incorporated environment. With the assistance of dispersed design, these substantial associations could conquer the issues of removing significant data from a gigantic information dump. One of the best open source apparatuses utilized as a part of the business sector to saddle the conveyed design with a specific end goal to take care of the information preparing issues is Apache Hadoop. Utilizing Apache Hadoop's different parts, for example, data clusters, map-reduce calculations, and distributed preparing, we will resolve different area-based complex information issues and give the applicable data again into the framework, along this it expanding the client experience [6].

Michael Frampton delineated that Hadoop MapReduce is a system for parallel treatment of huge data sets using disseminated fault tolerant limit over large clusters. The information set is isolated into pieces, which are the inputs to the Map capacities. The Map capacities then filter and sort these information lumps (whose size is configurable) on the Hadoop cluster information hubs. The yield of the Map systems is passed on to the Reduce forms, which adjust and consolidate the information to convey the subsequent output [7].

Zhanquan Sun illustrated feature selection which is an important research topic in machine learning and pattern recognition. It is useful in lessening dimensionality, removing unessential information, expanding learning accuracy, and upgrading result probability. Set up segment determination technique is out of work in dealing with significant scale data set as a result of absurd computational expense. For enhancing computational rate, parallel part choice is taken as the practical technique. MapReduce is a proficient distributional figuring model to plan wide scale data mining problems [8].

E. Sivaraman and Dr. R. Manickachezian illustrated that Hadoop is a rapidly growing biological system of taking into record Google's MapReduce calculation and record system work for realizing MapReduce calculations in an adaptable way and scattered on thing equipment. Hadoop engages customers to store and process

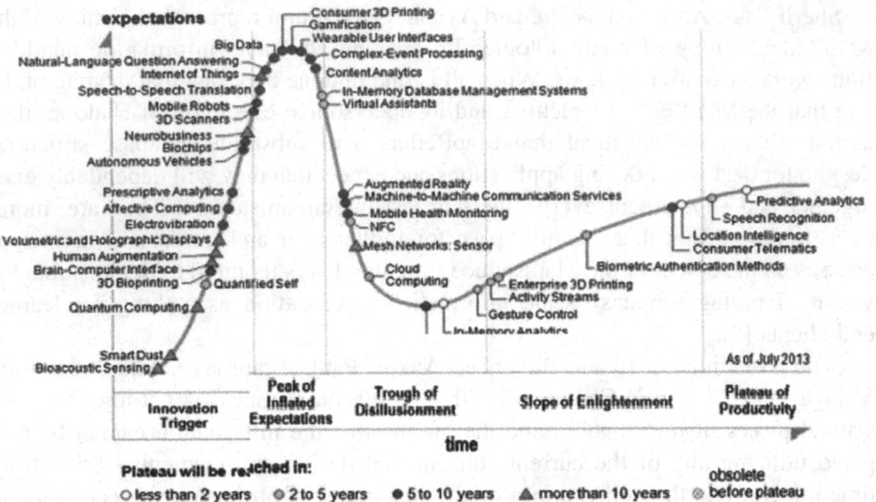

Fig. 2 Emerging technologies 2013 shows in Gartner's hype cycle

considerable volumes of data and separate it in ways not already conceivable with SQL-based philosophies or less versatile courses of action. Surprising upgrades in standard procedure and limit resources make Hadoop clusters achievable for the most part affiliations. They discussed the Big Data progression and the inevitable destiny of Big Data in perspective of Gartner's Hype Cycle [6] (Fig. 2).

Mohammed Muddasir N., Ranjitha H.C., and Meghana S. illustrated that data-preparing procedures are turning out to be more creative when the amount of data develops. They investigate such strategies like one is the conventional RDBMS approach and the other conveyed approach which handles the enormous data. Then, they found some favorable circumstances and weaknesses of these methodologies. Exceedingly utilized innovation for information handling by different associations is RDBMS, and it is replaced with new innovation because it has lot of difficulties. Hadoop, MapReduce, and so forth technology are the distributed processing techniques which are being utilized for preparing Big Data [9].

Robert J. Stewart, Phil W. Trinder, Hans-Wolfgang Loidl illustrated Hadoop MapReduce certification which contains diverse high-level query language (HLQL) like Pig, JAQL, and Hive on the top of it. They talked about a methodical execution associations and dialect examination of these three HLQLs based on different measurements like runtime, scale up, scale out, conciseness, and computational force. The development groups of HLQLs are occupied with the study and uncovered the restrictions that are affecting their development, and are depicted in this document.

Marissa Rae Hollingsworth, Amit jain illustrated that any association can give any product tools for client data administration and record the investigation for BI. With the goal that they can foresee the future installment design through past

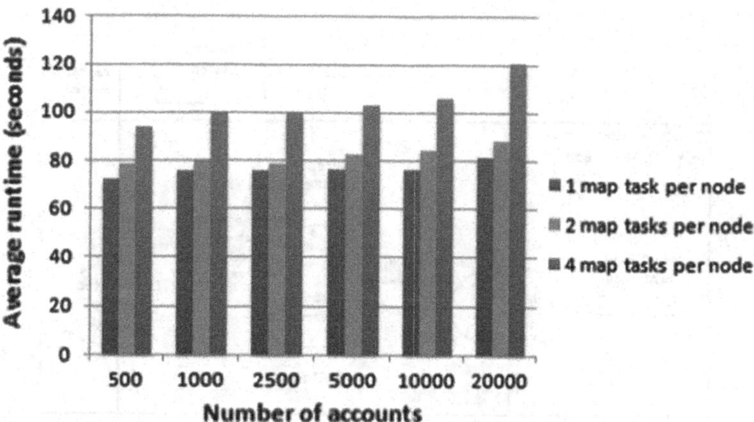

Fig. 3 Adaptability of MapReduce runtime (s) execution for a differed number of map assignments per data node

patterns, they found that in above situation the issue explanations are all information which is put away in one single machine and database access time increments as complexity of data increments. They check the correlation of MapReduce, Hive, and MYSQL. The fundamental objective is to decide when MapReduce and HIVE would beat MYSQL [10, 11] (Fig. 3).

Iman Elghandour and Ashraf Aboulnaga outline to investigating that large-scale data has ascended as a basic movement for a few relationship in the past couple of years. This huge scale information examination is energized by the MapReduce programming and execution model and its use, most strikingly Hadoop, and they use high-level inquiry dialects, for instance, Pig, Hive, or Jaql to express their complex assignments. The compilers of these dialects make an interpretation of queries into workflows of MapReduce organizations. Every job in these workflows examines its information from the coursed file structure utilized by the MapReduce framework and produces yield that is secured in this appropriated file structure and read as data by the accompanying job in the workflow. The present practice is to erase these broadly engaging results from the appropriated file framework toward the end of executing the workflow. One approach to manage enhance the execution of workflows of MapReduce program is to keep these moderate results and reuse them for future workflows submitted to the structure. They show ReStore, a system that arrangements with the limit and reuse of such in-termediate results. ReStore can reuse the yield of whole MapReduce occupations that are a bit of a workflow, and it can in like manner make extra reuse opportunities by creating and securing the yield of request execution regulators that are executed inside a MapReduce work [12] (Fig. 4).

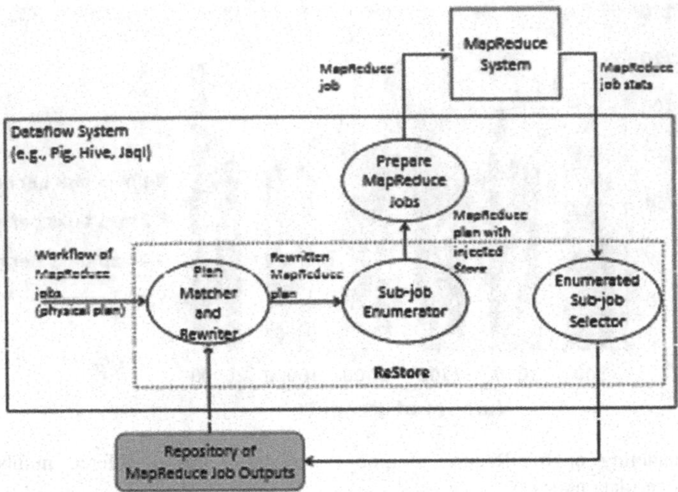

Fig. 4 ReStore system architecture

3 Analysis of Related Work

This section gives an insight into the working style, implementation, scalability criteria, access time, efficiency, and other parameters of the different systems under different case studies and shows the performance of each component of Hadoop.

S. No.	Title	Author	Year and publication	Major contribution
1	Comparative analysis of efficient methods for storing unstructured data into database with accent on performance	Jasmin Azemovic, Denis Music	2010, ICETC	In this paper, author began with theory that presents strategy for putting away and recovering unstructured information that is not efficient. Better approach for utilizing file system under database consistency was great chance to test new innovation in various zones of putting away data. Advantage is getting greatest execution taking into

(continued)

(continued)

S. No.	Title	Author	Year and publication	Major contribution
				account of hardware and software foundation. Additionally, frameworks where execution issue as of now exist, this model can recognize bottlenecks and find an approach to enhance it. Their research produces model for testing and benchmarking system for storing the unstructured data in hybrid way
2	Comparing high-level MapReduce query languages	Robert J. Stewart, Phil W. Trinder, Hans Wolfgang Loidl	2011, Springer Berlin Heidelberg	In this paper, they talked about a methodical execution associations and dialect examination of these three HLQLs based on different measurements like runtime, scale up, scale out, conciseness, and computational force The development groups of HLQLs are occupied with the study and uncovered the restrictions which are affecting their development that are depicted in this document
3	Hadoop and Hive as scalable alternatives to RDBMS: a case study	Marissa Rae Hollingsworth Amit Jain, Jyh-haw Yeh	2012, Boise State University	Payment historical analysis. Problem statements are all data stored in one single machine, and database access time increases as

(continued)

(continued)

S. No.	Title	Author	Year and publication	Major contribution
				complexity of data increases. RESULT is Map reduce is the best candidate for this study
4	ReStore: reusing results of MapReduce jobs in Pig	Iman Elghandour, Ashraf Aboulnaga	2012, ACM (ICMD)	They show ReStore, a system that arrangements with the limit and reuse of such intermediate results. ReStore can reuse the yield of whole MapReduce occupations that are a bit of a workflow, and it can in like manner make additional reuse opportunities by developing and securing the yield of request execution oversees that are executed inside a MapReduce work
5	The family of MapReduce and large-scale data processing systems	Sherif Sakr, Anna Liu, Nicta, Ayman G. Fayoumi	2013, ACM Computing Surveys	They give an outline of the MapReduce gathering of methodology for making flexible data get prepared systems and strategies. They in addition saw that the over straightforwardness of the MapReduce programming model has raised some key troubles on managing complex information models (e.g., settled models, XML and various leveled model, RDF, and graphs) capably. This constraint has

(continued)

(continued)

S. No.	Title	Author	Year and publication	Major contribution
				required the need of best in class time of tremendous data diagrams and structures that can give the required scale and execution qualities for these regions
6	Parallel feature selection based on MapReduce	Zhanquan Sun	2013, Springer International Publishing (CENet2013)	In this paper, a parallel segment decision framework considering MapReduce model is proposed. Liberal scale data set is allocated sub-data sets. Highlight decision is chipped away each computational focus point. Picked highlight variables are joined into one section vector in Reduce work. The parallel part determination system is versatile
7	High performance and fault tolerant distributed file system for big data storage and processing using Hadoop	E. Sivaraman, Dr. R. Manickachezian	2014, ICICA	They have highlighted the progression and move of tremendous data utilizing Gartner's Hype Cycle for rising advancements. They have talked about how HDFS produces distinctive copies of information pieces and appropriates them on method center points all through a gathering to connect with reliable, astoundingly

(continued)

(continued)

S. No.	Title	Author	Year and publication	Major contribution
				quick estimations. They understood, MapReduce imagined that scales to broad clusters of machines containing endless
8	Big data analysis using Apache Hadoop	Jyoti Nandimath; Ekata Banerjee; Ankur Patil; Pratima Kakade; Saumitra Vaidya; Divyansh Chaturvedi	2013, IRI	They illustrated the features of the Apache Hadoop, and how it is used in huge data sets. One of the best open source instruments utilized as a part of the business sector to bridle the conveyed engineering keeping in mind the end goal to take care of the information preparing issues is Apache Hadoop. Utilizing Apache Hadoop's different parts, for example, information bunches, map-decrease calculations, and dispersed handling, we will resolve different area-based complex information issues and give the significant data again into the framework, along these lines expanding the client experience
9	Comparing implementation features of MapReduce in RDBMS with distributed cluster	Mohammed Muddasir N., Ranjitha H.C., Meghana S.	2015, ICCTAC	Hadoop clusters and Oracle In-database Hadoop they discovered different pros and cones in both. These

(continued)

(continued)

S. No.	Title	Author	Year and publication	Major contribution
				technologies are used by the discretion of the individual or company on the basis of their need for preparing the data. At long last, they derive that traditional RDBMS could be sufficient if it is processing for medium-sized data, but if the size of data increases, then it will not be able to improve on the latency so they require the distributed processing infrastructure. Offering security to information on a distributed condition is a challenge. Moreover, the information is repeated and besides toughens the attempt. Encryption estimations are utilized at any rate they back off the strategy for limit and recovering
10	Processing data with MapReduce	Michael Frampton	2015, Apress	Hadoop Map Reduce is a structure for parallel get prepared of boundless information sets utilizing scattered inadequacy tolerant breaking point over huge clusters. The data set is confined into pieces, which are the inputs to the Map limits. The Map

(continued)

(continued)

S. No.	Title	Author	Year and publication	Major contribution
				limits then filter and sort these data pieces (whose size is configurable) on the Hadoop bunch data center points. The yield of the Map systems is gone on to the Reduce forms, which revise and pack the data to make the ensuing yield

4 Conclusion

There are various ways for Hadoop to run the job. The three programming approaches that are MapReduce, Hive, and Pig are used. Mapreduce, highly recommendable to process whole data, it's time consume and required project aptitudes like java (exceptionally recommendable), python, ruby and other programming languages. Total data aggregates and sorts by utilizing mapper and reducer capacities. Hadoop uses it by default. An alternative to Hadoop MapReduce Hive which is an open source was built so that Hadoop developers could do the same thing in Java in a less complex way by writing only fewer lines of code that will be very easy to understand. A Hadoop developer who knows about Hive does not require learning concepts of Java. Performing practically, we conclude that Pig takes less time as compared to both MapReduce and Hive. But all approaches have pros and cons so we have to choose according to our need and data.

Operations and functions	MapReduce	Hive	Pig
Implemented by	Apache	Facebook	Yahoo!
Basically for	Process large unstructured data sets in distributed manner by using large number of nodes	Processing large data sets without the users having to write Java-based MapReduce code	Processing large data sets without the users having to write Java-based MapReduce code

(continued)

(continued)

Operations and functions	MapReduce	Hive	Pig
Schemas	No need of schema	Yes (explicit)	Yes (implicit)
Partitions	No	Yes	No
Complex business logic	Composing complex business logic in MapReduce has more control on it	Composing complex business logic in Hive has less control	Composing complex business logic in Pig has less control
Run the job practically	Takes more time	Takes more time	Takes less time
Language	Complied language, which contains algorithms and can be implemented in C, Python, and Java	Has a declarative SQLish language (HiveQL)	Pig has a procedural data flow language (PigLatin)
Purpose	Process data in a cluster	Process structured data in Hadoop and creating reports	Executing data flow in parallel
Operates on	Client side	Hive operates on the server side of any cluster	Pig operates on the client side of any cluster
Designed for	Development of large-scale distributed and fault tolerant data processing application	Online transaction processing	Programmers and researchers
Supports AVRO	Yes	No	Yes
Web interface	Yes	Yes	No
Use of MapReduce in other technology	Based on a rigid procedural structure	It uses MapReduce for processing the queries	Platform for creating MapReduce programs
Technology is best for	Best for metapatterns problem involving job chaining and job merging	Hive is an excellent tool for analysts and business development types	MapReduce-based jobs are quickly developed in Pig, and it is best for prototyping
Data types	It is suited for all kinds of data, structured, semi-structured, and unstructured data	It is best suited for structured data	It is best suited for semi-structured data
Server	No	Optional (thrift)	No
COGROUP feature	No	Does not support	Perform outer joins
Communicate with ETL tools	Take more time	Less time	Takes more time

(continued)

(continued)

Operations and functions	MapReduce	Hive	Pig
Customization	Use custom partitioner	Ability to integrate custom programming, mappers, and reducers	Ability to create custom UDFs
Provide	Provide good scalability and fault tolerance	Good for ad hoc queries	Powerful, transformation, and processing capabilities
Abstraction	Lower level	Higher level	Higher level
Legacy code requirement in application	Yes	No	No
Advancement time	More development effort	Fast development	Fast development
Performance	MapReduce project would be speedier than Pig/Hive	Slower than MapReduce program, however, speedier than badly composed MapReduce code. Partitioning and bucketing enhance the execution	Slower MapReduce program, however, speedier than badly composed MapReduce code
Need	Expertise knowledge of programming is required to write a MapReduce program. Need of good amount of testability with large data sets	Eliminate tricky coding, managing the hive schema into embedded database, utilize the power of SQL. Unstructured data requires UDF like component	Since the entire program is based on Pig transformations, there is no much need of programming skills. Require more verbose coding
Lines of code	Huge line of code is required	Writing only fewer lines of code	As compared to MapReduce program, line of code is less. Equivalence ratio of MapReduce to Pig is 20:1
Recommended for	Functional programming concept to build data processing application	Data warehouse-oriented projects	Streaming data
Real-time analysis	No	It is not a language for real-time queries	Deployed in real-time scenario

(continued)

(continued)

Operations and functions	MapReduce	Hive	Pig
Complexity	High	Low	Low computational complexity
Productivity of data analysis	Fast approach	Relatively less	Relatively less
Code efficiency	High	Relatively less	Relatively less
Joins	This functionality is hard to achieve	Easy to achieve joins	It can be easily written in Pig
Advantage	Full control to minimize the number of jobs that your data processing flow require	It has engine that stores and partitions data. But its table can read from Pig or MapReduce	Ability to work with bags of data. More optimization and control on data flow
Best tool	Writing SQL	Ad hoc queries	Data loading
Turning complete	Yes	No	No
Hierarchal data	Yes	No	No
Connect from external application	No	Yes (using JDBC)	No

References

1. http://www.ijsr.net/archive/v3i10/T0NUMTQyNTE%3D.pdf.
2. Computer Science and Management Studies. *International Journal of Advance Research.* 2 (9), September 2014. www.ijarcsms.com. Big Data: Moving Forward with Emerging Technology and Challenge.
3. https://en.wikipedia.org/wiki/MapReduce.
4. 978-1-4244-6370-11$26.00 © 2010 IEEE.
5. http://dl.acm.org/citation.cfm?id=2522979&CFID=599320279&CFTOKEN=20228397.
6. ieeexplore.ieee.org/xpls/abs_all.jsp?arnumber=6965006.
7. http://link.springer.com/chapter/10.1007/978-1-4842-0094-0_4.
8. http://link.springer.com/chapter/10.1007%2F978-3-319-01766-2_35.
9. http://research.ijcaonline.org/icctac2015/number2/icctac2018.pdf.
10. http://scholarworks.boisestate.edu/cgi/viewcontent.cgi?article=1001&context=cs_gradproj.
11. cs.boisestate.edu.
12. vldb.org/pvldb/vol5/p 586_imanelghandour_vldb2012.pdf.
13. https://oraclesys.com/2013/04/03/difference-between-hadoop-and-rdbms/.

An Effective Speech Emotion Recognition Using Artificial Neural Networks

V. Anoop, P.V. Rao and S. Aruna

Abstract The speech signal has emerged as the quickest and the most normal mode of communication between the human beings. As an indispensable method of the human emotional behaviour comprehension, the speech emotion recognition (SER) has evinced zooming significance in the human-centred signal processing. In the current document, the speech signal is identified and distinguished by way of four distinct phases such as the feature extraction; modified cuckoo search-based generating finest weight, feature analysis and the artificial neural network (ANN) classification. The speech signal categorization, in turn, is performed in the working platform of MATLAB, and outcomes are assessed to ascertain the efficiency in the performance of the system. The performance metrics are analysed and made comparison for an improvement of the sensitivity 0.692%, specificity 24% and accuracy 5.88% with the existing method.

Keywords Speech emotion recognition · Modified cuckoo search
SNR · ANN classifier

1 Introduction

From times immemorial, the speech has established itself as the most important and usual method for the human beings to express their emotions, cognitive stages and intentions to one other [1, 2]. The speech signal has surfaced as the speediest and the most natural means of communication between the humans. This fact has

---

V. Anoop (✉)
Department of ECE, VTU Belgaum, Belagavi, Karnataka, India
e-mail: vanoop@gmail.com

P.V. Rao · S. Aruna
Rajarajeswari College of Engineering, Bengaluru, India
e-mail: pachararao@rediffmail.com

S. Aruna
e-mail: arunaraman6@gmail.com

M.S. Reddy et al. (eds.), *International Proceedings on Advances in Soft Computing, Intelligent Systems and Applications*, Advances in Intelligent Systems and Computing 628, https://doi.org/10.1007/978-981-10-5272-9_36

greatly influenced the inquisitive investigators to accept the speech as the quickest and effective means of interface between the human being and the machine. Nevertheless, it is highly essential that the machine possesses requisite skills to identify the human voices [3].

In this regard, the emotion detection segment of the speech, as a vital module of the affective human computer interface system, has thrown up several challenges to the speech-processing task [4, 5]. In a host of speech augmentation and sound decline methods, the decision is in accordance with the a priori SNR, and the distinctive approaches such as the spectral subtraction [6], Wiener filtering and maximum likelihood may be configured as a function of a priori SNR. In the concurrent applications, the a priori SNR evaluation is found to be fruitful, but for the sake of perfection, the restricted SNR is preferred rather than the a priori SNR [7, 8].

With a view to scaling up the excellence and transparency of the loud speech, the investigation in the SE [9] has riveted much attention on the superior designing of the speech and sound PDFs, the manner in which the sound contaminates the hygienic speech, the type of noise origin and so on [10]. The vital purpose of utilizing the speech emotion detection is to optimize the system feedback upon the identification of the irritation or annoyance in the voice of the speaker [5, 11]. The highly pervasive twist in the speech is caused by the additive noise, which is free from the clean speech [10]. The SE algorithms [4, 9] assessed may be categorized into two vital types such as the class dependent on the hidden Markov model (HMM), and that based on the modification of the signals, such as the MMSE evaluation, spectral subtraction and subspace-based approaches. Ephraim and Van Trees were instrumental in launching the subspace-based techniques. Prior to that, various noise reduction techniques were spearheaded [12]. The wavelet-based techniques employing the coefficient thresholding techniques were also employed for the purpose of the speech augmentation [12]. The rest of the paper is organized as follows: Sect. 2 gives a detailed description of proposed speech emotion recognition using ANNs. A summary of the simulation results and experimental set-up is given in Sect. 3. The obtained results and discussion on results are given in Sect. 4, and conclusion is given in Sect. 5.

2 Proposed Method for Speech Emotion Recognition Using ANN

In this document, an effective speech emotion recognition method is employed to destruct the noise as well as to improve the quality of the speech signal and to classify the speech utterances. The speech signal recognition system is home to four vital phases such as the feature extraction, modified CS-based generating finest weight, feature analysis and artificial neural network (ANN) classification. The original signal and noise speech signal are furnished as input to extract the features.

Thereafter, by means of the modified cuckoo search algorithm, the finest weight is evaluated. Further, the merging and ranking unit is effectively utilized to achieve the enhanced quality signal. Subsequently, several features such as the local invariant feature and the salient discriminative features are achieved. Thereafter, the procedure of testing and training so carried out by means of the ANN classifier. In the long run, the categorized signals are efficiently used to attain the important speech signals.

The speech emotion recognition technique proceeds through the following four vital phases:

- Feature extraction
- Optimization by modified cuckoo search algorithm
- Feature analysis
- Artificial neural network (ANN) classification

2.1 *Feature Extraction*

The input speech signal has to be pre-processed well-before the feature extraction phase. The input speech signal is distracted with noise signal. The input signals are processed by inserting noise signals at varying decibel levels such as the 0, 5 and 10 dB. The process of pre-processing is carried out in three phases such as the sampling, quantization and pre-emphasis.

The sampling represents the procedure of choosing the units from a region of interest. The sampling rate decides the spatial resolution of the digitized image, while the quantization level decides the number of grey levels in the digitized image. The quantization which is involved in the image processing task is a lossy compression method and is realized by compressing a range of values to a single quantum value. When the number of discrete symbols in a specified stream is decreased, the stream becomes highly compressible. The pre-emphasis functions as a robust high-frequency boost for the transmitter. The input speech signal is pre-processed, and the features are extracted with the help of the Amplitude Magnitude Spectrogram (AMS) method.

2.2 *Modified Cuckoo Search Algorithm*

The extracted features are adapted by means of a novel method employing the modified cuckoo search algorithm. The input noisy speech signal characterized by I is defined by

$$I = \{i_1, i_2, \ldots, i_n\}$$

where, n—denotes the total number of input signals.

In respect of the input signals, novel solutions are achieved by means of two methods such as the modified Levy flights and velocity update of particle swarm optimization techniques. Thereafter, the fitness solutions are assessed for both the methods and are subsequently integrated. Then the best fitness solution is substituted by the new solution by arbitrarily selecting the new solution. The substitution of the fittest solution is terminated till the new ones are generated from the section of the worst solutions. The finest quality solutions are combined and graded to ascertain the current best solution.

2.3 Feature Analysis

The feature analysis phase employs two distinct features such as the salient discriminative features and the local invariant features. In the current document, the feature learning algorithm is carried out by means of the convolutional neural network (CNN) in SER.

2.3.1 Local Invariant Analysis

The unsubstantiated feature learning (local invariant feature) has turned out amazing outcomes in a host of applications such as the image classification. An auto-encoder is a device habitually employed to learn a compressed depiction for a set of data. It is home to two segments such as the encoder and decoder. In the encoder side, the n inputs are mapped with a function x.

The encoder function in respect of the n inputs is expressed by the following Eq. 1.

$$n' = x(n) = s(Vn + \beta) \tag{1}$$

where, V represents the weight matrix, and β corresponds to the bias vector.

In Eq. (1), s represents a nonlinear activation function, which is essentially a sigmoid function. The sigmoid of p is furnished as per the following Eq. 2.

$$s(p) = \frac{1}{(1 + e^{-p})} \tag{2}$$

The decoder function y rebuilds n' in accordance with the feature vector n' and is represented as Eq. 3

$$y(n') = y(x(n)) = s'(V^E n' + \alpha) \quad (3)$$

where,

s' represents the decoder's activation function,
V^E denotes the weight matrix shared with the encoder and
α signifies the bias vector.

At last, the final local invariant feature vector is produced with the help of the CNN. The convolutional layer output is evaluated by means of Eq. 4.

$$h = [H(n), \ldots, H^r(n)] \quad (4)$$

Equation 4 is furnished as the input of salient discriminative feature analysis to segment the emotional salient factors from the others.

2.3.2 Salient Discriminative Features

The local invariant features achieved by means of the auto-encoder are furnished as the input to the fully connected layer, in which they are segmented into two distinct blocks of features as shown below.

(i) $\Psi^{(en)}(h)$ encodes the salient features of its input, and
(ii) $\Psi^{(o)}(h)$ encodes all other features.

Both of the two feature blocks are trained to lend a hand to rebuild their common input with a reconstruction loss function as illustrated in Eq. 5.

$$\begin{aligned} \hat{h} &= y\left(\left[\Psi^{(en)}(h),\ \Psi^{(o)}(h)\right]\right) \\ &= s'\left(V^C\left[\Psi^{(en)}(h),\ \Psi^{(o)}(h)\right] + \gamma\right) \end{aligned} \quad (5)$$

where V^C represents the weight matrix in the fully connected network, and γ_j signifies an offset to capture the mean value of h.

2.4 *Artificial Neural Network Classification*

The neural network characterizes an artificial intelligence approach which is effectively employed for the generation of the training data set and for testing the applied input data. Usually, the NN is home to three layers such as the input, hidden and output layers. The ANNs make an effort to replicate the brain functions as an algorithm and renovate its constraints by means of a learning technique based on the human conduct. The data achieved by means of test or replication outcomes are effectively employed to train and appraise the efficiency in the performance of the ANN algorithm. Hence, the entire data sets are segregated into two distinct parts such as the train data sets and test data sets. The neural network categorization is effectively carried out by means of the following steps.

2.4.1 The Bias Function of the Neural Network

The bias function represents a product of weights and inputs as illustrated in Eq. 6

$$f(n) = \beta + \sum_{n=1}^{h} (y_1 W_{1n} + y_2 W_{2n}) \tag{6}$$

where

β corresponds to a constant,
y_1, y_2 relate to the input function at nodes H_{11} and H_{12},
W_{1n}, W_{2n} signifies the weight function with value of n varying from $1, 2, \ldots,$ h.

2.4.2 Activation Functions for the Neural Network

The activation function signifies a nonlinear function which is characterized as per Eq. 7.

$$A = \frac{1}{1 + \exp^{-f(n)}} \tag{7}$$

3 Experimental Set-up and Simulation Results

The innovative approach for the speech improvement is performed in a system having 8 GB RAM with 32 bit operating system having i5 Processor using MATLAB Version 2011. The signal power is plotted for a frequency range between 0 and 2.5 kHz. The input signal, noisy signal and the de-noised signal are elegantly exhibited in Fig. 1.

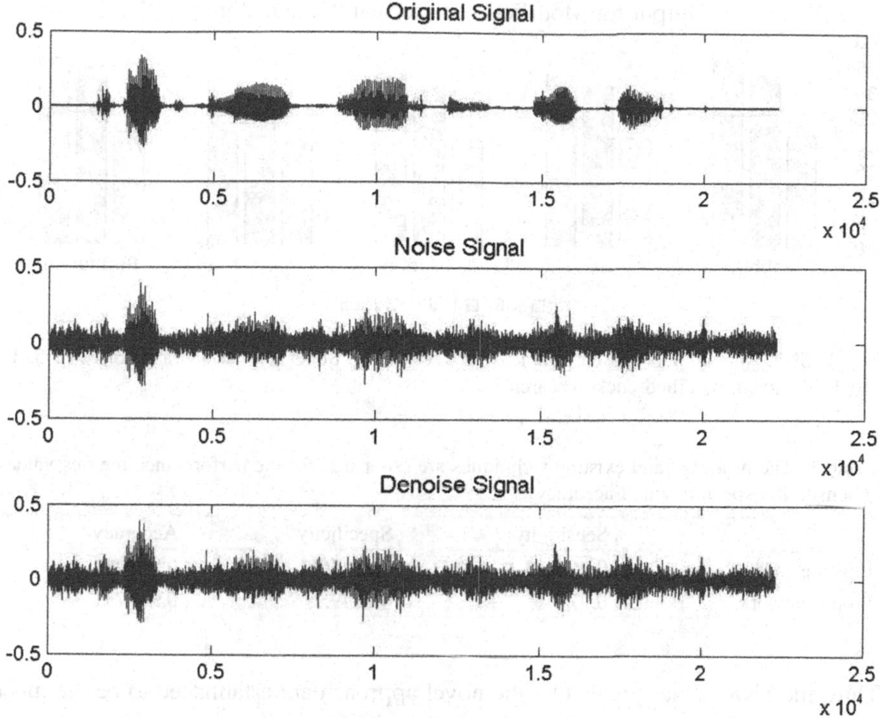

Fig. 1 Input signal, noise signal and de-noised signal for 5 dB Babble noise

4 Results and Discussion

Figure 2 shows the performance of modified cuckoo search algorithm for five different types of noises: Babble, Car, Exhibition, Street and Restaurant noises. *X*-axis of the figure indicates the SNR values, and *y*-axis shows different noise levels. The cheering outcomes emerge as the convincing credentials certifying the par excellence performance of the new-fangled approach which is well-geared to effectively decrease the noise and fine-tune the speech signal. The SNR values yielded by the innovative and the modern techniques in respect of the Babble, Car, Exhibition, Street and Restaurant noises added at the range of 5, 10 and 15 dB employing the modified CS are effectively exhibited in Table 1. Here for various types of noises, the quantities of noise added are 5, 10 and 15 dB, respectively, and in respect of these noise signals, diverse SNR values were evaluated. With an eye on validating the effectiveness of the novel technique viz. modified cuckoo search method, its performance is assessed and contrasted with that of the modern approach namely the basic cuckoo search technique. Table 1 authenticates the superior performance of the novel technique vis-à-vis that of the modern approach.

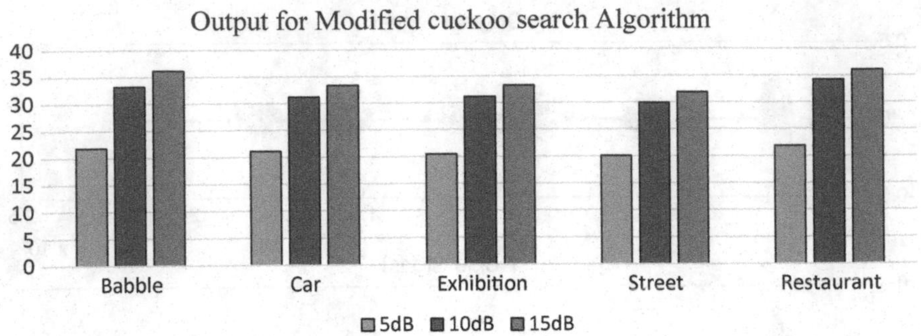

Fig. 2 SNR values for proposed Babble, Car, Exhibition, Street and Restaurant noises of 5, 10 and 15 dB using modified cuckoo search

Table 1 The proposed and existing techniques are compared for the performance metrics values of sensitivity, specificity and accuracy

	Sensitivity	Specificity	Accuracy
Existing method	0.963333	0.672566	0.891041
Proposed work	0.97	0.884956	0.946731

Thus, the SNR values yielded by the novel approach are established to be the most excellent for the augmentation of the speech signal quality.

The performance metrics such as the sensitivity, specificity and the accuracy are achieved and are analysed and contrasted with the values achieved by the novel technique and the modern method.

5 Conclusion

The learning of salient, discriminative features has emerged as a vital research issue for the SER. The significant contribution of the proposed work is towards the categorization of the speech signal together with the augmentation in the quality of the signal. The groundbreaking technique flows through four vital phases of processing such as the feature extraction, modified CS-based generating finest weight, feature analysis and artificial neural network (ANN) classification. The speech signal after optimization by means of the modified cuckoo search algorithm is furnished to the feature analysis phase in which the local invariant and the salient discriminative features are assessed and evaluated. In the long run, the speech signal is categorized by means of the ANN to achieve the significant speech signals. The speech signal categorization is carried out in the working platform of MATLAB, and results are assessed and estimated to ascertain the efficiency in execution of the epoch-making technique. The performance metrics are analysed and made comparison for an improvement of the sensitivity 0.692%, specificity 24% and accuracy 5.88% with the existing method.

References

1. Falk, Tiago H., Chenxi Zheng, and Wai-Yip Chan. 2010. A Non-Intrusive Quality and Intelligibility Measure of Reverberant and De-reverberated Speech. *IEEE Transactions on Audio, Speech, and Language Processing* 18 (7).
2. Wolfe, P.J., and S.J. Godsill. 2003. Efficient Alternatives to the Ephraim and Malah Suppression Rule for Audio Signal Enhancement. *EURASIP Journal on Applied Signal Processing* 2003 (10): 1043–1051.
3. El Ayadia, Moataz, Mohamed S. Kamel, and Fakhri Karray. 2011. Survey on Speech Emotion Recognition: Features, Classification Schemes, and Databases. *Pattern Recognition* 44: 572–587.
4. Anoop, V., and P.V. Rao. 2015. Performance Analysis of Speech Enhancement Methods Using Adaptive Algorithms and Optimization Techniques. *IEEE Digital Library—International Conference on Communication and Signal Processing, India,* 2–4 Apr 2015, 1322–1326. ISBN 978-1-4799-8080-2.
5. Ephraim, Y., and D. Malah. 1984. Speech Enhancement Using a Minimum Mean-Square Error Short-Time Spectral Amplitude Estimator. *IEEE Transactions on Acoustics, Speech, and Signal Processing* ASSP-32 (6): 1109–1121.
6. Huang, Zhengwei, Ming Dong, Qirong Mao, and Yongzhao Zhan. 2014. Speech Emotion Recognition Using CNN. IEEE Transactions on Audio, Speech, and Language Processing.
7. Mao, Qirong, Ming Dong, Member, Zhengwei Huang, and Yongzhao Zhan. 2014. Learning Salient Features for Speech Emotion Recognition Using Convolutional Neural Networks. *IEEE Transactions on Multimedia* 16 (8).
8. Hu, Y., and P. Loizou. 2006. Subjective Comparison of Speech Enhancement Algorithms. In *Proceedings of IEEE International Conference on Acoustics, Speech, Signal Processing*, vol. 1, 153–156.
9. Anoop, V., and P.V. Rao. 2014. Speech Signal Quality Improvement Using Cuckoo Search Algorithm. *International Journal of Engineering Innovation and Research* 2 (6): 201. ISSN 2277-5668.
10. Soon, I.Y., S.N. Koh, and C.K. Yeo. 1998. Noisy Speech Enhancement Using Discrete Cosine Transform. *Speech Communication* 24 (3): 249–257.
11. Laska, Brady N.M., Miodrag Bolic, and Rafik A. Goubran. 2010. Particle Filter Enhancement of Speech Spectral Amplitudes. IEEE Transactions on Audio, Speech, and Language Processing 18 (8).
12. Mao, Xia, Lijiang Chen, and Liqin Fu. 2009. Multi-Level Speech Emotion Recognition based on HMM and ANN. In *Proceedings of World Congress on Computer Science and Information Engineering*.

A New Pre-distorter for Linearizing Power Amplifiers Using Adaptive Genetic Algorithm

P.R. Bipin, P.V. Rao and S. Aruna

Abstract The power amplifier (PA) is naturally nonlinear in its operation. To get good energy efficiency, the PA is needed to function at its saturation level and results in the generation of the nonlinear outputs. To counter the nonlinearization in PA, a pre-distorter is appropriately designed and introduced in front of the PA. In this paper, an innovative pre-distorter is introduced by employing adaptive genetic algorithm (AGA) and their results are compared with that of genetic algorithm (GA) and particle swarm optimization (PSO) algorithm. The Wiener model is considered to model the PA, and the pre-distorter is built up by means of Hammerstein model. The new approach simulated using MATLAB and the outputs achieved are analyzed. The pre-distortion using AGA has produced better results in terms of MSE compared to that produced using PSO and GA optimization algorithms.

Keywords Digital pre-distorter · Particle swarm optimization
Adaptive genetic algorithm · Wiener model

1 Introduction

Power amplifiers (PAs) are the important subunits in almost all the wireless communication systems. PAs are designed to boost the power level of the signal before transmitting it through the antenna. They also show the memory effects [1], which is not desirable. Further, they tend to be invariably nonlinear. The amplifiers which

P.R. Bipin (✉)
Department of ECE, VTU Belgaum, Belgaum, Karnataka, India
e-mail: bipinpr@gmail.com

P.V. Rao · S. Aruna
Rajarajeswari College of Engineering, Bengaluru, India
e-mail: pachararao@rediffmail.com

S. Aruna
e-mail: arunaraman6@gmail.com

M.S. Reddy et al. (eds.), *International Proceedings on Advances in Soft Computing, Intelligent Systems and Applications*, Advances in Intelligent Systems and Computing 628, https://doi.org/10.1007/978-981-10-5272-9_37

are incredibly linear with good efficiency have become a rare specimen. The pre-distorter recompenses for the nonlinear distortion envisaged by the PA by working on the input signal. The theory of the digital pre-distorter (DPD) is easy to comprehend. Here, a nonlinear distortion function is generated within the digital horizon which represents the inverse of the amplifier function [2]. The DPD will be connected in front of the PA. In fact, it is very easy to devise an incredibly linear and inferior distortion system in principle, by connecting the two nonlinear systems (DPD and PA) in series.

The process followed in this paper offers the pre-distortion before the power amplifier with the help of the optimization method to achieve linearity in the combined system. The PA is modeled using Wiener model, and the pre-distorter is designed using Hammerstein model. At the output of Wiener HPA model, the authentic constraint vector is achieved and it is optimized by means of optimization approaches [3] such as particle swarm optimization (PSO) [4], genetic algorithm (GA) and adaptive genetic algorithms (AGA). In Sect. 2, a brief account of the Wiener HPA model and basics of PSO, GA, and AGA are given which is used for the optimization of Wiener HPA results to devise a pre-distorter. Test outcomes and consequential appraisal are presented in Sect. 3. Finally, the conclusions are effectively exhibited in Sect. 4.

2 Methodology

2.1 Power Amplifier Modeling

The PA model used here is the Wiener model which incorporates a memoryless nonlinearity preceded by a linear filter [2]. The inverse of Wiener model can be easily obtained by using Hammerstein model (a linear filter preceded by a memoryless nonlinearity).

The linear filter coefficient vector for the liner filter with order K_l can be denoted by

$$h = [h_0 h_1 \ldots h_{K_l}]^{\mathrm{T}} \tag{1}$$

The PA provides amplitude and phase distortion to the input signal applied to it [5], and this can be considered as the traveling wave tube (TWT) nonlinearity. Let $t = \left[\alpha_a \beta_a \alpha_\varnothing \beta_\varnothing\right]^{\mathrm{T}}$ gives the parameter vector for TWT nonlinearity [6] where $\alpha_{a,} \beta_a \alpha_\varnothing \beta_\varnothing$ are different parameters of the TWT nonlinearity.

2.2 Wiener Model Identification

For the Wiener parameter identification purpose, a normalized 64-QAM signal was generated and is then applied to Wiener model to construct training data set $\{x(k), y(k)\}$, where $x(k)$ is the input QAM and $y(k)$ is the output from the model, and the diagram is shown in Fig. 1. The true parameter of memory high-power amplifier is estimated using the training data. The true parameter vector is defined as

$$\beta = \left[\beta_1 \beta_2 \ldots \beta_{N_\beta}\right]^{\mathrm{T}} \tag{2}$$

where N_β represents the total number of parameter to be estimated, that is, the sum of number of linear filter coefficients and number of nonlinearity coefficients. The training data input $x(k)$ is given to the model and it produces an output $y(k)$. The output from the estimated Wiener model is indicated as $\hat{y}(k)$. The error results between the desired output $y(k)$ and the model output $\hat{y}(k)$ is $e(k) = y(k) - \hat{y}(k)$; thus, mean-square error cost function can be given by

$$J\left(\tilde{\beta}\right) = \frac{1}{K}\sum_{k=1}^{K} |e(k)|^2 \tag{3}$$

The true parameter vector β is estimated by obtaining the solution to the following optimization problem

$$\hat{\beta} = \arg\min_{\tilde{\beta} \in \varnothing} J\left(\tilde{\beta}\right) \tag{4}$$

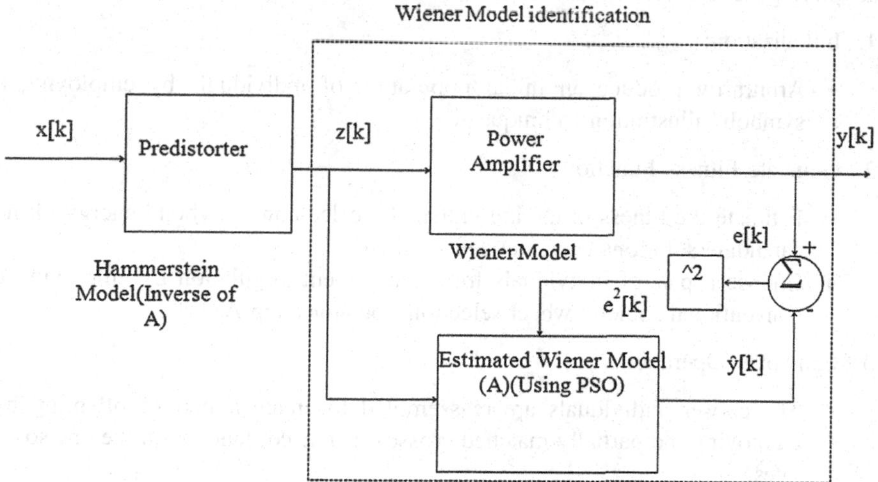

Fig. 1 Wiener model identification

where $\varnothing$ the search space is given as

$$\varnothing \triangleq \prod_{i=1}^{N_\beta} [\beta_{i,\min} \beta_{i,\max}] \tag{5}$$

The true parameter β is an element of the search space. The cost function (3) is a nonlinear function and has local minima. The above challenging identification problem is solved here using PSO algorithm, GA, and AGA.

2.3 Genetic Algorithm (GA)

The genetic algorithm [7] represents an adaptive global search technique in accordance with the evolutionary data of genetics. For the purpose of solving the optimization challenges, the genetic algorithm is effectively utilized as an arbitrary search technique. In the GA, the iterations are represented as the generation of modernized solutions and the population is characterized as the chromosomes.

2.4 Adaptive Genetic Algorithm (AGA)

The authentic constraint is achieved after the Wiener HPA model is adapted. The original population is allotted by the authentic constraint output. The traditional genetic algorithm is optimized by means of the mutation operator. The mutation function carried out in the innovative technique is the Cauchy mutation. The adaptive genetic algorithm is carried out by means of the steps detailed below:

1. Initialization:
   - Arbitrarily produce an initial population of individuals by employing a symbolic illustration technique.
2. Generate Fitness Function:
   - Estimate the fitness of the individuals by calculating the bond energy of the candidate solutions characterized by them.
   - Choose a pair of individuals from the present population by employing a conventional roulette wheel selection operator (step A).
3. Crossover Operation:
   - The chosen individuals are reassembled to create a pair of offspring by employing the partially matched crossover in accordance with the crossover rates.

- Appraise the fitness of the two offspring by evaluating the bond energy of the candidate solutions characterized by them.
- Place on record the percentage of perfection, or the percentage of deprivation, on their fitness values on account of the crossover function.

4. Mutation Operation:
   - Carry out the process of the conventional symbolic mutation of the two offspring and appraise their fitness by evaluating the bond energy of the candidate solutions characterized by them.
   - Place on record the percentage of perfection, or the percentage of deprivation, of their fitness values as a consequence of the mutation function.

Subsequent to the crossover function, the procedure of mutation is performed where the new chromosomes with the finest fitness values are mutated. In the novel technique, the Cauchy mutation is effectively employed for mutating the genes in the parent chromosomes.

In the Cauchy mutation, the optimal solution is achieved by moving each gene left in the parent chromosome and is replaced with the newer located genes. Now, the gene of the parent chromosome is moved one step left and the optimized new solution is achieved when the mutation procedure is concluded.

5. Updating Population:
   - Allot the consequential individuals into a fresh population pool. If the population size is not achieved, go back to step A.
   - Adapt the crossover and mutation rates according to the specified rules.
6. Termination:
   - Verify the stopping criterion.
   - End the genetic investigation procedure and select the best candidate solution over time as the ultimate solution if the stopping criterion is fulfilled.
   - Or else, move forward to the succeeding generation with the old population to be substituted by the new population, and go back to step A.

The AGA-optimized Wiener HPA pattern is effectively employed to devise a pre-distorter model which is exploited to scale down the nonlinear attributes.

2.5 Pre-distorter Design

The pre-distorter is implemented using Hammerstein model as it represents the inverse form of the Wiener model. The linear filter of Hammerstein model [8] is made to be inverse of linear filter of identified Wiener power amplifier model, and inverse nonlinearity [9] of estimated Wiener is used to implement the Hammerstein nonlinearity.

Consider the transfer function of the Hammerstein Pre-distorter's linear filter.

$$Q(z) = z^{-\tau} \sum_{i=0}^{N_h} q_i z^{-i} \tag{6}$$

where q_i represents the linear filter coefficient and τ is the delay. If $H(z)$ is the transfer function of linear filter of Wiener model and is a minimum phase filter, then $\tau = 0$. The filter coefficient of pre-distorter can be obtained by solving the linear equations derived from

$$Q(z).H(z) = z^{-\tau} \tag{7}$$

3 Results and Discussions

Figure 2 exhibits the output signal *y*(*k*) of the memory power amplifier when normalized 64-QAM signal, *x*(*k*), is given to its input for input back-off value of 5 dB. It is evident from the figure that output signal is spread around the input signal as a result of memory effect and nonlinearity of power amplifier. It will lead to larger bit-error rate and adjacent channel interference during transmission.

In this paper, identification process is done using both PSO algorithm and GA and AGA algorithms. The training data set taken contains 500 samples of normalized 64 QAM data. The noise with standard deviation 0.0 and 0.01 was added as input, and identification was done for different IBO values such as 5, 10, 15 dB. The results obtained were averaged over 100 runs.

The parameter vector for the estimated Wiener power amplifier model obtained for each case is given by

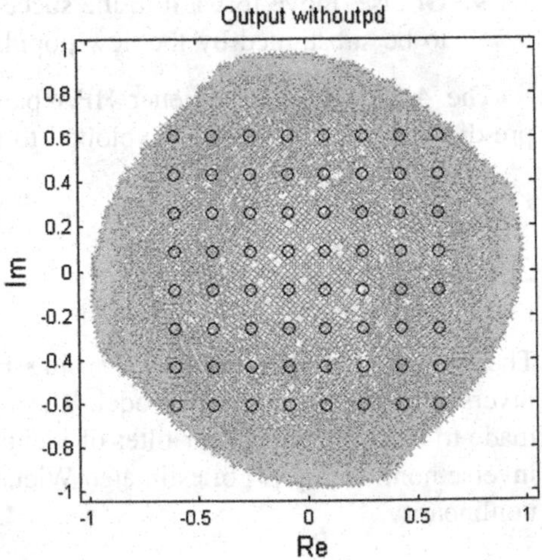

Fig. 2 Output without pre-distorter

h^T = [0.777822614 0.155516821 0.076451606]
t^T = [2.137008627 1.137645489 3.935456068 1.996214262] for PSO
h^T = [0.765382127 0.153215367 0.077231119]
t^T = [2.176933106 1.205994222 4.010059489 2.058612345] for GA and
h^T = [0.76918 0.15386 0.07672]
t^T = [2.15839 1.15415 3.99853 2.09696] for AGA.

The linear filter of memory length of eight has selected for compensating memory effect of power amplifier model. Then, solving expression (7) with the help of estimated Wiener model's linear filter coefficient, the resulting linear filter coefficients for each case are given as

h^T = [1.285640173 −0.257049189 −0.074970543 0.040254684 −0.000679669 −0.003820711 0.000830712 0.000209443] for PSO
h^T = [1.306536911 −0.261544561 −0.079480117 0.042301686 −0.000448051 −0.004178773 0.000881724 0.000245156] for GA
and
h^T = [1.300248917 −0.259939966 −0.078005391 0.041577799 −0.000514716 −0.004053171 0.000861742 0.000232875] for AGA.

The constellation diagrams of output signal from the combined pre-distorter using PSO, GA, and AGA algorithms combined with Wiener power amplifier model are shown in Figs. 3, 4, and 5 for IBO = 5 dB ('x' represents the output $y(k)$ and 'o' represents the input 64-QAM signal $x(k)$).

From Figs. 3, 4, and 5, it can be seen that designed pre-distorters almost completely cancel out the nonlinear distortions and memory effects caused by the Wiener memory high-power amplifier model. Compared to PSO and GA, the pre-distorter designed using AGA has produced better results.

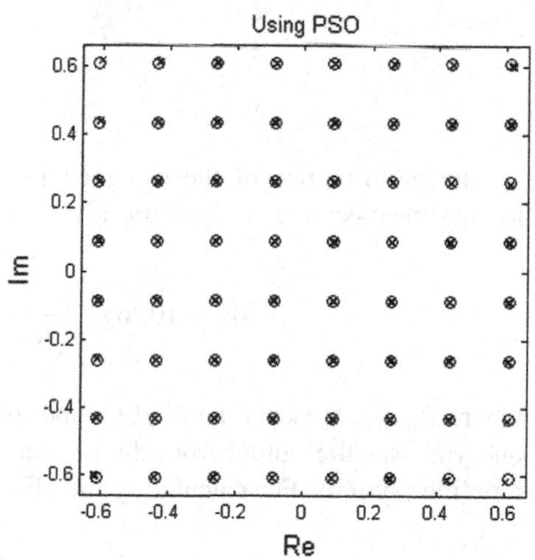

Fig. 3 Output with pre-distorter using GA

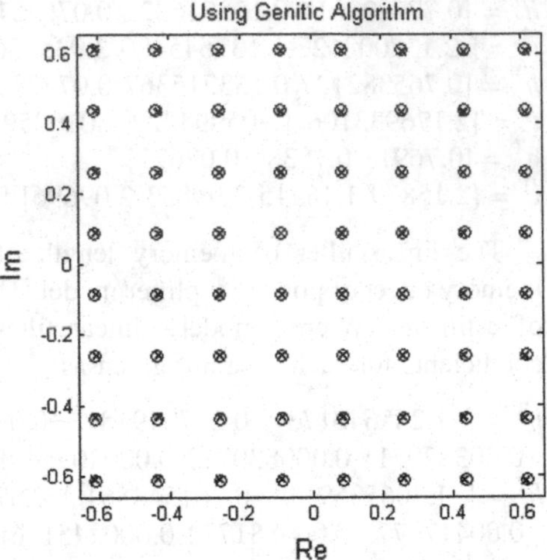

Fig. 4 Output with pre-distorter using PSO

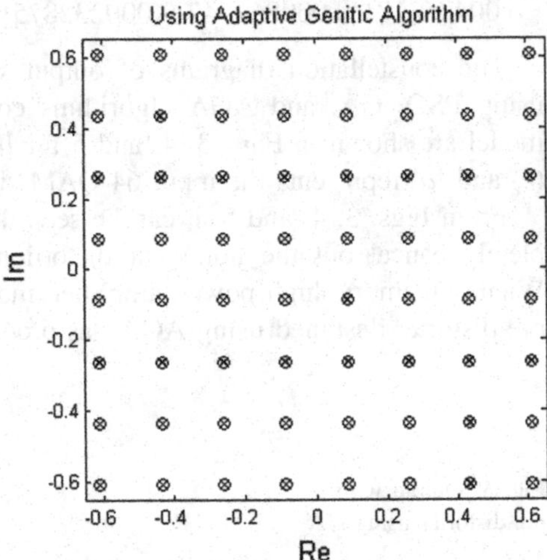

Fig. 5 Output with pre-distorter using AGA

The performance of the designed pre-distorters was evaluated using the following mean-square error metric given in Eq. (8),

$$\mathrm{MSE} = 10\,\log_{10}\left(\frac{1}{K_{\mathrm{total}}}\sum_{k=1}^{K_{\mathrm{total}}}|x(k) - y(k)|^2\right) \tag{8}$$

where K_{total} represents the total number of the test data, $x(k)$ was the input signal, and $y(k)$ was the output from the combined pre-distorter and memory high-power amplifier system. For calculating the effectiveness of pre-distorter, K_{total} = 20,000

Table 1 MSC values for different IBO values

IBO	Without PD	PSO	GA	AGA
1	−0.88	−18.6229	−18.24353093	−19.1624
2	−2.20	−23.7221	−23.47215939	−25.1326
3	−3.577	−29.2669	−30.40934644	−34.0816
4	−4.993	−33.0066	−39.62186106	−51.1595
5	−6.427	−35.4421	−44.51344478	−54.6613
6	−7.85	−37.9127	−46.86382147	−57.8648
7	−9.23	−40.3647	−48.1744839	−60.9684
8	−10.54	−42.7472	−49.2292658	−64.0139
9	−11.77	−45.0111	−50.25339467	−67.0112
10	−12.91	−47.1135	−51.28417344	−69.9547
11	−13.98	−49.0264	−52.31811731	−72.8268
12	−14.98	−50.7429	−53.34854158	−75.5979
13	−15.95	−52.2771	−54.37218999	−78.2279
14	−16.91	−53.6579	−55.38871141	−80.6707
15	−17.85	−54.9196	−56.39920396	−82.8851
16	−18.78	−56.0944	−57.40518485	−84.8483

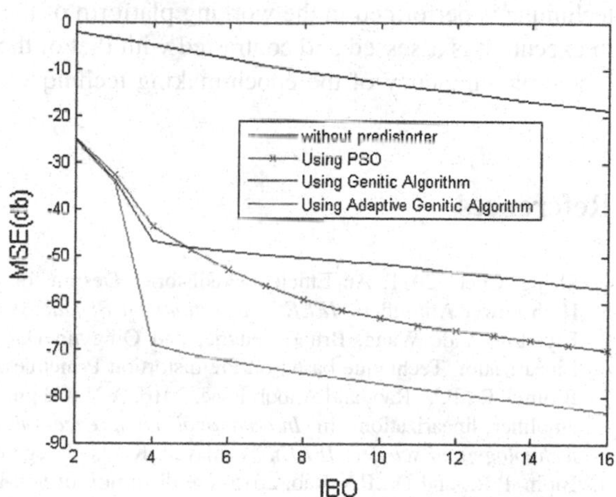

Fig. 6 MSE versus IBO plot

samples of normalized 64 QAM data were allowed to pass through the combination of pre-distorter and Wiener power amplifier.

The mean-square error metric (MSE) was computed for each pre-distorters by noting input and output data. The obtained MSE for the pre-distorters is shown in Table 1 as a function of IBO.

Figure 6 depicts the MSE versus IBO plot for pre-distorters designed using corresponding estimated parameter vectors, where Wiener power amplifier is implemented using h^{T} = [0.7692 0.1538 0.0769] and t^{T} = [2.1587 1.15 4.0 2.1].

From Fig. 6 and Table 1, it is clear that pre-distorters designed with AGA-based identification method have greater reduction in MSE value than pre-distorter designed using PSO algorithm at lower IBO values. Hence, pre-distorter designed using AGA can be considered as the best one that provides good linearization.

4 Conclusion

The power amplifiers have steadily surfaced as inevitable modules in the communication systems. In this regard, the HPAs have brilliantly played their role and are offered a red carpet welcome in the burgeoning gamut of applications, especially in the fascinating world of the wireless communications. However, the HPAs are habitually deformed and exhibit a tendency to generate the nonlinear outputs, thereby miserably failing to attain the saturation level. In the innovative technique, the Wiener HPA technique is employed to devise the high-power amplifier and the authentic constraint vector has to be adapted here. The optimization process is carried out by means of the adaptive genetic algorithm (AGA). Thereafter, the optimized Wiener HPA is employed to devise the pre-distorter. The innovative technique is performed in the working platform of the MATLAB and the efficiency in execution is assessed and contrasted with that of the GA and PSO to illustrate the incredible efficiency of the epoch-making technique.

References

1. Sheng, Chen. 2011. An Efficient Predistorter Design for Compensating Nonlinear Memory High Power Amplifiers. *IEEE Transactions on Broadcasting* 57 (4).
2. Biyi Lin, Yide Wang, Bruno Feuvrie, and Qingyun Dai. 2010. A Novel Power Amplifier Linearization Technique based on Pre-distortion Principle and Wiener model. *IJAST* 22.
3. Bipin, P.R., P.V. Rao, and Anoob Issac. 2016. A Novel predistorter based on MPSO for power amplifier linearization. In *International conference on emerging trends in Engineering technology and science* (*IEEE*). Thanjavur: Kings college of engineering. February 24–26.
4. Bipin, P R., and Dr. P.V Rao. 2015. Pre-distortion of non-linear power amplifiers using ABC and PSO algorithms. In *IEEE 4th International Conference on Communication and Signal Processing*. Tamilnadu: Adiparasakthi Engineering college.
5. Bipin, P.R., P.V Rao. 2013. Artificial Bee Colony for Pre-Distortion of Nonlinear Power Amplifiers. *IJEIR*-2013.
6. Ding, L., G.T. Zhou, D.R. Morgan, Z. Ma, J.S. Kenney, J. Kim, and C.R. Giardina. 2004. A Robust Digital Baseband Predistorter Constructed Using Memory Polynomials. *IEEE Transactions Communications* 52 (1): 159–165.
7. Goodman, Joel, Matthew Herman, Bradley Bondand, and Benjamin Miller. 2009. A Log-Frequency Approach to the Identification of the Wiener–Hammerstein Model. IEEE Signal Processing Letters 16 (10).

8. Bipin, P.R., and P.V Rao. 2013. Artificial Bee Colony for Pre-distortion of Non-linear Power Amplifiers. *International Journal of Engineering Innovation and research* 2 (6) (ISSN: 2277 – 5668).
9. Saleh, A.A.M. 1981. Frequency-Independent and Frequency-Dependent Nonlinear Models of TWT Amplifiers. *IEEE Transactions Communications* 29 (11): 1715–1720.

Design and Implementation of ECC-Based RFID Tag for Wireless Communications on FPGAs

Neelappa and N.G. Kurahatti

Abstract This paper proposes an elliptic curve cryptography (ECC) and direct spread spectrum (DSS) to implement Radio-Frequency Identification (RFID) tag chip for highly secure wireless communication. Digital baseband controller (DBC) and nonlinear feedback shift register (NLFSR) have been used to generate control signal, data transfer, and random number sequences for ECC processor and DSS compatible with ISO/IEC 14443. In order to achieve optimized resources in field programmable gate array (FPGA) for ECC, point multiplication, reusable registers, and asynchronous counter are adopted. The proposed work has been implemented on two Spartan 6 FPGAs. Wireless communication between them has been established via Zigbee modules. Single-user DS-SS system using pseudo-chaotic sequence as spreading sequence and RFID transmitter and receiver in FPGA development kit targeted to Xilinx's Spartan 6 device XC6S45-2tq324 has been implemented. The proposed elliptic curve processor (ECP) in digital baseband controller (DBC) needs only 8.58 K gate area and has a delay of 7.509 ns. The synthesis results show that the power consumption of DBC including ECP and other units in transmitter and receiver is only 381.58 μW at 35,810 kHz. Considerable improvement in power dissipation, area, and delay has been achieved. Security of the data has been ensured by using ECC.

Keywords Elliptic curve cryptography · Direct spread spectrum
Nonlinear feedback shift register · Digital baseband controller · Protocol and detector · Radio-frequency identification tag

Neelappa (✉)
Government of Engineering College, Kushal nagar, Karnataka, India
e-mail: neel_m_d@yahoo.co.in; neel.m.dy@gmail.com

N.G. Kurahatti
East Point College of Engineering, Bengaluru, Karnataka, India
e-mail: ngkurahatti@yahoo.co.in

M.S. Reddy et al. (eds.), *International Proceedings on Advances in Soft Computing, Intelligent Systems and Applications*, Advances in Intelligent Systems and Computing 628, https://doi.org/10.1007/978-981-10-5272-9_38

1 Introduction

One of the features of improvements of RFID is an auto proof of identity expertise. It is widely used for proof of identity, control chain organization, wireless sensor networks (WSNs), ad hoc wireless communication, and other applications. With the development of the Internet of Things (IOT) and WSN, the demand on security-related RFID systems has expanded [1]. RFID applications require low-power and low-cost implementation with high data security. To satisfy security, a suitable public key cryptography scheme is required. RFID passive tags obtain energy from radio frequency signals transmitted from the reader and have limited power supply. Therefore, these tags cannot use the energy-demanding algorithms such Revert-Shamir-Adleman (RSA) cryptography. Elliptic curve cryptography (ECC), proposed by Koblitz [2], has been employed, the main advantages being by employing ECC are smaller key sizes and offers comparable security level as RSA [3]. This feature has been incorporated in the implementation of RFID tag chips. The essential operation in the elliptic curve cryptosystems is scalar point multiplication. The point multiplication is achieved by finite field arithmetic computations such as field addition, field multiplication, field squaring, and inversion. Tawallebh et al. [4] proposed processor architecture for elliptic curves over prime fields. The speed of point multiplication is increased by proper selection of the coordinate system [5]. A number of hardware implementations for elliptic curve cryptography have been proposed in the literature, but only a few of them are aimed for low-end devices. The proposed implementation is emphasis on speed, area, and power which are based on FPGA technology [6].

In this paper, we propose an ECC algorithm based on projective coordinates, which can be adopted for both binary and prime fields for resource limited RFID tags. The data is encoded and decoded by chaos-based direct spread spectrum/binary phase shift keying (DSS/BPSK) modulation. The proposed RFID system with DSS/ECC offers improved security feature and makes analysis of the RFID tag in terms of area, speed, and power. The proposed work has been implemented and tested on Spartan-6 FPGA boards.

The paper proceeds as follows: Sect. 2 describes mathematical foundation. Section 3 describes RFID transmitter, and Sect. 4 describes RFID receiver. Section 5 describes experimental results and comparison and conclusion.

2 Mathematical Foundation

In cryptography, two of the most studied fields are finite fields of characteristic two, denoted by $GF(2^m)$ and prime fields. Advantage of $GF(2^m)$ fields is the simple hardware required for computation of commonly used arithmetic operations such as addition and squaring. Addition and squaring in $GF(2^m)$ can be performed by a simple XOR operation. These operations are simpler than the addition and squaring circuits of a *GF* (*P*) field. In the proposed paper, ECC operations are performed on binary field.

2.1 *Finite Field Operations (2^m)*

EC over field $F(2^m)$ includes arithmetic of integer with length m bits. The binary string can be declared as polynomial:

Binary string: $(b_{m-1} \ldots b_1\, b_0)$

Polynomial: $b_{m-1}\, y^m - 1 + b_{m-1}\, y^{m-2} + \ldots + b_2 y^2 + b_1 y + b_0$ where $b_i = 0$. For example, $y^3 + y^2 + y$ is polynomial for a four-bit binary number 1110.

In addition

If $A = y^3 + x^2 + 1$ and $B = y^2 + y$ are two polynomial, then $A + B$ is called polynomial addition that returns $y^3 + 2y^2 + y + 1$ after taking mod 2 over coefficients as $A + B = y^3 + y + 1$. On binary representation, polynomial addition can be achieved by simple XOR of two numbers. If $A = 1101$ and $B = 0110$, then $A + B = A$ XOR $B = A + B = (1011)_2$.

In multiplication

If $A = y^3 + y^2 + 1$ and $B = y^2 + x$ are two polynomials, then $A * B$ is called polynomial multiplication that returns $y^5 + y^3 + y^2 + y$, $m = 4$. The result should be reduced to a degree less than 4 by irreducible polynomial $y^4 + y + 1$.

$y^5 + y^3 + y^2 + y \pmod{f(y)} = (y^4 + y + 1)(y + y^5 + y^3 + y^2 + y) = 2y^5 + y^3 + 2y^2 + 2y = y^3$ (after reducing the coefficient on mod 2).

If $A = (1101)_2$ and $B = (0110)_2$, then $A * B = (1000)_2$.

3 Methodology

In this proposed work, the approach is divided into two sections: (i) block diagrams of DSS and ECC-based RFID tag chip at transmitter and (ii) block diagram of DSS and ECC-based RFID tag chip at receiver.

3.1 *Transmitter*

Figure 1 shows the block diagram of DSS and ECC-based RFID tag at transmitter section. A typical transmitter-embedded RFID tag chip can be divided into six parts: analog front end (AFE), NLFSR, EEPROM, ECC processor, wireless Zigbee transmitter, and digital baseband controller. AFE realizes the comprehensive functions of physical layer according to the RFID protocol, tag id stored in reuse register. NLFSR generates sequences of numbers for pseudo-chaotic sequence

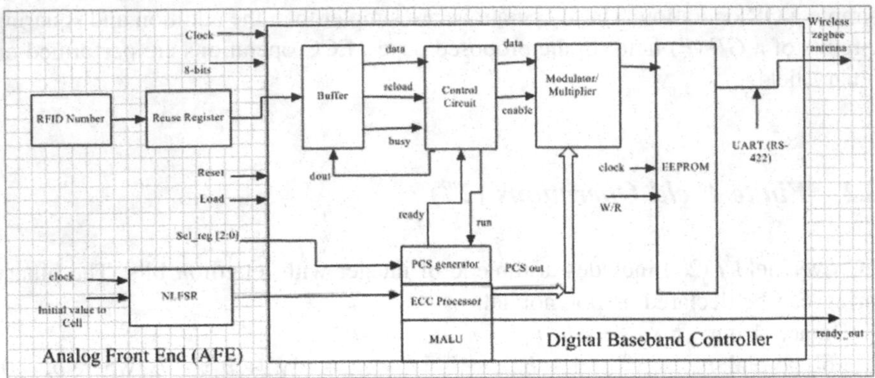

Fig. 1 Proposed block diagram of DSS and ECC-based RFID tag chip at transmitter section

(PCS) generator and ECC processor, AFE including carrier signal, clock generation, and reset signal generation. PCS creates randomness for each authentication so that the data in the authentication is highly secure. EEPROM is used for storing private or public information, such as the private key, base point of elliptic curve (EC), and the EC equation parameters. Utilizing the stream line bus structure, baseband controller integrates system controller, memory interface, buffer, multiplier, PCS, and ECC processor into one unit.

3.2 *Direct Spread Spectrum (DSS)*

In DSS system, the data bits are generated using PCS generator are spread. To generate PCS sequence, initially all 8-bit registers from R_1 to R_8 have to be initialized based on two sets of initial values. To load eight registers of the PCS generator, set load = 1 and using three signal pins of sel_reg registers R_1–R_8 are selected. The 8-bit initial values to each of these eight registers are loaded using eight signal pin reg_init internally [5]. After loading the initial values to all the eight registers, ready out signal pin gives an indication to the user. At the same time, signal ready is set to high and gives an indication to the control circuit to start its operation. Prior to loading the initial values to the registers R_1–R_8, the busy and done signal should be high and low, respectively. After receiving the 8-bit tag id data frame from RFID tag, the control circuit permits the PCS generator to generate PCS sequence by setting signal run = 1. It also enables the multiplier by setting enable = 1 and indicates the buffer that it is busy by setting the signal busy = 1. During this time, the PCS generator starts generating the 64 bits of PCS sequence. The control circuit then transfers one bit at a time serially to the multiplier, where it is multiplied by the 64 bits of generated PCS sequence resulting in a 64 bits of spread sequence, and the same is transmitted. After the first RFID number is serialized into bit spread by 64 bits of PCS sequence, the second data bit is received in the multiplier and is multiplied by the next 64 bits of the PCS sequence. Thus, the PCS generator generates a total of 512 bits to spread all the 8 bits of RFID data. After transmitting all the 512 bits with respect to one data (8 bits), the control circuit makes signal done = 1 and busy = 0 and then it accepts next frame of 8 bits of RFID id data.

The transmitter operation is carried out using direct spectrum sequence by multiplying the RFID tag number $b(t)$ with PCS sequence $p(t)$ which acts as a carrier signal for modulator. The spreading signal $s(t)$ is then modulated with $p(t)$ by means of BPSK, and the resultant of DSS/BPSK signal is given by

$$e(t) = Ap(t)b(t)\sqrt{2p}\cos(2\pi f_0 t) \tag{1}$$

where A and f_0 are the amplitude and frequency of the carrier. The modulated signal $e(t)$ is transmitted through wireless Zigbee via serial communication protocol universal asynchronous receiver/transmitter (UART) shown in Fig. 1.

3.3 ECC Processor

Prime field-based ECC processor with high-speed operating frequency of 50 MHz and scalar multiplication to perform both point addition and point doubling in affine coordination is adopted in this work. Figure 2 shows the overall ECC dual-field architecture with input/output buffers, control unit, data selector, register file, and ECC scalar multiplication. The data is fed into an input buffer and read the output buffer through I/O interfacing. ECC parameters are written into the buffer before the computation. All operations are controlled by the control unit. The control instructions are stored in the control register and decoded by the main controller architecture of ECC arithmetic unit. The Karatsuba multiplier [7] is used to perform point addition and doubling for both fields. Results are stored in the register files.

3.3.1 Scalar Multiplication

The ECC scheme requires the point and scalar multiplication defined as follows:

$$Q = kP = P + P + \cdots + P\,(k \text{ times})$$

Here, P denotes a point on the elliptic curve and k is a random integer. Point addition and point doubling play a key role in scalar multiplication.

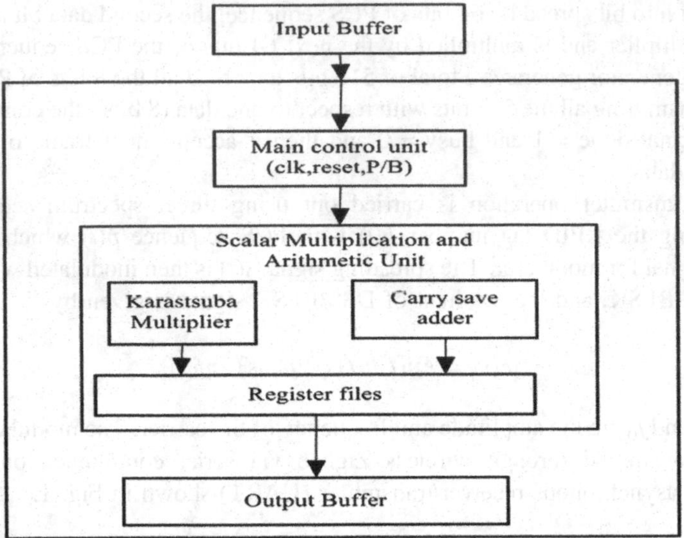

Fig. 2 Proposed architecture of the ECC processor (GF^{163})

The scalar multiplication algorithm is as follows:

Input : $k = (k_{n-1}, k_{n-2} \text{-------} k_1, k_0), P$;
Output $= [k]\,P$;
$R_0 = 0$; $R_1 = P$;
For $i = n - 1$ down to 0
do
$b = k_i$; $R_{1-b} = R_{1-b} + R_b$;
$R_b = 2R_b$;
end for;
return R_0

An elliptic curve equation over prime field is given by

$$y^2 \bmod p = (x^3 + ax^2 + b) \bmod P \quad (2)$$

where a and b are the parameters and x and y are the points on curves.

Binary field equation is given by

$$y^2 + xy = x^3 + ax^2 + b \quad (3)$$

ECC over binary field achieves high performance without considering carry and modular reduction. These fields are optimal for the use in hardware in terms of area and speed.

3.3.2 Binary Field

The most important elliptic curve equation is: $y^2 + xy = x^3 + ax^2 + b$ [8].

In binary field, addition is performed by an EX-OR operation and multiplication is polynomial based. The result is reduced by using the irreducible polynomial. Squaring is achieved by shift operation.

Point Addition Over Binary Field

In this work, one point is in projective coordinate and another point is an affine coordinate. The resulting point will be in projective coordinate which avoids the inversion operation.

Algorithm for addition is as follows:

Inputs: A(x_2, y_2), $Q(X_4,Y_4,Z_4)$.
Outputs: $R(X_3,Y_3,Z_3)$.

$$
\begin{aligned}
&A = Y_4 + y_2 * Z_4^2;\\
&B = X_4 + x_2 * Z_4;\\
&C = B * Z_4;\\
&Z_3 = C * C;\\
&D = x_2 * Z_3;\\
&E = A + B * B + aC;\\
&X_3 = A * A + C * E;\\
&I = D + X_3;\\
&J = A * C + Z_3;\\
&F = I * J;\\
&K = Z_3 * Z_3;\\
&Y_3 = F + x_2 * K + y_2 * K
\end{aligned}
$$

The point doubling operation is to add a point on the elliptic curve with itself. In these equations, 'a' and 'b' are considered as parameters of an elliptic curve.

Algorithm for point doubling is as follows:

Inputs: (x_1, y_1, z_1);
Outputs: (x_4, y_4, z_4);

$$
\begin{aligned}
&z_4 = z_1^2 * x_1^2,\\
&x_4 = x_1^4 + bz_1^4,\\
&y_4 = \left(y_1^2 + az_4 + bz_1^4\right) * x_4 + z_4 * bz_1^4.
\end{aligned}
$$

3.3.3 Prime Field

In prime field, each elliptic curve addition and doubling requires a fixed number of modular multiplications, square, additions, shifts, and similar basic arithmetic operations. The actual number of these operations depends on the progression of the curve. The operation of the multiplication and square operations that dominate the running time, which scales exactly as the number of arithmetic operations to be performed.

Point Addition Over Prime Field

For an elliptic curve defined over *GF*(*P*), the normal elliptic point (x, y) is projected to (X_1, Y_1, Z_1), where $x = X/Z^2$, and $y = Y/Z^3$, and the second point considered is affine point that is (x_2, y_2).

Algorithm for a point addition is as follows:

Inputs: $Q = (X_4, Y_4, Z_4)$, $A = (x_2, y_2)$
Output: $R = (X_3, Y_3, Z_3) = P + Q$;

$$
\begin{aligned}
A &= X_4;\\
B &= x_2 * Z_1^2;\\
C &= A - B;\\
D &= Y_1;\\
E &= y_2 * Z_1^3;\\
F &= D - E;\\
G &= A + B;\\
H &= D + E;\\
Z_3 &= Z_1 * C;\\
X_3 &= F^2 - G * C^2;\\
I &= G * C^2 - 2 * X_3;\\
Y_3 &= \left(I * F - H * C^2\right)/2;
\end{aligned}
$$

In the *GF*(*P*), the point doubling algorithm is as follows:

Inputs: $P = (X_1, Y_1, Z_1)$, a;
Output: $Q = (X_4, Y_4, Z_4) = 2P$;

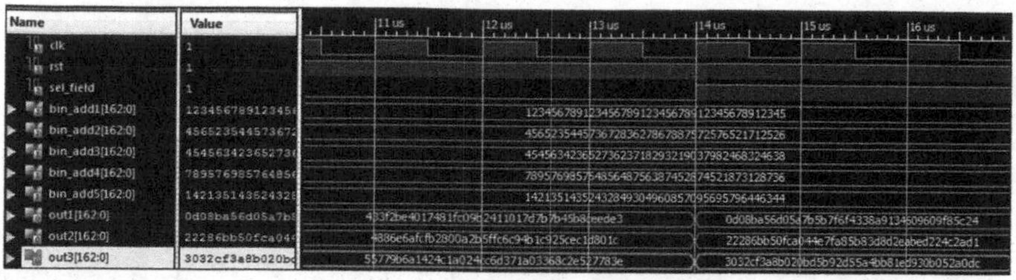

Fig. 3 Simulated results of ECC processor in GF (2^{163}) for prime and binary fields

$$A = 3 * X_1^2 + a * Z_1^4;$$
$$B = 4 * X_1 * Y_1^2;$$
$$X_4 = A^2 - 2 * B;$$
$$Z_4 = 2 * Y_1 * Z_1;$$
$$C = 8 * Y_1^4;$$
$$Y_4 = A * (B - X_4) - C;$$

ECC processor has been designed for 163 bits for both binary and prime fields, and to select particular field, sel_field control signal selects either binary field or prime field, when it is '1' binary field is selected else prime field. The clock frequency for ECC processor is 100 MHz which is generated from Spartan 6 FPGA. Reset clears all internal registers and memories. Out1, Out2, and Out3 are the keys generated from ECC based on field selection and are shown in Fig. 3.

4 Receiver

Receiver block diagram is as shown in Fig. 4. It consists of detector, control circuit, PCS generator, and demodulator. The initial 64-bit data is loaded into eight registers from R_1 to R_8 which are used in the transmitter, and after the initial data loaded, ready control signal will become high (Logic 1) and it is control signal to PCS generator through control circuit. During the process phase, the control circuit maintains the signal ld = 1, enable = 0, and run = 1. Now, the detector block is initialized with a process sequence, the first 512 bits of the PCS sequence are loaded into the eight lower registers, each length is 64 bits of the detector. The synchronization between transmitter and receiver is important in a DSS system when the receiver block uses pseudo-chaotic sequence. When the control circuit of receiver generates ld = 1, then the 64 bits of the PCS producer stored into the shift register 1 of 64-bit detector. After loading all the 512 bits into shift register 1 of eight 64-bit detectors, control circuit resets ld = 0, so that the receiver bits are moving to the shift register 2. After receiving each bit from channel, the detector

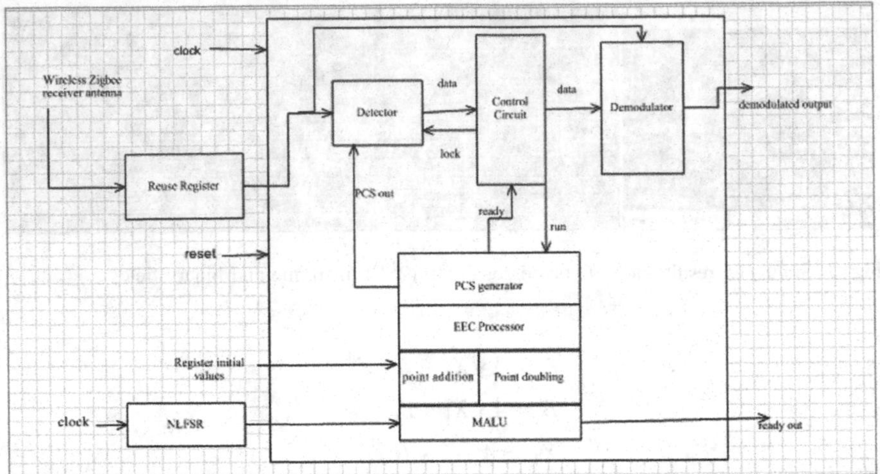

Fig. 4 Proposed block diagram of the DSS and ECC-based RFID tag chip at receiver

block multiplies each bit with shift register 1 and shift register 2 and sums up the output of the bit-wise multiplier, if the sum exceeds the selected threshold value, detector block decides the received bit is logic 1 or 0. The mathematical analysis of the receiver is as follows.

The received signal $r(t)$ at the input of the receiver is the sum of the transmitted signal $e(t)$ and channel noise $n(t)$. Firstly, the received signal is multiplied with the pseudorandom number (PN) sequence and resulting signal $g(t)$ is given as follows:

$$\begin{aligned} g(t) &= r(t).p(t) = (e(t) + n(t))p(t) \\ &= \mathrm{Ab}(t).p^2(t)\cos(2\pi f_0 t) + n(t)p(t) \\ &= \mathrm{Ab}(t).\cos(2\pi f_0 t) + n(t).p(t). \end{aligned} \tag{4}$$

Signal (t) is multiplied with the sinusoidal carrier, and resulting signal $s(t)$ is given is as follows:

$$\begin{aligned} s(t) &= Ag(t)\cos(2\pi f_0 t) \\ &= Ab(t))\cos^2(2\pi f_0 t) + An(t)p(t)\cos(2\pi f_0 t) \\ &= A^2/b(t) + A^2/b(t)\cos(4\pi f_0 t) + An(t)p(t)\cos(2\pi f_0 t). \end{aligned} \tag{5}$$

The signal $s(t)$ is fed to the integrator whose output is reset to zero by the trigger of each pulse of the train $p(t)$. It means that the integration period of each bit is equal to the corresponding interpulse interval which is also the corresponding bit duration. Before each reset instance, the output signal of the integrator $i(t)$ is sampled. The output value of the sampler at the instance t_{n+1} is given by.

$$\begin{aligned} i(t_{n+1}) &= \int_{t_n}^{t_{n+1}} s(t)\mathrm{d}t \\ &= \int_0^{T_{bn}} \frac{A^2}{2} b(t)\mathrm{d}t + \int_0^{T_{bn}} \frac{A^2}{2} b(t)\cos(4\pi f_0 t)dt + \int_0^{T_{bn}} An(t)p(t)\cos(2\pi f_0 t)\mathrm{d}t \\ &= \frac{A^2}{2} b_n T_{bn} + 0 + \int_0^{T_{bn}} An(t)p(t)\cos(2\pi f_0 t)\mathrm{d}t \end{aligned} \tag{6}$$

where ($A^2/2bn$) is the energy of the desired signal. $\int_0^{T_{bn}} \frac{A^2}{2} b(t)\cos(4\pi f_0 t)\mathrm{d}t$ equal to zero because the period T_{bn} is a multiple of the carrier cycle. $\int_0^{T_{bn}} An(t)c(t)$ $\cos(2\pi f_0 t)\mathrm{d}t$ is the energy produced by the channel noise. It is noted that the correlation between $p(t)$ and $An(t)$ $\cos(2\pi f_0 t)$ is very low, and hence the noise energy is much less than the signal energy. The resulting sample is fed to the decision device to recover the binary value of the nth bit as follows:

$$\begin{aligned} b_n &= \{1, i(t_{n+1}) \geq 0, \\ &= \{0, i(t_{n+1}) < 0. \end{aligned} \tag{7}$$

5 Results and Comparison with Methods Proposed in the Literature

In the proposed work, the results obtained from simulation are verified on two Spartan 6 FPGAs (XCS6LX45). At the transmitter section, one FPGA along with Zigbee module and RFID tag has been used to process and transmit the modulated signal. At the receiver section, another FPGA along with Zigbee module and RFID tag has been used to retrieve the RFID tag number which is displayed on LCD of receiver FPGA. The 163-bit ECC processor using scalar multiplication (both binary and prime fields) has been incorporated in transmitter section to secure tag number. Figure 5 shows loading of initial values into eight registers from R_1 to R_8 using sel_reg control signal.

Before transmitting and after modulation, all 64-bit data is concatenated to get 512 bits and transmitted along with RFID tag number, The output of multiplier with 512 bits at RFID transmitter is shown in Fig. 6.

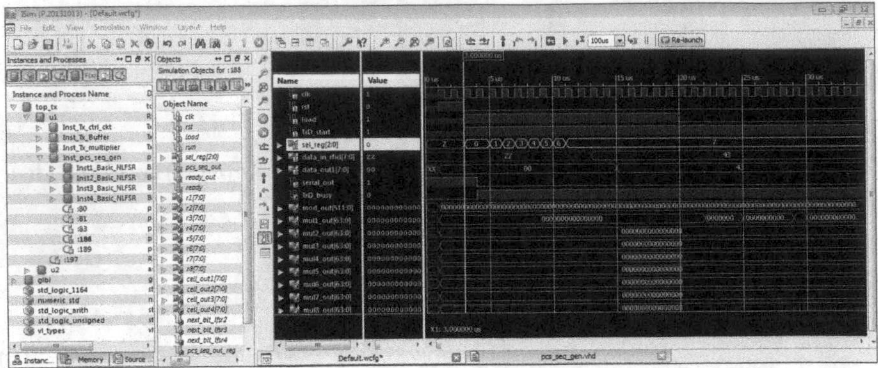

Fig. 5 Loading initial values into R_1–R_8 registers at RFID transmitter

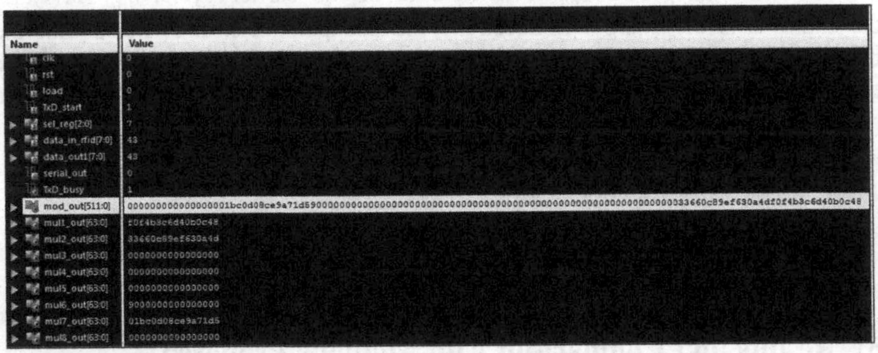

Fig. 6 Concatenation of all 64-bit multiplier outputs to get 512 bits at RFID transmitter

PCS generator generates spread spectrum with each size of 64 bits stored in different registers and then applied to both modulator and multiplier; these two modules generate the final modulated signals. These modulated signals are transmitted through Zigbee transmitter using UART protocol with the baud rate of 11,520 bits per second. The simulated results of tag number and multiplier output are shown in Fig. 7.

In receiver section, Zigbee receives the data serially through R_x pin and stores in 8-bit register to get 512-bit data which contains both RFID tag number and carrier signal and then sends to detector to decode tag number. After demodulation, RFID tag number is displayed on LCD of FPGA board. The demodulated tag number at the receiver section is as shown in Fig. 8.

The proposed work has implemented using Cadence software tools to measure the area, power, and delay. Table 1 shows the comparison of present work and existing work. The proposed ECP in DBC needs only 8.58 K gate area and has a

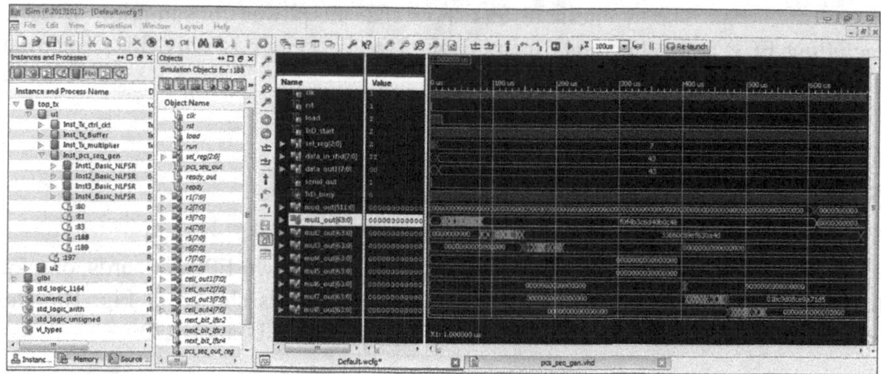

Fig. 7 RFID tag number and 64-bit multiplier outputs at RFID transmitter

Fig. 8 Demodulated tag number at receiver section

Table 1 Comparison of performance characteristics of power consumption, frequency, delay, and area

Ref. No.	Digit size	Field size	Frequency (kHz)	Power (μW)	ECP area (gates) K	Delay (ns)	Technology
This work	8	GF(2^{163})	35,810	383.17	8.58	7.50	FPGA
01	8	GF(2^{163})	13,536	253	14.1	13.2	Umc 0.13 μm
06	8	GF(2^{163})	2454	320.3	12.5	13.1	Umc 0.13 μm
07	8	GF(2^{163})	13560	NA	12.5	244	Umc 0.13 μm
08	8	GF(2^{163})	400	7.3	56.7	31.8	Umc 0.13 μm
10	8	GF(2^{163})	1059	148.5	13.2	547.8	Umc 0.13 μm
15	8	GF(2^{163})	13,536	208.4	21.8	12.5	Umc 0 .13 μm
16	8	GF(2^{100})	500	400	18.7	NA	TSMC 0.18 μm
17	8	GF(2^{163})	200	15	6.1	NA	TSMC 0.18 μm
18	8	GF(2^{163})	76	4.88	13.78	NA	65 nm CMOS

delay of 7.50 ns. The synthesis results show that the power consumption of DBC including ECP and other units in transmitter and receiver is only 381.58 μW at 35,810 kHz.

5.1 Hardware Implemented Results

The proposed work has been implemented and tested on two FPGA boards, one for transmission is shown in Fig. 9 and other for receiver is shown in Fig. 10. The receiver module receives tag number and displays on LCD and LED.

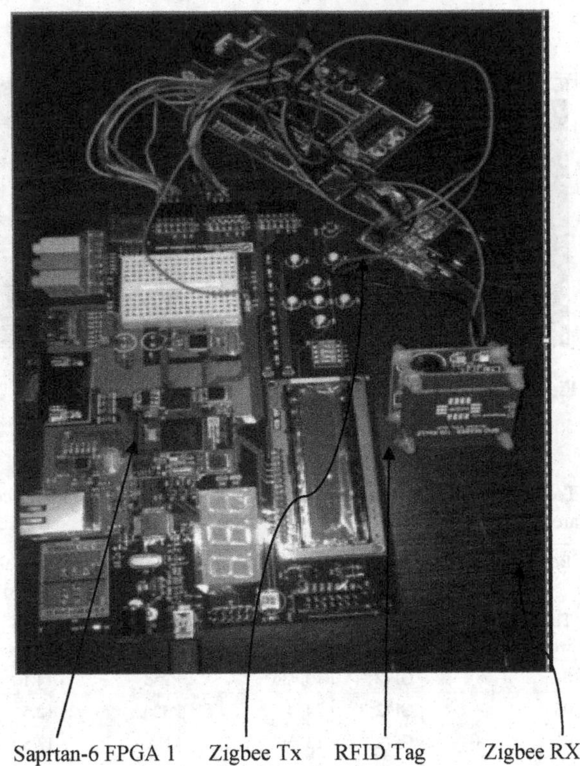

Fig. 9 Hardware implementation for transmission

Fig. 10 Hardware implementation for receiver

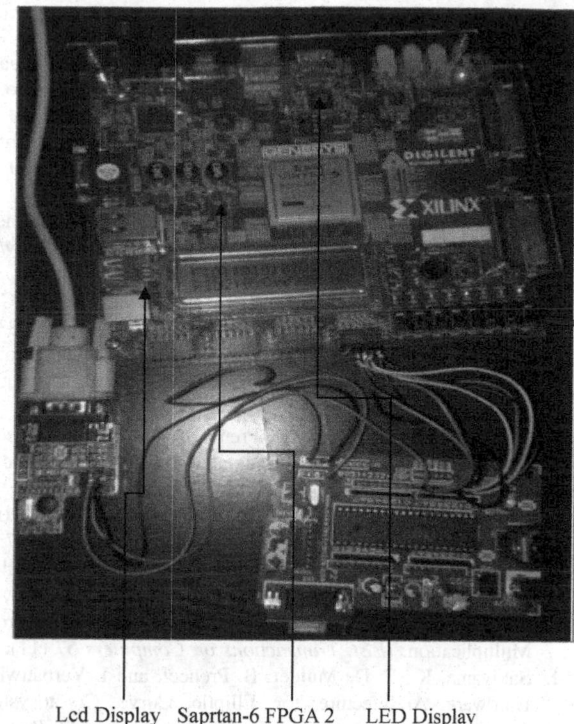

5.2 Conclusion

Single-user DS-SS system using pseudo-chaotic sequence as spreading sequence and RFID transmitter and receiver in FPGA development kit targeted to Xilinx's Spartan 6 device XC6S45-2tq324 has been implemented. Transmission has been achieved through Zigbee. Considerable improvement in power dissipation, area, and delay has been achieved. Security of the data has been assured using ECC.

References

1. Zilong, Liu , Dongsheng Liu, Xuecheng Zou, Hui Lin, and Jian Cheng. 2014. Design of an Elliptic Curve Cryptography Processor for RFID Tag Chips *Sensors* 14: 17883–17904
2. Koblitz, N. 1987. Elliptic Curve Cryptosystems. *Mathematics of Computation* 48: 203–220.
3. Lai, Jyu-Yuan, and Chih-Tsun Huang. 2011. Energy-Adaptive Dual-Field Processor for High-Performance Elliptic Curve Cryptographic Applications. *IEEE Transactions on VLSI* 19 (8).
4. Kocabas, U., J. Fan, and I. Verbauwhede. 2010. Implementation of binary Edwards curves for very-constrained devices. In *Proceedings of 2010 the 21st International Conference on*

Application-Specific Systems Architectures and Processor (*ASAP*), Rennes, France, 7–9 July 185–191.

5. Hein, D., J. Wolkerstorfer, and N. Felber. 2009. ECC is Ready for RFID-a proof in silicon. In *Selected Areas in Cryptography*, vol. 5381 ed. Roberto, M.A., and K. Liam, 401–413. Heidelberg, Germany: Springer.
6. Azarderakhsh, R., and A.R. Masoleh. 2013. High-Performance Implementation of Point Multiplication on Koblitz Curves. *IEEE Transactions on Circuits and Systema II Express Briefs* 60: 41–45.
7. National Institute of Standards and Technology. Recommended Elliptic Curves for Federal Government Use. Available online: http://csrc.nist.gov/groups/ST/toolkit/documents/dss/NISTReCur.pdf (Accessed on 15 Sept 2014).
8. Ting, Hsin-Yu, and Chih-Tsun Huang. Design of Low-Cost Elliptic Curve Cryptographic Engines for Ubiquitous Security. 978-1-4799-2776-0/14/$31.00 ©2014 IEEE.
9. Lai, Jyu-Yuan, and Chih-Tsun Huang. 2008. High-Throughput Cost-Effective Dual-Field Processors and the Design Framework for Elliptic Curve Cryptography. *IEEE Transactions on VLSI* 16 (11).
10. Sakiyama, K., L. Batina, B. Preneel, and I. Verbauwhede. 2007. Multi-Corecurve-Based Cryptoprocessor with Reconfiguarable Modular Arithmetic Logic Units Over GF(2^{m}). *IEEE Transacions on Computers* 56 (9): 1269–1282.
11. Lee, Y.K., K. Sakiyama, L. Batina, and I. Verbauwhede. 2008. Elliptic-Curve-Based Security Processor for RFID. *IEEE Transactions on Computers* 57: 1514–1527.
12. Kumar, S., and C. Paar. 2006. Are standards compliant elliptic curve cryptosystems feasible on RFID? In *Proceedings of Workshop on RFID Security, Graz, Austria*, 12–14, July 2006.
13. Ansari, B., and M.A. Hasan. 2008. High-Performance Architecture of Elliptic Curve Scalar Multiplication. *IEEE Transactions on Computers* 57 (11): 1143–1153.
14. Sakiyama, K., E. De Mulder, B. Preneel, and I. Verbauwhede. 2006. A Parallel Processing Hardware Architecture for Elliptic Curve Cryptosystems. *In Proceedings of IEEE International Conference Acoustic, Speech Signal Process* (*ICASSP*), vol. 3. Toulouse, France 904–907.
15. Shylashree, N., A. Deepika, and V. Sridhar. 2012. High-Speed FPGA-Based Elliptic Curve Cryptography Using Mixed Co-ordinates. *Journal of Discrete Mathematics and Cryptography* 46: 8511–8516 (Elixir).
16. Liu, Dongsheng, Zilong Liu, Zhenqiang Yong, Xuecheng Zou, and Jian Cheng. 2015. Design and Implementation of an ECC-Based Digital Baseband Controller for RFID Tag Chip. *IEEE Transactions on Industrial Electronics* 62 (7).
17. G Gaubatz et al. 2005. State of the Art in Ultra-Low Power Public Key Cryptography for Wireless Sensor Networks. In *3rd IEEE Per Com 2005 Workshops*, 146–150.
18. L Batina et al. 2007. Public-Key Cryptography on the Top of a Needle. *ISCAS* 2007: I831–1834.

Handling Imbalanced Data: A Survey

Neelam Rout, Debahuti Mishra and Manas Kumar Mallick

Abstract Nowadays, handling of the imbalance data is a major challenge. Imbalanced data set means the instances of one class are much more than the instances of another class where the majority and minority class or classes are taken as negative and positive, respectively. In this paper, the meaning of the imbalanced data, examples of the imbalanced data, different challenges of handling the imbalanced data, imbalance class problems and performance analysis metrics for the imbalanced data are discussed. Then different methods are summarized with their pros and cons. Finally, the examples of the imbalanced data sets having low-to-high imbalance ratio (IR) values are shown.

Keywords Imbalanced data · Performance analysis metrics · Different methods

1 Introduction

In recent years, the imbalanced data sets problem plays a key role in machine learning. The imbalance problem means the instances of one of the classes (majority class) are much more than the other class (minority class). The ratio between the majority and minority classes may be 100:1, 1000:1 and 10000:1; in short, the instances of majority class outnumber the amount of minority class instances. This problem is not only in binary class data but also in multi-class data. Usually, negative examples are defined as majority class and positive examples are called minority class [1, 2] Some applications utilizing the imbalanced data sets are the medical diagnosis, detection of oil spills and financial industry [3, 4]. Nowadays,

N. Rout (✉) · D. Mishra · M.K. Mallick
Siksha 'O' Anusandhan University, Bhubaneswar, India
e-mail: neelamrout@soauniversity.ac.in

D. Mishra
e-mail: debahutimishra@soauniversity.ac.in

M.K. Mallick
e-mail: mkmallick@soauniversity.ac.in

M.S. Reddy et al. (eds.), *International Proceedings on Advances in Soft Computing, Intelligent Systems and Applications*, Advances in Intelligent Systems and Computing 628, https://doi.org/10.1007/978-981-10-5272-9_39

this is a very challenging topic of research [5] where lots of points need focus at a time like the binary class problem, class overlapping, cost of misclassified class, multiple class problem, small size and small disjuncts of the imbalanced data sets. There are lots of efforts given to binary class imbalance problem, but the different types of issues related to the multi-class imbalance problem are not solved yet. In the multiple class imbalance problems, the number of the majority class may be one or more and for the minority class, the situation is also same. The multi-class problem might be solved by using decomposition or any other techniques, but it still needs focus. So, when the data are skewed in nature, then it is really very challenging to work with the minority class [6].

There are several solutions existed from earlier like data-level solutions and algorithmic-level solutions [7]. The accurate extraction of knowledge from the skewed data sets with different levels of noise is very difficult [8]. When the learning of instances for some classes is costly than other classes, then the misclassified costs are varied between classes and the degree of hardness between the classes is another issued [9, 10] which can be solved by using different techniques. Feature selection is a major challenge in the imbalanced data sets to deal with specific decision [11]. There are lots of publicly available sites from which imbalanced data sets might be found, e.g., UCI machine learning repository, Broad Institute, KEEL data sets repository (http://sci2s.ugr.es/keel/datasets.php), etc. Here, no specific area of the data set is used, and those which are imbalanced might be taken for the experiment, for example, cancers data, glass data and yeast data [3]. For the performance evaluation, the area under the curve (AUC), the receiver operating characteristic (ROC) graph, F-measure, G-mean, sensitivity, specificity are used instead of measuring accuracy using confusion matrix [12–16].

The rest of the paper is structured as follows. In Sect. 2, the imbalanced class problem in classification and performance evaluation is elaborated. In Sect. 3, the summary of different existing methods based on binary and multi-class imbalance problem is discussed. Section 4 introduces examples of the imbalanced data sets. Finally, we make our concluding remarks.

2 Imbalanced Class Problem in Classification and Performance Evaluation

In this section, the problem of imbalanced data sets, the different approaches for dealing with class imbalance problem and performance evaluation in the imbalanced domains are discussed. There are several techniques by which the class imbalance problem may be solved. To solve the class imbalance problem, the concept of supervised classification should be known. The objective of classification is to predict categorical labels (e.g., loan application data; "yes" or "no"), where input and output are known. Data classification in which knowledge extraction is done by j features $b_1, b_2, \ldots, b_j \in B$ of n input instances, $x = (x_1, x_2, \ldots, x_n)$. The output class labels (e.g., $Y_k \in C = \{c_1, \ldots, c_m\}$) of the supervised classification is known before and the mapping function is defined over the pattern $B^j \rightarrow C$.

The class imbalance problems are small sample size, class overlapping and small disjuncts. To deal with the class imbalance problem, different approaches are categorized like algorithm-level approaches, data-level approaches and cost-sensitive learning [3, 17]. Some of the existing techniques are random undersampling, oversampling, synthetic minority oversampling technique (SMOTE), selective pre-processing of imbalanced data (SPIDER) [3, 18], and some of the filtering-based methods are SMOTE-TL and SMOTE-EL [17] to eliminate noise in the imbalanced data sets. The ensemble-based method is another technique which is used to deal with the imbalanced data sets, and the ensemble technique is combined the result or performance of several classifiers to improve the performance of single classifier. The most common and effective ensemble algorithms are bagging and boosting (AdaBoost) [3]. The multiple class imbalanced data sets problems are difficult to handle than the binary class imbalanced data sets problems. Multi-class problem might be solved by one-versus-all (one-against-all) approach [19]. Till now, there are so many techniques, and modification of existing techniques is implemented for dealing with the imbalanced data sets. Some of the useful and powerful data-level techniques are SMOTE [18], modification of SMOTE, extension of SMOTE, B1-SMOTE and B2-SMOTE [20] to deal with imbalanced data. Ensemble learning [21] approaches are also very useful and fruitful.

Another important point is that to choose appropriately performance metrics for performance evaluation and analysis. Normally, confusion matrix [22] is used to calculate accuracy rate. But by using this method, the minority class data has not shown good result or ignored. This performance metrics has not given importance to the data of minority class and is usually reduced the global quantities such as error rate to give the best result. So, due to this reason for the performance evaluation of the imbalanced data, other metrics should be used rather than this one. Receiver operating characteristic (ROC) curves is a very workable visual tool which shows the trade-off in-between benefits or True Positive (TP rate) and costs or False Positive (FP rate) [12, 22–24]. Other metrics may be used like geometric-mean (Gm), F-measure (Fm), sensitivity, and specificity [22, 25].

3 Solutions of the Imbalanced Data Sets

3.1 Data-Level Approaches

To deal with the minority class of the imbalanced data, the authors [26] developed majority weighted minority oversampling technique (MWMOTE) method, and for experimental purposes 20 real-world data sets are used where G-mean, ROC, and AUC are taken as performance metrics. In [17], modification of the original sampling technique is used for handling the imbalanced data with neighbourhood-balanced bagging (NBBag). In [16], the imbalanced data set is divided into two parts, i.e. training and testing sets; then different types of classification algorithms are taken for experiment, and finally the performance analysis is done by using different performance measures and ANOVA test is also done. As per their conclusion, SMOTE +PSO+C5 is the best classifier for 5 year survivability of breast cancer patients. In [19], the authors used one-versus-one binarization technique to decompose multi-class data into the two classes, and SMOTE+OVA algorithm is used to handle these data sets. Though Random Forests or Decision Trees is a successful and fast algorithm, it is used for classification purpose [19]. In [18] for handling the imbalanced data, Radial Basis Function Networks (RBFN) weights training methodology is used, and local and global terms are taken for designing of this method. In local weights training methods, the higher value of imbalance ratio (IR) gives better results and lower value of IR should be balanced with SMOTE or any other methods.

In [20], it is stated that the noisy and borderline examples might create problems in the imbalanced data sets and the solution of this problem is re-sampling method with some filtering techniques, and the authors used the extension of SMOTE and also Iterative–Partitioning Filter (IPF) to tackle noisy and borderline examples problem. For the binary class imbalance problems, the authors [27] used RBF classifier (combined SMOTE and PSO) and to analyse results different types of performance metrics are taken, i.e. TP%, FP%, precision, G-mean and F-measures having different $\beta\%$ and ρ. In [28], the researchers tried to develop the solutions for the both imbalanced and noisy data by using 7 different sampling techniques. In [29], novel inverse random under-sampling (IRUS) method is proposed to tackle the problem due to the imbalanced data sets and the idea inverse (ratio of imbalance class cardinality) is used. It has also advantages on the multi-label classification; the IRSU is also compared with several other imbalance techniques, and for each method decision tree (C4.5) is taken as the base classifier. In Table 1, the sampling methods and its variant for handling the imbalanced data are summarized.

3.2 Algorithm-Level Approaches

Support Vector Machines (SVM): To balance the imbalanced data set, only cancer data sets (breast and colon) are taken [30] because nowadays this type of cancer-related diseases are very common. Here the input is protein sequences data having different dimensions of feature space (amino acid composition, split amino acid composition, pseudo-amino acid composition—series, and pseudo-amino acid composition—parallel). In this paper, the majority class is kept aside and megatrend diffusion (MTD) technique is used to synthesize some more data for the minority class for balancing the imbalance data. By developing SVM/KNN models and analysing computational cost, it is concluded that hybrid MTD-AVM is the best model as compared to RF, QPDR, NB, and KNN. The SVMPseAAC-S model has

Table 1 Theme: sampling method and its variants for handling the imbalanced data

S. No.	Algorithms/Methods/Approaches used	Findings	Limitations
1	SMOTE+PSO+C5 [16]	Estimate 5-year survivability of breast cancer patients	Breast cancer patients
2	Modification of sampling technique, i.e., Neighbourhood Balanced Bagging (NBBag) [17]	Better than existing over-sampling bagging extension and competitive to roughly balance bagging	Costly
3	SMOTE+OVA, random forest algorithm [19]	Useful for classification multi-class imbalanced data	Fix up the oversampling rate
4	SMOTE, radial basis function networks (Ribbons) [18]	Higher the dataset imbalance ratio tends to better result	More storage space
5	Extension of SMOTE, i.e., SMOTE-IPF [20]	Addressing the problem due to noisy and borderline examples in imbalanced datasets	Small sample size
6	Combined SMOTE and PSO-based RBF classifiers (SMOTE+PSO-RBF) [27]	To generate synthetic instances for the positive class, and RBF performs well	More storage space
7	Majority weighted minority over-sampling technique (MWMOTE) [26]	Generate the synthetic minority class samples according to euclidean distance	Multi-class problem
8	Data sampling techniques [28]	Robust with noisy imbalance data	Cannot compete to advance noise handling techniques
9	Inverse random under-sampling (IRUS) [29]	Beneficial for disproportionate training datasets sizes and improve the accuracy of the multi-label classification	Other different applications to multi-label classification

given accuracy 96.71% (C/NC data set), 96.50% (B/NBC data set) and 95.18% (B/NBC data set). From [31], it is known that Ant Colony Optimization (ACO) algorithm (where '0' indicates sample is eliminated and '1' indicates sample is kept for further use) is suitable to use with under-sampling method for classifying DNA microarray imbalance data and the classification technique support vector machine (SVM) is used for the imbalanced data sets. To tackle the problem due to the majority and minority class in the imbalanced data, in [32] the authors have experimented with the kernel scaling method to improve support vector machine and adjusted F-measure performance matrix is used to evaluate the performance of the classifier. Finally, the comparison of performance results of classifiers (C4.5, L2 Loss SVM, etc.) is seen in this research paper. Though SVM might not able to handle the imbalanced data properly, the authors have experimented with the base model EnSVM and selective ensemble EnSVM+ with additional re-sampling method [33].

To handle the imbalance class, some researchers worked with feature extraction and selection methods like wrapper method [34]. In this paper, second-order cone programming support vector machines (SOCP-SVM) method is taken where linear programming support vector machine is used as formulation principle and the area under the curve (AUC) metric is used for performance analysis which worked well on six benchmark data sets. When parallel selective sampling (PSS) is combined with support vector machine (SVM), PSS-SVM method is produced and its performance is excellent than support vector machine (SVM) because it has no convergence [35]. In [36], the trained support vector machine (SVM) is taken as the preprocessor to improve the performance of MLP, LR, and RF intelligent algorithm. In this paper, two-phase balancing approach is taken, wherein phase one, the available unbalanced data is trained with SVM to get modified balance data; in the second phase, this modified data is the input to MLP, LR and RF. After that, the performance analysis is done. In [37], a new variant of SVM Near-Bayesian Support Vector Machine (NBSVM) is proposed and to reduce Bayes error, the authors developed NBSVM. This technique has many advantages over existing methods like it does not over-sample the minority class. This method is able to solve the problem due to overlapping between the target and non-target classes, small disjuncts in the imbalance data sets and noisy data sets. In Table 2, these methods are summarized in a tabular manner.

Clustering: The similarity-based hierarchical decomposition method is based on the outliers' detection and clustering techniques. In hierarchy construction, there are two parts: one is perfectly classified clusters and another one has misclassified clusters; the researchers have used this method to solve the problem classes overlapping and varieties of the majority and minority classes [38]. Fuzzy is the another useful technique for handling the imbalanced data to improve the performance of FRBCSs, and the authors have used 2-tuple genetic tuning where they have taken high and low imbalance ratio data sets [39], and also the pre-processing technique (SMOTE algorithm) is used to balance the imbalanced data. The summary of these methods are shown in Table 3.

3.3 Ensemble and Hybrid Methods

A new ensemble method is proposed in [40] to overcome the problem due to other different methods, that is bagging-based ensemble method. This method does not change the original class distribution like sampling methods, cost-sensitive learning methods. Firstly, the proposed method converts the imbalanced data (binary) into multiple balanced binary class data after that specific classification algorithm is applied to build multiple classifiers, and finally a specific ensemble rule is used like max distance, min distance, etc. and the max distance rule is performed the best among others. The hybrid method is the combination of modified back-propagation (MBP) and Gabriel graph editing (GGE) and used to handle the class imbalance and class overlapping for the multi-class problem [41]. This method is examined over

Table 2 Theme: support vector machine (SVM) and its variants for handling the imbalanced data

S. No.	Algorithms/Methods/Approaches used	Findings	Limitations
1	MTD-SVM and MTD-SVM [30]	To increase the sample of the minority class to predict human breast and colon cancers	Synthetic data generation is slightly expensive, and mega-trend diffusion (MTD) technique works well on small size data
2	ACO sampling, support vector machine [31]	To address imbalanced data classification problem by sample selection procedure based on ACO algorithm	Excessive computational and storage cost
3	Support vector machine with suitable kernel transformation, adjusted F-measure [32]	To handle the imbalance data with kernel scaling and also manage cost function	Efficient estimation strategy for parameters and different kernels
4	EnSVM and EnSVM$^+$: selective ensembles [33]	Effective than normal SVM	*K* value is not automatically determined
5	SOCP-SVM [34]	To improve classification performance and robust due to LP-SVM formulation	Only designed for imbalanced data
6	PSS-SVM [35]	Accurate statistical predictions and low computational complexity	For parallel and distributed computing
7	Support vector machine (SVM) [36]	Effectively balance the data and creates more number of instances for the minority class	Not so simpler and faster
8	Near-Bayesian support vector machine (NBSVM) [37]	Reduce misclassification cost due to rare class	Performance metrics

Table 3 Theme: clustering and its variants for handling the imbalanced data

S. No.	Algorithms/Methods/Approaches used	Findings	Limitations
1	Similarity-based hierarchical decomposition technique based on clustering and outlier detection [38]	Works effectively when data is highly overlapping, and classes are more imbalance	Computational complexity is more during training of this method
2	2-tuple-based genetic tuning, fuzzy rule-based classification systems (FRBCSs), genetic algorithms, genetic fuzzy systems [39]	To increase the performance of simple FRBCSs	Highly imbalanced datasets

Table 4 Theme: ensemble or hybrid methods and its variants for handling the imbalanced data

S. No.	Algorithms/Methods/Approaches used	Findings	Limitations
1	A new ensemble method includes three components, i.e., data balancing, modelling and classifying [40]	Prevent information loss or unexpected mistakes	Can handle only binary class imbalanced problem
2	Hybrid method (MBP+GGE) [41]	To deal with the class imbalance and class overlapping for the multi-class problems	Speed of the neural network convergence

seven strategies (SBP, MBP, SBP+GGE, MBP+GGE, SMOTE, SMOTE+GGE and RUS), and the authors have concluded that it is a very effective technique. The gist of these methods is shown in Table 4.

3.4 *Other Different Techniques*

According to [42], density-based feature selection (DBFS) is a simple technique but effective for the high-dimensional and small sample size classes of the imbalanced data sets. For studying the problem of customer churn prediction, the authors [43] carried out the following steps.

1. First, the useless features are discarded from the original data sets, then missing values are replaced, and nominal to numeric conversion is done.
2. The under-sampling method is applied which is based on RUS and PSO.
3. The PCA, Fisher's Ratio, F-Measure and Mr techniques are used for the dimension reduction.
4. The KNN and RF classifiers are used for the model building, and finally the performance evaluation is done using the metrics like sensitivity, specificity and AUC.

The proposed Che-PmRF approach is a combination of PSO-based sampling, Mr-based feature selection and RF classifier, and it is very effective for the competitive telecommunication industry to tackle the problem of customer churn prediction. To get the best performance result for the imbalanced data sets, the LSI-based feature extraction is implemented to the information granulation-based data mining model, and the main advantage of this technique is that it could save storage space [44]. Distribution optimally balanced–stratified cross-validation (DOB-SCV) is used to deal with the imbalanced data sets to obtain better performance [45]. In [46], the authors proved that the weighted extreme learning machine (ELM) has the following advantages.

1. Simple and fast in implementation.
2. Directly apply to the multi-class classification tasks.
3. Different types of feature mapping functions or kernel methods are available.

Lastly, for the performance analysis, G-mean performance metric is considered for the unweighted ELM, weighted ELM W1 and weighted ELM W2 techniques.

In [47], the authors proposed two different approaches, i.e., decomposition-based and Hellinger distance-based methods, to solve the issues related to feature selection in the imbalance data sets. In the first method, large classes are decomposed into the pseudo-subclasses and later different strategies are used for the classification. The second one is used to measure the distributional divergence for handling the challenges of the imbalance class distribution. After that both methods are compared with the three traditional feature rank methods, which are correlation, Fisher and mutual information methods using the F-measure, AUC and ROC metrics. The basic information of these methods is given in Table 5.

Table 5 Theme: other different techniques for handling the imbalanced data

S. No.	Algorithms/Methods/Approaches used	Findings	Limitations
1	Density-based feature selection (DBFS) [42]	To handle the small sample size and high-dimensional problem in imbalanced datasets	Problems due to more than two classes
2	Chr-PmRF [43]	PSO-based sampling for balancing the datasets, mRMR for feature selection and RF classifier are used to handle customer churn prediction problem	High computational cost
3	Information granulation-based data mining approach, latent semantic indexing [44]	Accuracy is slightly better and faster than the numerical computing method	Overlapping class
4	Distribution optimally balanced stratified cross-validation (DOB–SCV) [45]	Obtaining a better performance estimation result for the classifier	Storage space
5	Weighted extreme learning machine (ELM) [46]	Simple theory, fast in implementation and apply directly to the multi-class classification tasks	Large variety in class distribution
6	Decomposition-based and Hellinger distance-based methods [47]	To solve the feature selection issues in the imbalanced datasets	Comparison with only three traditional feature selection methods

Table 6 Statistic summary of the 21 imbalanced datasets

ID	Dataset	#Attributes (R/I/N)	#Examples	%class (min., maj.)	Imbalance ratio (IR)
1	glass1	9 (9/0/0)	214	(35.46, 64.54)	1.82
2	glass0	9 (9/0/0)	214	(32.68, 67.32)	2.06
3	Ecoli1	7 (7/0/0)	336	(22.94, 77.06)	3.36
4	New-thyroid2	5 (4/1/0)	215	(16.29, 83.71)	5.14
5	Yeast 3	8 (8/0/0)	1484	(10.99, 89.01)	8.1
6	yeast-2_vs_4	8 (8/0/0)	514	(9.92, 90.08)	9.08
7	Glass 2	9 (9/0/0)	214	(7.94, 92.06)	11.59
8	ecoli-0-1-4-6_vs_5	6 (6/0/0)	280	(7.14, 92.86)	13
9	yeast-1_vs_7	7 (7/0/0)	459	(6.54, 93.46)	14.3
10	glass4	9 (9/0/0)	214	(6.07, 93.93)	15.47
11	Abalone9-18	8 (7/0/1)	731	(5.75, 94.25)	16.4
12	glass-0-1-6_vs_5	9 (9/0/0)	184	(4.89, 95.11)	19.44
13	shuttle-c2-vs-c4	9 (0/9/0)	129	(4.65, 95.35)	20.5
14	Glass 5	9 (9/0/0)	214	(4.21, 95.79)	22.78
15	yeast-2_vs_8	8 (8/0/0)	482	(4.15, 95.85)	23.1
16	Car-good	6 (0/0/6)	1728	(3.99, 96.01)	24.04
17	winequality-red-4	11 (11/0/0)	1599	(3.31, 96.69)	29.17
18	Winequality_red_8_vs_6	11 (11/0/0)	656	(2.74, 97.26)	35.44
19	Abalone_9_vs_10_11-12_13	8 (7/0/1)	1622	(1.97, 98.03)	49.69
20	kddcup-rootkit-imap_vs_back	41 (26/0/15)	2225	(0.99, 99.01)	100.14
21	abalone19	8 (7/0/1)	4174	(0.77, 99.23)	129.44

4 Examples of the Imbalanced Data Sets

In Table 6, different types of data sets are mentioned with their IR values in the ascending order. The data sets are taken from KEEL data set repository [48]. For each imbalanced data set, the number of attributes, number of examples, percentage of minority and majority of each class, and imbalance ratio (IR) are given in the table.

5 Conclusion

From the study, it is concluded that in the presence of the imbalance data sets, most of the standard classifier learning algorithms, such as nearest neighbour, decision tree, back-propagation neural networks, failed to give good results. In this paper, different types of existing techniques are discussed for tackling the imbalance class problems but still improvement techniques are needed, necessarily. It is also known that the ensemble learning algorithms are the useful and powerful methods to deal with the imbalance class problem. Some of the imbalanced data sets have been shown with different IR values in the tabular manner. It is very important to balance the imbalance data with effective techniques and at the same time, cost factor should be given attention. The correct classifier techniques and performance evaluation metrics must be applied to achieve good results. There are many methods for handling the imbalanced data, but the main focus is to use the appropriate technique from the existing techniques or develop the new methods according to the need because it is not necessary that if one technique is worked well on an imbalanced data set, then the same method is worked for an another imbalanced data set.

References

1. He, Habib, and Edwardo Garcia. 2009. Learning from Imbalanced Data. *IEEE Transactions on Knowledge and Data Engineering* 21 (9): 1263–1284.
2. Van Pulse, Jason, and Tag hi Jehoshaphat. 2009. Knowledge Discovery from Imbalanced and Noisy Data. *Data and Knowledge Engineering* 68 (12): 1513–1542.
3. Galar, M., A. Fernandez, E. Barrenechea, H. Bustince, and F. Herrera. 2012. A Review on Ensembles for the Class Imbalance Problem: Bagging-, Boosting-, and Hybrid-Based Approaches. *IEEE Transactions on Systems, Man, and Cybernetics, Part C: Applications and Reviews* 42 (4): 463–484.
4. He, Habib, and Yunnan Ma. (eds.). 2013. *Imbalanced Learning: Foundations, Algorithms, and Applications*. Wiley.
5. Yang, Anglia, and Wu Donning. 2006. 10 Challenging Problems in Data Mining Research. *International Journal of Information Technology and Decision Making* 5 (04): 597–604.
6. Wang, Shu, and In Tao. 2012. Multi Class Imbalance Problems: Analysis and Potential Solutions. *IEEE Transactions on Systems, Man, and Cybernetics, Part B: Cybernetics* 42 (4): 1119–1130.
7. Lakshmi, T. Jay, and C. Pradesh. 2014. A Study on Classifying Imbalanced Datasets. In *First International Conference on Networks and Soft Computing (ICNSC)*, IEEE.
8. Neapolitan, Ami. 2009. *Classification Techniques for Noisy and Imbalanced Data*. Dis. Florida Atlantic University.
9. Org Mennicke, J. 2006. Classifier Learning for Imbalanced Data with Varying Misclassification Costs.
10. Scrupulousness's, M.G., D.S. Antifascist, S.B. Konstantin, and P.E. Intelsat. Local Cost Sensitive Learning for Handling Imbalanced Data Sets. In *Mediterranean Conference on Control and Automation, 2007, MED'07*, 1–6. IEEE.
11. Yin, Lithium, et al. Feature Selection for High-Dimensional Imbalanced Data. *Supercomputing* 105 (2013): 3–11.
12. Y, Wenona, Yuan-chin Ivan Chang, and Eunice Park. 2014. A Modified Area Under the ROC Curve and its Application to Marker Selection and Classification. *Journal of the Korean Statistical Society* 43 (2): 161–175.
13. Lou, Zen, Ruy Wang, Ming Tao, and Xian fa CAI. 2015. A Class-Oriented Feature Selection Approach for Multi-Class Imbalanced Network Traffic Datasets Based on Local and Global Metrics Fusion. *Supercomputing* 168: 365–381.

14. Mahmoud, Shani, Par ham Moravia, Cardin Highland, and Rasoul Moradi. 2014. Diversity and Separable Metrics in Over-Sampling Technique for Imbalanced Data Classification. In *4th International eConference on Computer and Knowledge Engineering (ICCKE)*, IEEE, 152–158.
15. Ghanavati, Mojgan, Raymond K. Wong, Fang Chen, Yang Wang, and Chang-Shing Perng. 2014. An Effective Integrated Method for Learning Big Imbalanced Data. In *IEEE International Congress on Big Data (Big Data Congress)*, IEEE, 691–698.
16. Wang, Kung-Jeng, Bunjira Makond, Kun-Huang Chen, and Kung-Min Wang. 2014. A Hybrid Classifier Combining SMOTE with PSO to Estimate 5 year Survivability of Breast Cancer Patients. *Applied Soft Computing* 20: 15–24.
17. Błaszczynski, Jerzy, and Jerzy Stefanowski. 2015. Neighbourhood Sampling in Bagging for Imbalanced Data. *Neurocomputing* 150: 529–542.
18. Perez-Godoy, M.D., A.J. Rivera, C.J. Carmona, and M.J. del Jesus. 2014. Training Algorithms for Radial Basis Function Networks to Tackle Learning Processes with Imbalanced Datasets. *Applied Soft Computing* 25: 26–39.
19. Bhagat, Reshma C., and Sachin S. Patil. 2015. Enhanced SMOTE Algorithm for Classification of Imbalanced Big-Data Using Random Forest. *IEEE International Conference on Advance Computing (IACC), 2015*, IEEE.
20. Saez, J.A., J. Luengo, J. Stefanowski, and F. Herrera. 2015. SMOTE–IPF: Addressing the Noisy and Borderline Examples Problem in Imbalanced Classification by a Re-Sampling Method with Filtering. *Information Sciences* 291: 184–203.
21. Hu, Xiao-Sheng, and Run-Jing Zhang. 2013. Clustering-Based Subset Ensemble Learning Method for Imbalanced Data. *International Conference on Machine Learning and Cybernetics (ICMLC), 2013*, vol. 1. IEEE.
22. Han, Jiawei, and Micheline Kamber. 2001. Data Mining: Concepts and Techniques.
23. Subtil, Fabien, and Muriel Rabilloud. 2015. An Enhancement of ROC Curves Made Them Clinically Relevant for Diagnostic-Test Comparison and Optimal-Threshold Determination. *Journal of clinical epidemiology*.
24. Wang, Qihua, Lili Yao, and Peng Lai. 2009. Estimation of the Area Under ROC Curve with Censored Data. *Journal of Statistical Planning and Inference* 139 (3): 1033–1044.
25. Batuwita, Rukshan, and Vasile Palade. 2009. A New Performance Measure for Class Imbalance Learning. Application to Bioinformatics Problems. In *International Conference on Machine Learning and Applications, ICMLA'09, 2009*, IEEE.
26. Barua, S., M.M. Islam, X. Yao, and K. Murase. 2014. MWMOTE–majority Weighted Minority Oversampling Technique for Imbalanced Data Set Learning. *IEEE Transactions on Knowledge and Data Engineering* 26 (2): 405–425.
27. Gao, Ming, Xia Hong, Sheng Chen, and Chris J. Harris. 2011. A Combined SMOTE and PSO Based RBF Classifier for Two-Class Imbalanced Problems. *Neurocomputing* 74 (17): 3456–3466.
28. Seiffert, C., T.M. Khoshgoftaar, J. Van Hulse, and A. Folleco. 2014. An Empirical Study of the Classification Performance of Learners on Imbalanced and Noisy Software Quality Data. *Information Sciences* 259: 571–595.
29. Tahir, Muhammad Atif, Josef Kittler, and Fei Yan. 2012. Inverse Random Under sampling for Class Imbalance Problem and Its Application to Multi-label Classification. *Pattern Recognition* 45 (10): 3738–3750.
30. Majid, A., S. Ali, M. Iqbal, and N. Kausar. 2014. Prediction of Human Breast and Colon Cancers from Imbalanced Data Using Nearest Neighbor and Support Vector Machines. *Computer Methods and Programs in Biomedicine* 113 (3): 792–808.
31. Yu, Hualong, Jun Ni, and Jing Zhao. 2013. ACO Sampling: An Ant Colony Optimization-based Undersampling Method for Classifying Imbalanced DNA Microarray Data. *Neurocomputing* 101: 309–318.
32. Maratea, Antonio, Alfredo Petrosino, and Mario Manzo. 2014. Adjusted F-Measure and Kernel Scaling for Imbalanced Data Learning. *Information Sciences* 257: 331–341.

33. Liu, Y., X. Yu, J.X. Huang, and A. An. 2011. Combining Integrated Sampling with SVM Ensembles for Learning from Imbalanced Datasets. *Information Processing and Management* 47 (4): 617–631.
34. Maldonado, Sebastian, and Julio Lopez. 2014. Imbalanced Data Classification Using Second-Order Cone Programming Support Vector Machines. *Pattern Recognition* 47 (5): 2070–2079.
35. D'Addabbo, Annarita, and Rosalia Maglietta. 2015. Parallel Selective Sampling Method for Imbalanced and Large Data Classification. *Pattern Recognition Letters* 62: 61–67.
36. Farquad, M.A.H., and Indranil Bose. 2012. Preprocessing Unbalanced Data Using Support Vector Machine. *Decision Support Systems* 53 (1): 226–233.
37. Datta, Shounak, and Swagatam Das. 2015. Near-Bayesian Support Vector Machines for Imbalanced Data Classification with Equal or Unequal Misclassification Costs. *Neural Networks* 70: 39–52.
38. Beyan, Cigdem, and Robert Fisher. 2015. Classifying Imbalanced Data Sets Using Similarity Based Hierarchical Decomposition. *Pattern Recognition* 48 (5): 1653–1672.
39. Fernandez, Alberto, Maria Jose del Jesus, and Francisco Herrera. 2010. On the 2-Tuples Based Genetic Tuning Performance for Fuzzy Rule Based Classification Systems in Imbalanced Datasets. *Information Sciences* 180 (8): 1268–1291.
40. Sun, Z., Q. Song, X. Zhu, H. Sun, B. Xu, and Y. Zhou. 2015. A NOVEL Ensemble Method for Classifying Imbalanced Data. *Pattern Recognition* 48 (5): 1623–1637.
41. Alejo, R., R.M. Valdovinos, V. García, and J.H. Pacheco-Sanchez. 2013. A Hybrid Method to Face Class Overlap and Class Imbalance on Neural Networks and Multi-class Scenarios. *Pattern Recognition Letters* 34 (4): 380–388.
42. Alibeigi, Mina, Sattar Hashemi, and Ali Hamzeh. 2012. DBFS: An Effective Density Based Feature Selection Scheme for Small Sample Size and High Dimensional Imbalanced Data Sets. *Data and Knowledge Engineering* 81: 67–103.
43. Idris, Adnan, Muhammad Rizwan, and Asifullah Khan. 2012. Churn Prediction in Telecom Using RANDOM Forest and PSO Based Data Balancing in Combination with Various Feature Selection Strategies. *Computers and Electrical Engineering* 38 (6): 1808–1819.
44. Chen, M.C., L.S. Chen, C.C. Hsu, and W.R. Zeng. 2008. An Information Granulation Based Data Mining Approach for Classifying Imbalanced Data. *Information Sciences* 178 (16): 3214–3227.
45. Lopez, Victoria, Alberto Fernandez, and Francisco Herrera. 2014. On the Importance of the Validation Technique for Classification WITH Imbalanced Datasets: Addressing Covariate Shift When Data is Skewed. *Information Sciences* 257: 1–13.
46. Zong, Weiwei, Guang-Bin Huang, and Yiqiang Chen. 2013. Weighted Extreme Learning Machine for Imbalance Learning. *Neurocomputing* 101: 229–242.
47. Yin, L., Y. Ge, K. Xiao, X. Wang, and X. Quan. 2013. Feature selection for High-Dimensional Imbalanced Data. *Neurocomputing* 105: 3–11.
48. http://sci2s.ugr.es/keel/datasets.php.

Author Index

M.S. Reddy et al. (eds.), *International Proceedings on Advances in Soft Computing, Intelligent Systems and Applications*, Advances in Intelligent Systems and Computing 628, https://doi.org/10.1007/978-981-10-5272-9

Printed in the United States
By Bookmasters